T0331772

ORDINARY DIFFERENTIAL EQUATIONS WITH APPLICATIONS

Third Edition

SERIES ON APPLIED MATHEMATICS

Editor-in-Chief: Zhong-ci Shi

*For the complete list of the volumes in this series, please visit
http://www.worldscientific.com/series/sam

Series on
Applied Mathematics
Volume 23

ORDINARY DIFFERENTIAL EQUATIONS WITH APPLICATIONS

Third Edition

Sze-Bi Hsu
Kuo-Chang Chen

National Tsing Hua University, Taiwan

World Scientific

NEW JERSEY · LONDON · SINGAPORE · BEIJING · SHANGHAI · HONG KONG · TAIPEI · CHENNAI

Published by

World Scientific Publishing Co. Pte. Ltd.

5 Toh Tuck Link, Singapore 596224

USA office: 27 Warren Street, Suite 401-402, Hackensack, NJ 07601

UK office: 57 Shelton Street, Covent Garden, London WC2H 9HE

Library of Congress Cataloging-in-Publication Data
Names: Hsu, Sze-Bi, 1948–   author. | Chen, Guozhan, 1937–   author.
Title: Ordinary differential equations with applications / Sze-Bi Hsu,
    Kuo-Chang Chen, National Tsing Hua University, Taiwan.
Description: Third edition. | New Jersey : World Scientific, [2022] |
    Series: Series on applied mathematics, 1793-0928 ; vol. 23 |
    Includes bibliographical references and index.
Identifiers: LCCN 2022013691 | ISBN 9789811250743 (hardcover) |
    ISBN 9789811250750 (ebook for institutions) | ISBN 9789811250767 (ebook for individuals)
Subjects: LCSH: Differential equations.
Classification: LCC QA372 .H85 2022 | DDC 515/.352--dc23/eng20220621
LC record available at https://lccn.loc.gov/2022013691

British Library Cataloguing-in-Publication Data
A catalogue record for this book is available from the British Library.

For any available supplementary material, please visit
https://www.worldscientific.com/worldscibooks/10.1142/12682#t=suppl

Printed in Singapore

Preface to the First Edition

Ordinary Differential Equations (ODE) is a well-developed discipline. It is a classical subject for most graduate students to study if they want to learn analysis and applied mathematics. There are many standard texts of ODE, for instance, the books, Coddington and Levinson [CL], Hartman [Ha], J. Hale [H1], Miller and Michel [MM]. Based on my past twenty years of teaching experience of the graduate course in ODE, I find that for most graduate students these books are either too difficult to understand or too abstract to follow. This is the reason why I have decided to write my own ODE lecture notes. This book is aimed to present the classical ODE materials in an easy way with rigorous proofs. In this text many examples from mathematical biology and physical science to interpret the meanings and the applications of the theorems have been given. Besides the classical materials, Brouwer degree in finite dimensional space following the index theory has been discussed in Chapter 8. Brouwer degree is a powerful topological tool for studying the nonlinear analysis. Introducing Brouwer degree paves the way to the Leray-Schauder degree in infinite dimensional space. In Chapter 9 we introduce perturbation method, the regular perturbation and singular perturbation methods which are very important in applied mathematics.

This book is designed for a one-year graduate course. Let's briefly state the contents of each chapter in this book.

Chapter 1: Introduction. In this chapter important examples of nonlinear systems of ODEs from physical science and mathematical biology have been introduced.

Chapter 2: Fundamental Theory. In this chapter I first prove the local existence and uniqueness of solutions of initial value problem of ODE. Then the continuation of the solutions and the global existence of the solutions are studied. The continuous dependence on initial conditions and parameters is proved. Applying continuous dependence on initial conditions and parameters gives the results of scalar differential inequalities. The differential inequalities for systems are also discussed.

Chapter 3: Linear Systems. In this chapter we discuss the fundamental matrices of general linear systems $x' = A(t)x$. Then for the linear system with constant coefficients $x' = Ax$ we introduce exponential matrix e^{At} as a fundamental matrix. For periodic linear system $x' = A(t)x$, $A(t + \omega) = A(t)$, $\omega > 0$, we study the structure of the solutions by proving Floque's Theorem. For two-dimensional linear autonomous $\begin{pmatrix} x \\ y \end{pmatrix}' = \begin{pmatrix} a & b \\ c & d \end{pmatrix} \begin{pmatrix} x \\ y \end{pmatrix}$, we introduce the important notions of node, focus, center and saddle. Similar to linear algebra, we introduce adjoint systems of $x' = Ax$. We obtain important results similar to the fundamental theorem of linear algebra.

Chapter 4: Stability of Nonlinear Systems. In this chapter we introduce concepts of stability, asymptotic stability and instability of an equilibrium solution. To verify these stabilities, we introduce the method of linearization by checking the eigenvalues of variational matrix. When the equilibrium is a saddle, we prove the existence of stable and unstable manifolds. For periodic orbit, we introduce the concept of orbit stability which is different from the asymptotic stability. The results related to Floquet multipliers have been discussed in Chapter 3. In Section 4.5 we study the existence of travelling wave solutions of the well-known partial differential equations, the Fisher's equation and the bistable equation by applying the Stable Manifold Theorem.

Chapter 5: Method of Lyapunov Functions. In this chapter we first introduce the concept of dynamical system, α-limit set, ω-limit set and their properties in abstract setting. In fact, autonomous ODE system is an important example of dynamical system. Then we discuss Lyapunov functions and their use in determining the global asymptotic stability or the domain of attraction of a locally stable equilibrium. We prove the LaSalle's invariance principle which explains how to apply Lyapunov function to locate ω-limit set.

Chapter 6: Two-Dimensional Systems. In this chapter we prove one of the most important theorems in nonlinear dynamics, Poincaré-Bendixson Theorem which states that a bounded trajectory of a two-dimensional system either converges to an equilibrium or approaches a limit cycle. We also discuss Dulac Criterion which provides a method to eliminate the existence of periodic solutions. In Section 6.2 we discuss the Levinson-Smith Theorem which proves the uniqueness of limit cycle for Lienard equation. An important example of Lienard equation is van der Pol equation. We discuss the relaxation property of the van der Pol oscillators. In Section 6.3, we introduce Hopf Bifurcation Theorem which is an important tool to detect the existence of periodic solution for autonomous system $x' = f(\lambda, x)$ where λ is a bifurcation parameter.

Chapter 7: Second Order Linear Equations. In this chapter we first discuss the Sturm's Comparison Theorem which is an important tool to study oscillation properties of a solution of a second order linear equation. Similar to the eigenvalue problems $Ax = \lambda x$, we consider the Sturm-Liouville boundary value problem and prove the existence of infinitely many discrete eigenvalues and show that the corresponding eigenfunctions satisfy the "node" properties. For second order nonlinear equation $x'' = f(t, x, x')$, $x(a) = x_0$, $x(b) = x_1$, we introduce Green's function so that we convert the boundary value problem into an integral equation.

Chapter 8: The Index Theory and Brouwer Degree. In this chapter we first introduce the Index $I_f(C)$ for a vector field f in the plane and a simple closed curve C. Then we prove various properties of index and apply index theory to prove fundamental theorem of algebra and Brouwer fixed point theorem. In Section 8.2 we generalize the index $I_f(C)$ to the topological degree in finite dimensional space, the Brouwer degree in $\mathbb{R}^n$. Topological degree is an important tool in studying nonlinear problems.

Chapter 9: Introduction to Regular and Singular Perturbation Methods. Perturbation methods are very important in applied mathematics. In this chapter we explain why regular perturbation method works by using Implicit Function Theorem and Fredholm Alternatives. Then we discuss singular perturbation methods for both boundary value problems and initial value problems.

Chapter 10: Introduction to Monotone Dynamical System. In this chapter we state without proof the main results of Monotone Dynamical Systems which was developed by Morris Hirsch in 1980. In the first section we apply the theory to dynamics of cooperative systems of n-species. Then we prove Poincaré-Bendixson Theorem holds for 3-species competition system. In section two we discuss the uniform persistence for n-species population interaction model. We prove Butler-McGehee Lemma which is a basic ingredient in proving the Theorem of Uniform Persistence. In section three we introduce my joint work with Paul Waltman, the competition of two species for a single nutrient with inhibition. In this model, we illustrate the application of Poincaré-Bendixson Theorem and Theorem of Persistence to the model. In section four, we state three abstract theorems in the frame of ordered Banach space for two species competition models. The theorems can be applied for the cases of competitive exclusion, stable coexistence and bistability. In summary Monotone Dynamics was developed in the late 20th century. It is a powerful tool to prove the global asymptotic behavior for some important models in population biology.

Chapter 11: Introduction to Hamiltonian Systems. We begin with classic examples from classical mechanics. Then we discuss fundamental matrices and stability of linear Hamiltonian systems. As preparation, we discuss properties of skew-symmetric matrices and symplectic matrices. For investigations of nonlinear Hamiltonian systems, we introduce first integrals, Poisson bracket, Poisson theorem, together with several applications to mechanical systems. Next we introduce symplectic transformations and their connections with Hamiltonian systems. We show that a system is locally Hamiltonian if and only if the induced local flow is symplectic. Action-angle variables were introduced, brief remarks on perturbations of integrable systems were made. The last section introduces generating functions and the Hamilton-Jacobi method. We use simple examples to illustrate how Hamilton-Jacobi's method can be used to find suitable generating functions so as to solve nonlinear Hamiltonian systems.

Acknowledgments: I would like to express my gratitude to my wife Taily for her support in the past ten years; to my former Ph.D. thesis adviser Professor Paul Waltman for his encouragement to write this book. I want to thank my colleagues Professor Wen-Wei Lin, Professor Shin-Hwa Wang, Professor Shuh-Jye Chern of National Tsing Hua University, Professor Chin-An Wang, Professor Dong-Ho Tsai of National Chung-Cheng University for using this text in their graduate ODE courses and for giving me some suggestions to improve the text. I also want to thank my post doctor Dr. Cheng-Che Li and my Ph.D. student Yun-Huei Tzeng for their patience in proofreading the draft. Finally I especially want to thank Miss Alice Feng for typing this text in the past three years. Without her help, it is impossible to have this book published.

S. B. Hsu
National Tsing Hua University
Hsinchu, Taiwan
Aug. 30, 2005

Preface to the Second Edition

This new Second Edition contains corrections, additional materials and suggestions from various readers and users. I have added several new exercises for each chapter. A new chapter, Chapter 10, on Monotone Dynamical Systems is added to take into account of new developments in ordinary differential equations and dynamical systems. The description of Chapter 10 is presented in the preface of the 1st edition.

I would like to take this opportunity to thank my colleague Prof. Shin-Hwa Wang and my Ph.D. student Chiu-Ju Lin who have pointed out errors in the first edition and make useful suggestions.

The author is indebted to the Publisher as well as the readers for their invaluable assistance in the publication of the new edition.

S. B. Hsu
National Tsing Hua University
Hsinchu, Taiwan
Feb. 27, 2013

Preface to the Third Edition

This new Third Edition contains corrections, additional materials and suggestions from various readers and users. We have added several new exercises to each chapter; moving several complicated proofs to Appendices. A new chapter, Chapter 11, "An Introduction to Hamiltonian Systems", written by Professor Kuo-Chang Chen is added to take celestial mechanics into the material of ordinary differential equations and dynamical systems. The description of Chapter 11 is presented in the preface of the 1st edition.

The authors are indebted to the Publisher as well as the readers for their invaluable assistance in the publication of the new edition.

S. B. Hsu and K. C. Chen
National Tsing Hua University
Hsinchu, Taiwan
Mar. 31, 2021

Contents

Chapter 1

INTRODUCTION

1.1 Where do ODEs arise

The theory of ordinary differential equations deals with the large time behavior of the solution $x(t, x_0)$ of the initial value problem (I.V.P.) of the first order system of differential equations:

$$\frac{dx_1}{dt} = f_1(t, x_1, x_2, \cdots, x_n)$$

$$\vdots$$

$$\frac{dx_n}{dt} = f_n(t, x_1, x_2, \cdots, x_n)$$

$$x_i(0) = x_{i0}, \quad i = 1, 2, \cdots, n$$

or in vector form

$$\frac{dx}{dt} = f(t, x), \tag{1.1}$$

$$x(0) = x_0$$

where $x = (x_1, \cdots, x_n)$, $f = (f_1, \cdots, f_n)$, $f : D \to \mathbb{R}^n$, D is open in $\mathbb{R} \times \mathbb{R}^n$.

If the right-hand side of (1.1) is independent of time t, i.e.,

$$\frac{dx}{dt} = f(x), \quad x \in \Omega \subseteq \mathbb{R}^n, \tag{1.2}$$

then we say that (1.2) is an autonomous system. In this case, we call f a vector field on its domain Ω. If the right-hand side depends on time t, then we say that (1.1) is a nonautonomous system. The most important nonautonomous system is the periodic system, i.e., $f(t, x)$ satisfies

$$f(t + w, x) = f(t, x),$$

1

for some $w > 0$ (w is called the period). If $f(t, x) = A(t)x$ where $A(t) \in \mathbb{R}^{n \times n}$, then we say that

$$\frac{dx}{dt} = A(t)x \tag{1.3}$$

is a linear system of differential equations. It is easy to verify that if $\varphi(t), \psi(t)$ are solutions of (1.3), then $\alpha\varphi(t) + \beta\psi(t)$ is also a solution of the linear system (1.3) for $\alpha, \beta \in \mathbb{R}$. The system

$$\frac{dx}{dt} = A(t)x + g(t) \tag{1.4}$$

is called a linear system with nonhomogeneous part $g(t)$. If $A(t) \equiv A$, then

$$\frac{dx}{dt} = Ax \tag{1.5}$$

is a linear system with constant coefficients. We say that system (1.1) is nonlinear if it is not linear. It is usually much harder to analyze nonlinear systems than the linear ones. The main difference between linear systems and nonlinear systems is the superposition principle. The superposition principle states that the linear combination of solutions is also a solution. For linear systems (1.3), (1.4), (1.5) as we can see in Chapter 3 that we have nice solution structures. However, nonlinear systems arise in many areas of science and engineering. It is still a great challenge to understand the nonlinear phenomena. In the following we present some important examples of differential equations from physics, chemistry and biology.

Example 1.1.1 $m\ddot{x} + c\dot{x} + kx = 0$.

The equation describes the motion of a spring with damping and restoring forces. Applying Newton's law, $F = ma$, we have

$$ma = m\ddot{x} = F = -c\dot{x} - kx = \text{Friction} + \text{restoring force}.$$

Let $y = \dot{x}$. Then we convert the equation into a first order system of two equations

$$\begin{cases} \dot{x} = y, \\ \dot{y} = -\frac{c}{m}y - \frac{k}{m}x. \end{cases}$$

The analysis of the model can be found in Example 5.2.3, Chapter 5.

Example 1.1.2 $m\ddot{x} + c\dot{x} + kx = F\cos wt$. The equation describes the motion of a spring with external periodic force. It can be rewritten as

$$\begin{pmatrix} \dot{x} \\ \dot{y} \end{pmatrix} = \begin{pmatrix} y \\ -\frac{c}{m}y - \frac{k}{m}x + F\cos wt \end{pmatrix}.$$

If $c = 0$ and $w = \sqrt{k/m}$, then we have "resonance" ([BDiP], p. 184). The resonance is discussed in Example 3.6.2 and Exercise 3.25, Chapter 3.

Example 1.1.3 Electrical Networks ([BDiP], p. 184).

Let $Q(t)$ be the charge in the RLC circuit at time t. Use the Kirchoff's 2nd law: In a closed circuit, the impressed voltage equals the sum of the voltage drops in the rest of the circuit.

Fig. 1.1

(1) The voltage drop across a resistance of R(ohms) equals RI (Ohm's law).
(2) The voltage drop across an inductance of L(henrys) equals $L\frac{dI}{dt}$.
(3) The voltage drop across a capacitance of C(farads) equals Q/C.

Hence

$$E(t) = L\frac{dI}{dt} + RI + \frac{Q}{C}.$$

Since $I(t) = \frac{dQ}{dt}$, it follows that

$$L\frac{d^2Q}{dt^2} + R\frac{dQ}{dt} + \frac{1}{C}Q = E(t).$$

The equation of electric network is similar to that of mechanical vibration with $L \to m$, $R \to c$ and $\frac{1}{C} \to k$.

Example 1.1.4 Van der Pol Oscillator ([Kee1], p. 481, [HK], p. 172)

$$u'' + \epsilon u'(u^2 - 1) + u = 0, \quad 0 < \epsilon \ll 1.$$

Let $E(t) = \frac{u'^2}{2} + \frac{u^2}{2}$ be the energy. Then

$$E'(t) = u'u'' + uu' = u'(-\epsilon u'(u^2 - 1) - u) + uu',$$

$$= -\epsilon(u')^2(u^2 - 1) = \begin{cases} < 0, & |u| > 1, \\ > 0, & |u| < 1. \end{cases}$$

Hence the oscillator is "self-excited".

Example 1.1.5 Van der Pol Oscillator with periodic forcing

$$u'' + \epsilon u'(u^2 - 1) + u = A \cos wt.$$

This is the equation Cartwright and Littlewood studied in 1945 and it led Smale to construct Smale's horseshoe in 1960. It is one of the model equations in chaotic dynamical systems [Sma]. The analysis related to the model can be found in Section 8.3, Chapter 8.

Example 1.1.6 Second order conservative system

$$\ddot{x} + g(x) = 0,$$

or equivalently

$$\dot{x}_1 = x_2,$$
$$\dot{x}_2 = -g(x_1).$$

The energy $E(x_1, x_2) = \frac{1}{2}x_2^2 + V(x_1)$, where $V(x_1) = \int_0^{x_1} g(s)ds$, is the potential. Then the energy E satisfies

$$\frac{d}{dt}E = 0.$$

The analysis of the equation can be found in Example 5.2.2, Chapter 5.

Example 1.1.7 Duffing's equation

$$\ddot{x} + (x^3 - x) = 0.$$

The potential $V(x) = -(1/2)x^2 + (1/4)x^4$ is a double-well potential (see Fig. 1.2).

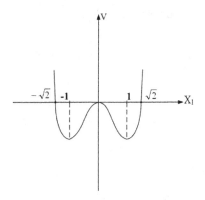

Fig. 1.2

Example 1.1.8 Duffing's equation with damping and periodic forcing.

$$\ddot{x} + \beta \dot{x} + (x^3 - x) = A \cos wt.$$

This is also a typical model equation in chaotic dynamical systems.

Example 1.1.9 Simple pendulum equation (see Fig. 1.3)

$$\frac{d^2\theta}{dt^2} + \frac{g}{\ell} \sin \theta = 0.$$

$$F = ma,$$
$$-mg \sin \theta = m\ell \cdot \frac{d^2\theta}{dt^2}.$$

We analyze the simple pendulum equation in Example 5.3.1, Chapter 5.

Fig. 1.3

Example 1.1.10 Gradient Vector Fields

$$\dot{x} = -\nabla U(x),$$

where $U : \mathbb{R}^n \to \mathbb{R}$ is a C^2 function. In classical mechanics, U is the potential energy. We note that $U(x, t)$ satisfies

$$\frac{d}{dt} U(x(t)) = -|\nabla U(x(t))|^2 \le 0.$$

We analyze the gradient vector field in Section 5.4, Chapter 5.

Example 1.1.11 N-Body Problem
By Newton's law, the equation is

$$m_i \ddot{\vec{q}}_i = -G \sum_{j \ne i}^{n} \frac{m_i m_j (\vec{q}_i - \vec{q}_j)}{r_{ij}^3}, \ r_{ij} := |\vec{q}_i - \vec{q}_j|.$$

Example 1.1.12 Lorentz equation ([Str], p. 301, [V]) (also see Example 6.3.4, Chapter 6)

$$\dot{x} = \sigma(y - x),$$
$$\dot{y} = rx - y - xz \text{ where } \sigma, r, b > 0,$$
$$\dot{z} = xy - bz.$$

When $\sigma = 10$, $b = \frac{8}{3}$, $r = 28$, $(x(0), y(0), z(0)) \approx (0, 0, 0)$, we have the "butterfly effect" phenomenon.

Example 1.1.13 Michaelis-Menten Enzyme Kinetics ([Kee1] p. 511, [LS] p. 302).
Consider the conversion of a chemical substrate S to a product P by enzyme catalysis. The reaction scheme

$$E + S \underset{k_{-1}}{\overset{k_1}{\rightleftharpoons}} ES \overset{k_2}{\to} E + P$$

was proposed by Michaelis and Menten in 1913. The law of Mass action states that the rate of reaction is proportional to the concentrations of reactants. Then by the law of mass action, we have the following equations:

$$\frac{d}{dt}[E] = -k_1[E][S] + k_{-1}[ES] + k_2[ES],$$
$$\frac{d}{dt}[S] = -k_1[E][S] + k_{-1}[ES],$$
$$\frac{d}{dt}[ES] = k_1[E][S] - k_{-1}[ES] - k_2[ES],$$
$$\frac{d}{dt}[P] = k_2[ES],$$

with initial concentrations

$$[E](0) = E_0, \ [S](0) = S_0, \ [ES](0) = [P](0) = 0.$$

Since

$$\frac{d}{dt}[ES] + \frac{d}{dt}[E] = 0,$$

then

$$[ES] + [E] \equiv E_0.$$

Let

$$u = [ES]/C_0, \ v = [S]/S_0, \ \tau = k_1 E_0 t,$$

and

$$\kappa = (k_{-1} + k_2)/k_1 S_0, \ \epsilon = C_0/S_0, \ \lambda = \frac{k-1}{k_{-1} + k_2}, \ C_0 = E_0/1 + \kappa.$$

Then we have the following equations

$$\epsilon \frac{du}{d\tau} = v - \frac{(v + \kappa)u}{1 + \kappa}, \quad 0 < \epsilon \ll 1,$$

$$\frac{dv}{d\tau} = -v + \frac{(v + \kappa\lambda)u}{1 + \kappa},$$

$$u(0) = 0, \ v(0) = 1.$$

We shall introduce method of singular perturbation to study this system in Chapter 9.

Example 1.1.14 Belousov-Zhabotinskii Reaction [Murr]

$$\epsilon \frac{dx}{dt} = qy - xy + x(1 - x),$$

$$\delta \frac{dy}{dt} = -qy - xy + 2fz,$$

$$\frac{dz}{dt} = x - z,$$

where ϵ, δ, q are small, $f \approx 0.5$. This is an important model of chemical oscillator. The analytic work can be found in Section 6, Chapter 3 in [Smi].

Example 1.1.15 Logistic equation.
Let $x(t)$ be the population of a species. Then $x(t)$ satisfies

$$\frac{dx}{dt} = rx - bx^2 = rx\left(1 - \frac{x}{K}\right), \quad r, b > 0,$$

where K and r are the carrying capacity and the intrinsic growth rate of the species respectively.

Example 1.1.16 The Lotka-Volterra model for Predator-Prey interaction. Let $x(t), y(t)$ be the population sizes of prey and predator at time t respectively ([Murr] pp. 124 and 62). Then we have the following two predator-prey models of Lotka-Volterra type:

$$\frac{dx}{dt} = ax - bxy,$$
$$a, b, c, d > 0,$$
$$\frac{dy}{dt} = cxy - dy,$$

(the analysis of above model, see Example 5.2.4, Chapter 5) and

$$\begin{cases} \frac{dx}{dt} = rx(1 - \frac{x}{K}) - bxy, \\ \\ \frac{dy}{dt} = cxy - dy. \end{cases}$$

(The analysis of above model, see Example 5.2.7, Chapter 5.)
The first model assumes the prey species grows exponentially in the absence of predation while the second assumes the prey species grows logistically with carrying capacity K.

Example 1.1.17 The Lotka-Volterra two species competition model ([Murr]).
Let $x_i(t)$, $i = 1, 2$, be the population of i-th competing species. We assume that the i-th species grows logistically in the absence of competition with intrinsic rate r_i and carrying capacity K_i. In the following model, α, $\beta > 0$ are called the competition coefficients. The model takes the form:

$$\frac{dx_1}{dt} = r_1 x_1 \left(1 - \frac{x_1}{K_1}\right) - \alpha x_1 x_2,$$
$$\frac{dx_2}{dt} = r_2 x_2 \left(1 - \frac{x_2}{K_2}\right) - \beta x_1 x_2,$$
$$x_1(0) > 0, \ x_2(0) > 0.$$

We analyze the model in Example 4.2.4, Chapter 4 and Example 5.2.8, Chapter 5.

Example 1.1.18 Rosenzweig-McArthur Predator-Prey model [M-S].
Let $x(t)$, $y(t)$ be population density of prey and predator at time t respectively. Assume that $x(t)$ grows logistically in the absence of predation with intrinsic growth rate r and carrying capacity K. Predator's growth rate $\frac{mx}{a+x}$ is of Holling-type II functional response where m is the maximal growth rate and a is the half-saturation constant. The positive constants d, c represent predator's death rate and conversion constant. The model takes the form

$$\frac{dx}{dt} = rx\left(1 - \frac{x}{K}\right) - c\frac{mx}{a+x}y,$$

$$\frac{dy}{dt} = \left(\frac{mx}{a+x} - d\right)y,$$

$$x(0) > 0, \ y(0) > 0.$$

The analysis of the Rosenzweig-McArthur model can be found in Example 4.2.3, Chapter 4.

Example 1.1.19 Food chain model with Holling-type II functional responses [KH].
Let $x(t)$, $y(t)$, $z(t)$ be population densities of prey, predator and top predator respectively. The model takes the form

$$\frac{dx}{dt} = rx\left(1 - \frac{x}{K}\right) - c_1\frac{m_1 x}{a_1 + x}y,$$

$$\frac{dy}{dt} = \left(\frac{m_1 x}{a_1 + x} - d_1\right)y - c_2\frac{m_2 y}{a_2 + y}z,$$

$$\frac{dz}{dt} = \left(\frac{m_2 y}{a_2 + y} - d_2\right)z,$$

$$x(0) > 0, \ y(0) > 0, \ z(0) > 0.$$

This is a well-known model in mathematical ecology with chaotic dynamics.

Example 1.1.20 SIR model in Epidemics [Murr] p. 612.
Let $S(t)$, $I(t)$, $R(t)$ be the population densities of susceptible, infective and removed respectively. The model takes the form

$$\frac{dS}{dt} = -rSI,$$

$$\frac{dI}{dt} = rSI - aI,$$

$$\frac{dR}{dt} = aI,$$

$S(0) = S_0 > 0$, $I(0) = I_0 > 0$, $R(0) = 0$ where $r > 0$ is the infection rate and $a > 0$ is the removable rate of infectives. For analytic results, see Exercise 4.23.

Example 1.1.21 Model of two predators competing for a prey [HHW].

$$\frac{dS}{dt} = rS\left(1 - \frac{S}{K}\right) - \frac{1}{y_1}\frac{m_1 S}{a_1 + S}x_1 - \frac{1}{y_2}\frac{m_2 S}{a_2 + S}x_2,$$

$$\frac{dx_1}{dt} = \left(\frac{m_1 S}{a_1 + S} - d_1\right)x_1,$$

$$\frac{dx_2}{dt} = \left(\frac{m_2 S}{a_2 + S} - d_2\right)x_2,$$

$$S(0) > 0, \ x_1(0) > 0, \ x_2(0) > 0.$$

Chapter 2

FUNDAMENTAL THEORY

2.1 Introduction and Preliminaries

In this chapter we shall study the fundamental properties of the initial value problem (I.V.P.)

$$\begin{cases} \frac{dx}{dt} = f(t,x), & f : D \subseteq \mathbb{R} \times \mathbb{R}^n \to \mathbb{R}^n, \\ x(t_0) = x_0, \end{cases} \tag{2.1}$$

where D is an open set of $\mathbb{R} \times \mathbb{R}^n$ containing (t_0, x_0). We shall answer the following questions:

(Q1) What is the least condition on $f(t,x)$ to ensure the local existence of a solution of I.V.P. (2.1)?

(Q2) When is the solution of (2.1) unique?

(Q3) When does the solution of (2.1) exist globally? That is, the solution's maximal interval of existence is the whole real line $\mathbb{R}$?

(Q4) Is the initial value problem well-posed? That is, the solutions of (2.1) continuously depend on the initial conditions and parameters.

Before we study these questions, we need the following preliminaries: Let $x \in \mathbb{R}^n$. $|\ |$ is a norm if the function $|\ | : \mathbb{R}^n \longrightarrow \mathbb{R}_+$ satisfies:

(i) $|x| \geq 0$ and $|x| = 0$ iff $x = 0$,

(ii) $|\alpha x| = |\alpha||x|$, $\alpha \in \mathbb{R}$, $x \in \mathbb{R}^n$,

(iii) $|x + y| \leq |x| + |y|$.

Three most commonly used norms of a vector $x = (x_1, \cdots, x_n)$ are (from [IK])

$$|x|_\infty = \sup_{1 \leq i \leq n} |x_i|,$$
$$|x|_1 = \sum_{i=1}^{n} |x_i|,$$
$$|x|_2 = \left(\sum_{i=1}^{n} |x_i|^2 \right)^{1/2}.$$

11

It is easy to verify that these norms satisfy

$$|x|_2 \leq |x|_1 \leq \sqrt{n}|x|_2,$$
$$|x|_\infty \leq |x|_2 \leq \sqrt{n}|x|_\infty,$$
$$|x|_\infty \leq |x|_1 \leq n|x|_\infty,$$

i.e., these norms are equivalent. For $A \in \mathbb{R}^{n \times n}$, given norm $|\ |$, the norm of A is defined as follows:

$$\|A\| = \sup \frac{|Ax|}{|x|}.$$

It is well known from linear algebra ([IK] p. 9) that for $A, B \in \mathbb{R}^{n \times n}$, $\alpha \in \mathbb{R}$

$$|Ax| \leq \|A\||x|,$$
$$\|AB\| \leq \|A\|\|B\|,$$
$$\|A + B\| \leq \|A\| + \|B\|,$$
$$\|\alpha A\| = |\alpha|\|A\|.$$

From [IK], it follows that

$$\|A\|_2 = \left(\max_{1 \leq i \leq n} \left\{ |\lambda_i| : \lambda_i \text{ is an eigenvalue of } A^T A \right\} \right)^{\frac{1}{2}},$$

$$\|A\|_1 = \sup_k \sum_{i=1}^n |a_{ik}|,$$

$$\|A\|_\infty = \sup_i \sum_{k=1}^n |a_{ik}|.$$

From (iii) we obtain by induction

$$|x_1 + x_2 + \cdots + x_m| \leq |x_1| + |x_2| + \cdots + |x_m|.$$

From this inequality and (ii) we can deduce the inequality

$$\left| \int_a^b x(t)dt \right| \leq \int_a^b |x(t)|dt,$$

for any vector function $x(t)$ which is continuous on $[a, b]$. In fact, let $\delta = \frac{b-a}{m}$ and $t_k = a + k\delta$, $k = 1, \cdots, m$. Then

$$\left| \int_a^b x(t)dt \right| = |\lim_{m \to \infty} \sum_{k=1}^m x(t_k)\delta|$$
$$\leq \lim_{m \to \infty} \sum_{k=1}^m |x(t_k)|\delta = \int_a^b |x(t)|dt.$$

A vector space X, not necessarily a finite dimensional vector space, is called a normed space if a real-valued function $|x|$ is defined for all $x \in X$ having properties (i)–(iii). A Banach space is a *complete* normed space, i.e., any Cauchy sequence $\{x_n\}$ is a convergent sequence.

Let $I = [a,b]$ be a bounded, closed interval and $C(I) = \{f \,|\, f : I \to \mathbb{R}^n$ is continuous $\}$ with the norm

$$\| f \|_\infty = \sup_{a \leq t \leq b} |f(t)|,$$

then $\| \ \|_\infty$ is a norm of $C(I)$. We note that $f_m \to f$ in $C(I)$ means $\| f_m - f \|_\infty \to 0$, i.e., $f_m \to f$ uniformly on I.

Theorem 2.1.1 ([Apo] p. 222).
$C(I)$ *is a Banach space.*

In the following, we need the notion of equicontinuity. A family of functions $F = \{f\}$ defined on an interval I is said to be equicontinuous on I if given $\varepsilon > 0$ there exists a $\delta = \delta(\epsilon) > 0$, independent of $f \in F$, and such that for any $t, \bar{t} \in I$

$$|f(t) - f(\bar{t})| < \epsilon \quad \text{whenever} \quad |t - \bar{t}| < \delta, \ f \in F.$$

We say that F is uniformly bounded if there exists $M > 0$ such that $\| f \|_\infty < M$ for all $f \in F$. The following Ascoli-Arzela's Theorem is a generalization of Bolzano-Weierstrass Theorem in $\mathbb{R}^n$, which deals the property of compactness in the Banach space $C(I)$.

Theorem 2.1.2 (Ascoli-Arzela) [Cop].
Let $f_m \in C(I), I = [a,b]$. *If* $\{f_m\}_{m=1}^\infty$ *is equicontinuous and uniformly bounded, then there exists a convergent subsequence* $\{f_{m_k}\}$.

Proof. Let $\{\gamma_k\}$ be the set of rational numbers in I. Since $\{f_m(\gamma_1)\}$ is a bounded sequence, we extract a subsequence $\{f_{1m}\}_{m=1}^\infty$ of $\{f_m\}$ such that $f_{1m}(\gamma_1)$ converges to a point denoted as $f(\gamma_1)$. Similarly $\{f_{1m}(\gamma_2)\}_{m=1}^\infty$ is a bounded sequence, we can extract a subsequence $\{f_{2m}\}$ of $\{f_{1m}\}$ such that $f_{2m}(\gamma_2)$ converges to $f(\gamma_2)$. Continue this process, inductively, we can extract a subsequence $\{f_{km}\}$ of $\{f_{k-1,m}\}$ such that $f_{k,m}(\gamma_k)$ converges to a point, denoted as $f(\gamma_k)$, by the diagonal process,

$$f_{11}, f_{12}, f_{13}, f_{14}, \cdots$$
$$\searrow$$
$$f_{21}, f_{22}, f_{23}, f_{24}, \cdots$$
$$\vdots \quad \searrow$$
$$\vdots \qquad \searrow$$
$$f_{k1}, f_{k2}, \cdots f_{kk}, \cdots$$

$f_{mm}(\gamma_k) \to f(\gamma_k)$ as $m \to \infty$ for all $k = 1, 2, \cdots$. Let $g_m \stackrel{\text{def}}{=} f_{mm}$. Claim: $\{g_m\}$ is a Cauchy sequence. Then $\{g_m\}$ is a desired convergent sequence. Since $\{g_m(\gamma_j)\}$ converges for each j, given $\varepsilon > 0$ there exists $M_j(\varepsilon) > 0$ such that

$$|g_m(\gamma_j) - g_i(\gamma_j)| < \varepsilon \text{ for } m, i \geq M_j(\varepsilon).$$

By the equicontinuity of $\{g_m\}$, there exists $\delta > 0$ such that for each i

$$|g_i(x) - g_i(y)| < \varepsilon \text{ for } |x - y| < \delta.$$

From Heine-Borel's Theorem, the open covering $\{B(\gamma_j, \delta)\}_{j=1}^{\infty}$ of $[a, b]$, has a finite subcovering $\{B(\gamma_{j_k}, \delta)\}_{k=1}^{L}$. Let $M(\varepsilon) = \max\{M_{j_1}(\varepsilon), \cdots, M_{j_L}(\varepsilon)\}$. Then $x \in [a, b]$ implies $x \in B(\gamma_{j_\ell}, \delta)$, for some ℓ, $1 \leq \ell \leq L$. For $m, i \geq M(\varepsilon)$,

$$|g_m(x) - g_i(x)| \leq |g_m(x) - g_m(\gamma_{j\ell})|$$
$$+|g_m(\gamma_{j\ell}) - g_i(\gamma_{j\ell})| + |g_i(\gamma_{j\ell}) - g_i(x)|$$
$$< \varepsilon + \varepsilon + \varepsilon = 3\varepsilon.$$

Thus $\{g_m\}$ is a Cauchy sequence and we have completed the proof of the theorem by the completeness of $C(I)$. □

2.2 Local Existence and Uniqueness of Solutions of I.V.P.

Before we prove the theorems for the local existence of solutions of I.V.P. (2.1), we present some examples to illustrate local existence.

Example 2.2.1 The system

$$\begin{cases} \frac{dx}{dt} = 1 + x^2, \\ x(0) = 0, \end{cases}$$

has a unique solution $x(t) = \tan t$ defined on $(-\pi/2, \pi/2)$. The solution $x(t)$ is not defined outside the interval $(-\pi/2, \pi/2)$.

Example 2.2.2 Consider the system

$$\begin{cases} \frac{dx}{dt} = x^2, \\ x(0) = x_0. \end{cases}$$

By separation of variables, the solution $x(t) = \frac{1}{\frac{1}{x_0}-t}$. If $x_0 > 0$ then $x(t)$ is defined on the interval $(-\infty, \frac{1}{x_0})$. If $x_0 < 0$ then $x(t)$ is defined on the interval $(\frac{1}{x_0}, \infty)$. If $x_0 = 0$ then $x(t) = 0$ for all t.

Example 2.2.3 Consider the I.V.P. $\frac{dx}{dt} = f(t,x) = \begin{cases} \sqrt{x}, & x \geq 0 \\ 0, & x < 0 \end{cases}$, $x(0) = 0$, then there are infinitely many solutions

$$x(t) = \begin{cases} \frac{(t-c)^2}{4}, & t \geq c \geq 0 \\ 0, & t \leq c \end{cases}$$

$x(t) \equiv 0$ is also a solution. We note that $f(t,x)$ is not Lipschitz at $x = 0$. (Prove it!)

Remark 2.2.1 From the application's viewpoint, most of I.V.P.'s from the physical world should have existence and uniqueness properties and the solution is defined globally, i.e., the interval of existence of the solution is $(-\infty, \infty)$. A nice thing about the solution of an ODE system is that the time can be "reversed", i.e., $x(t)$ is defined for $t < t_0$, where t_0 is the initial time.

Lemma 2.2.1 *If the function $f(t,x)$ and $x(t)$ are continuous, then the initial value problem (2.1) is equivalent to the integral equation.*

$$x(t) = x_0 + \int_{t_0}^{t} f(s, x(s)) ds. \tag{2.2}$$

Proof. Obviously, a solution $x(t)$ of (2.1) satisfies (2.2). Conversely, let $x(t)$ be a solution of (2.2). Substituting $t = t_0$ in (2.2), we obtain $x(t_0) = x_0$. Moreover, from the assumptions we have that $f(t, x(t))$ is continuous. From (2.2) and fundamental theorem of Calculus it follows that $x(t)$ is differentiable and $x'(t) = f(t, x(t))$. $\qquad \square$

Next we state a theorem for the local existence of initial value problem (2.1) and defer its proof to Appendix A.1.

Theorem 2.2.1 *For any $\varepsilon > 0$, there exists an ε-approximate solution of I.V.P. (2.1) on the interval $I = \{t : |t - t_0| \leq c\}$.*

Theorem 2.2.2 *Let $f \in C(D)$, $(t_0, x_0) \in D$. Then the I.V.P. (2.1) has a solution on the interval $I = [t_0 - c, t_0 + c]$ for some $c > 0$.*

To prove the uniqueness of the solutions of I.V.P., we need the following Gronwall inequality.

Theorem 2.2.3 *Let $\lambda(t)$ be a real value continuous function and $\mu(t)$ a nonnegative continuous function on $[a, b]$. If a continuous function $y(t)$ has the property that*

$$y(t) \leq \lambda(t) + \int_a^t \mu(s)y(s)ds, \quad a \leq t \leq b, \tag{2.3}$$

then on $[a, b]$ we have

$$y(t) \leq \lambda(t) + \int_a^t \lambda(s)\mu(s) \exp\left(\int_s^t \mu(\tau)d\tau\right) ds.$$

In particular if $\lambda(t) \equiv \lambda$ is a constant,

$$y(t) \leq \lambda \exp\left(\int_a^t \mu(s)ds\right).$$

Proof. Let

$$z(t) = \int_a^t \mu(s)y(s)ds.$$

Then $z(t)$ is differentiable and from (2.3) we have

$$z'(t) - \mu(t)z(t) \leq \lambda(t)\mu(t). \tag{2.4}$$

Let

$$w(t) = z(t) \exp\left(-\int_a^t \mu(\tau)d\tau\right).$$

Then from (2.4), it follows that

$$w'(t) \leq \lambda(t)\mu(t) \exp\left(-\int_a^t \mu(\tau)d\tau\right).$$

Since $w(a) = 0$, integrating above from a to t yields

$$w(t) \leq \int_a^t \lambda(s)\mu(s) \exp\left(-\int_a^s \mu(\tau)d\tau\right) ds.$$

By the definition of $w(t)$, it follows that

$$z(t) \leq \int_a^t \lambda(s)\mu(s) \exp\left(\int_s^t \mu(\tau)d\tau\right) ds.$$

Then from (2.3) and above inequality we complete the proof.   □

Remark 2.2.2 Gronwall inequality is frequently used to estimate a bound on the solutions of an ordinary differential equation.

The next theorem concerns the uniqueness of the solutions of I.V.P.

Theorem 2.2.4 *Let $x_1(t), x_2(t)$ be differentiable functions such that $|x_1(a) - x_2(a)| \leq \delta$ and $|x_i'(t) - f(t, x_i(t))| \leq \epsilon_i$, $i = 1, 2$, for $a \leq t \leq b$. If the function $f(t, x)$ satisfies a Lipschitz condition in x,*

$$|f(t, x_1) - f(t, x_2)| \leq L|x_1 - x_2|,$$

then

$$|x_1(t) - x_2(t)| \leq \delta e^{L(t-a)} + (\epsilon_1 + \epsilon_2)\left[e^{L(t-a)} - 1\right]\Big/L$$

for $a \leq t \leq b$.

Proof. Put $\epsilon = \epsilon_1 + \epsilon_2$, $\gamma(t) = x_1(t) - x_2(t)$. Then

$$|\gamma'(t)| \leq |f(t, x_1(t)) - f(t, x_2(t))| + \epsilon$$
$$\leq L|\gamma(t)| + \epsilon.$$

Since

$$\int_a^t \gamma'(s)ds = \gamma(t) - \gamma(a),$$

it follows that

$$|\gamma(t)| \leq |\gamma(a)| + \left| \int_a^t \gamma'(s)ds \right|$$

$$\leq \delta + \int_a^t |\gamma'(s)|ds$$

$$\leq \delta + \int_a^t (L|\gamma(t)| + \epsilon)ds$$

$$= \delta + \epsilon(t-a) + \int_a^t L|\gamma(s)|ds.$$

Therefore, by Gronwall's inequality,

$$|\gamma(t)| \leq \delta + \epsilon(t-a) + \int_a^t L\{\delta + \epsilon(s-a)\} e^{L(t-s)}ds$$

or, after integrating by parts the right-hand side,

$$|\gamma(t)| \leq \delta e^{L(t-a)} + \epsilon \left[e^{L(t-a)} - 1 \right] \Big/ L.$$

$\square$

Corollary 2.2.1 *Let $\epsilon_1 = \epsilon_2 = \delta = 0$ in Theorem 2.2.4. Then the uniqueness of solutions of I.V.P. follows if $f(t,x)$ satisfies Lipschitz condition in x.*

In the following Corollary 2.2.2, the definition of continuous dependence of solutions on initial conditions is stated as follows:
For fixed compact interval $[t_0, t_0+T]$, given $\varepsilon > 0$, there exists $\delta = \delta(\varepsilon, T) > 0$ such that

$$|x_1(t_0) - x_2(t_0)| < \delta \quad \text{implies} \quad |x_1(t) - x_2(t)| < \varepsilon$$

for all $t_0 \leq t \leq t_0 + T$.

Corollary 2.2.2 *The continuous dependence of the solutions on the initial conditions holds when f satisfies a global Lipschitz condition.*

Proof. Let $x_1(t), x_2(t)$ be two solutions of I.V.P. (2.1) with $|x_1(t_0) - x_2(t_0)| \leq \delta$, then applying Theorem 2.2.4 with $\epsilon_1 = \epsilon_2 = 0$ yields

$$|x_1(t) - x_2(t)| \leq \delta e^{L(t-t_0)} \quad \text{for} \quad t \geq t_0. \tag{2.5}$$

Choose $\delta > 0$ satisfying $\delta < \varepsilon e^{-LT}$. Then from (2.5) we obtain the continuous dependence on initial conditions. $\square$

Corollary 2.2.3 *If $f \in C^1(D)$, then we have local existence and uniqueness of the solutions of I.V.P.*

Proof. It suffices to show that if $f \in C^1(D)$ then f is locally Lipschitz. Since $f \in C^1(D)$, $(t_0, x_0) \in D$ then $D_x f(t, x)$ is continuous on $R = \{(t, x) : |x - x_0| \leq \delta_1, \ |t - t_0| \leq \delta_2\}$ for some $\delta_1, \delta_2 > 0$. Claim: $f(t, x)$ satisfies Lipschitz condition in x on the rectangle R. Since

$$f(t, x_1) - f(t, x_2) = \int_0^1 \frac{d}{ds} f(t, sx_1 + (1 - s)x_2) ds$$
$$= \int_0^1 D_x f(t, sx_1 + (1 - s)x_2) \cdot (x_1 - x_2) ds,$$

we have

$$|f(t, x_1) - f(t, x_2)| \leq M|x_1 - x_2|,$$

where

$$M = \sup_{\substack{0 \leq s \leq 1 \\ |t - t_0| \leq \delta_2}} \| D_x f(t, sx_1 + (1 - s)x_2) \| .$$

$\square$

Let X be a Banach space and $F \subseteq X$ be a closed subset of X. We say $T : F \to F$ is a contraction if $|Tx_1 - Tx_2| \leq \theta|x_1 - x_2|$ for some $0 < \theta < 1$ and for any $x_1, x_2 \in F$. It is easy to verify that a contraction map has a unique fixed point.

Theorem 2.2.5 (Contraction mapping principle).
There exists a unique fixed point for a contraction map $T : F \to F$.

Proof. Given any $x_0 \in F$, defined a sequence $\{x_n\}_{n=0}^\infty$ by $x_{n+1} = Tx_n$. Then

$$|x_{n+1} - x_n| \leq \theta|x_n - x_{n-1}| \leq \cdots \leq \theta^n|x_1 - x_0|.$$

<u>Claim:</u> $\{x_n\}$ is a Cauchy sequence. Let $m > n$. Then

$$|x_m - x_n| \leq |x_m - x_{m-1}| + |x_{m-1} - x_{m-2}| + \cdots + |x_{n+1} - x_n|$$
$$\leq \theta^n \left(1 + \theta + \cdots + \theta^{m-n-1}\right)|x_1 - x_0|$$
$$\leq \frac{\theta^n}{1 - \theta}|x_1 - x_0| \to 0 \quad \text{as} \quad n \to \infty.$$

Since F is a closed set, then $x_n \to x \in F$ as $n \to \infty$. Claim: x is the desired fixed point of T. Since

$$\begin{aligned}
|Tx - x| &\le |Tx - x_{n+1}| + |x_{n+1} - x| \\
&= |Tx - Tx_n| + |x_{n+1} - x| \\
&\le \theta|x - x_{n+1}| + |x - x_{n+1}| \to 0, \quad \text{as } n \to \infty.
\end{aligned}$$

It follows that $Tx = x$.   □

Now we shall apply contraction principle to show the existence and uniqueness of solutions of I.V.P. (2.1).

Theorem 2.2.6 *Let $f(t,x)$ be continuous on $S = \{(t,x) : |t - t_0| \le a,$ $|x - x_0| \le b\}$ and let $f(t,x)$ satisfy a Lipschitz condition in x with Lipschitz constant L. Let $M = \max\{|f(t,x)| : (t,x) \in S\}$. Then there exists a unique solution of I.V.P.*

$$\frac{dx}{dt} = f(t,x)$$
$$x(t_0) = x_0$$

on $I = \{t : |t - t_0| \le \alpha\}$, where $\alpha < \min\{a, b/M, 1/L\}$.

Proof. Let B be the closed subset of $C(I)$ defined by

$$B = \{\varphi \in C(I) : |\varphi(t) - x_0| \le b, \quad t \in I\}.$$

Define a mapping on B by

$$(T\varphi)(t) = x_0 + \int_{t_0}^{t} f(s, \varphi(s))ds. \tag{2.6}$$

First we show that T maps B into B. Since $f(t, \varphi(t))$ is continuous, T certainly maps B into $C(I)$. If $\varphi \in B$, then

$$\begin{aligned}
|(T\varphi)(t) - x_0| &= \left| \int_{t_0}^{t} f(s, \varphi(s))ds \right| \\
&\le \int_{t_0}^{t} |f(s, \varphi(s))|ds \le M|t - t_0| \le M\alpha < b.
\end{aligned}$$

Hence $T\varphi \in B$. Next we prove that T is a contraction mapping. Let $\| x \|$ be the sup norm of x. Then

$$
\begin{aligned}
|Tx(t) - Ty(t)| &= \left| \int_{t_0}^{t} (f(s, x(s)) - f(s, y(s))) ds \right| \\
&\leq \int_{t_0}^{t} |f(s, x(s)) - f(s, y(s))| ds \\
&\leq L \int_{t_0}^{t} |x(s) - y(s)| ds \leq L \| x - y \| |t - t_0| \\
&\leq L\alpha \| x - y \| = \theta \| x - y \|
\end{aligned}
$$

where $\theta = L\alpha < 1$. Hence $\| Tx - Ty \| \leq \theta \| x - y \|$.

Thus we complete the proof by the contraction mapping principle. $\square$

Remark 2.2.3 From (2.6), $\{x_{n+1}(t)\}$ is a successive approximation of $x(t)$. It is used for theoretical purpose, not for numerical computation. If

$$
x_{n+1}(t) = x_0 + \int_{t_0}^{t} f(s, x_n(s)) ds,
$$

$$
x_0(t) \equiv x_0, \quad |t - t_0| \leq \alpha,
$$

then $\{x_n(t)\}$ is called Picard iterations of I.V.P. (2.1).

2.3 Continuation of Solutions

In this section we discuss the properties of the solution on the maximal interval of existence and the global existence of the solution of the I.V.P. (2.1).

Theorem 2.3.1 Let $f \in C(D)$ and $|f| \leq M$ on D. Suppose φ is a solution of (2.1) on the interval $J = (a, b)$. Then

(i) the two limits $\lim_{t \to a^+} \varphi(t) = \varphi(a^+)$ and $\lim_{t \to b^-} \varphi(t) = \varphi(b^-)$ exist;

(ii) if $(a, \varphi(a^+))$ (respectively, $(b, \varphi(b^-))$) is in D, then the solution φ can be continued to the left passing through the point $t = a$ (respectively, to the right passing through the point $t = b$).

Proof. Consider the right endpoint b. The proof for the left endpoint a is similar. Fix $\tau \in J$ and set $\xi = \varphi(\tau)$. Then for $\tau < t < u < b$, we have

$$
|\varphi(u) - \varphi(t)| = \left| \int_{t}^{u} f(s, \varphi(s)) ds \right| \leq M(u - t). \tag{2.7}
$$

Given any sequence $\{t_m\} \uparrow b$, from (2.7) we see that $\{\varphi(t_m)\}$ is a Cauchy sequence. Thus, the limit $\varphi(b^-)$ exists.

If $(b, \varphi(b^-))$ is in D, then by local existence theorem we can extend the solution φ to $b + \delta$ for some $\delta > 0$ by the following: Let $\tilde{\varphi}(t)$ be the solution of I.V.P. for $b \leq t \leq b + \delta$,

$$\frac{dx}{dt} = f(t, x),$$

$$x(b) = \varphi(b^-).$$

Then we verify that the function

$$\varphi^*(t) = \begin{cases} \varphi(t), & a \leq t \leq b \\ \tilde{\varphi}(t), & b \leq t \leq b + \delta \end{cases}$$

is a solution of the I.V.P. (2.1).   □

Remark 2.3.1 The assumption $|f| \leq M$ on D is too strong for the theorem. In fact from (2.7) we need $|f(t, \varphi(t))| \leq M$ for t closed to b^-.

Definition 2.3.1 *We say $(t, \varphi(t)) \to \partial D$ as $t \to b^-$ if for any compact set $K \subseteq D$, there exists $t^* \in (a, b)$ such that $(t, \varphi(t)) \notin K$ for all $t \in (t^*, b)$.*

Example 2.3.1 $\frac{dx}{dt} = 1 + x^2$, $x(0) = 0$. Then the solution is $x(t) = \tan(t)$, $\frac{-\pi}{2} < t < \frac{\pi}{2}$.

In this case $D = \mathbb{R} \times \mathbb{R}$, ∂D is an empty set.

Corollary 2.3.1 *If $f \in C(D)$ and φ is a solution of (2.1) on an interval J, then φ can be continued to a maximal interval $J^* \supset J$ in such a way that $(t, \varphi(t)) \to \partial D$ as $t \to \partial J^*$ (and $|t| + |\varphi(t)| \to \infty$ if ∂D is empty). The extended solution φ^* on J^* is noncontinuable.*

Proof. By Zorn's Lemma, there exists a noncontinuable solution φ^* on a maximal interval J^*. By Theorem 2.3.1, $J^* = (a, b)$ must be open. If $b = \infty$, then obviously from Definition 2.3.1 $(t, \varphi^*(t)) \to \partial D$ as $t \to \infty$. Assume $b < \infty$. Let c be any number in (a, b).

Claim: For any compact set $K \subseteq D$, $\{(t, \varphi^*(t)) : t \in [c, b)\}$ is not contained in K.

If not, then $f(t, \varphi^*(t))$ is bounded on $[c, b)$ and $(b, \varphi^*(b-)) \in K \subseteq D$. By Theorem 2.3.1, φ^* can be continued to pass b. This leads to a contradiction. Now we prove $(t, \varphi^*(t)) \to \partial D$ as $t \to b^-$. Let K be any compact set in D, we claim that there exists $\delta > 0$ such that $(t, \varphi^*(t)) \notin K$ for all $t \in (b - \delta, b)$.

If not, then there exists a sequence $\{t_m\} \uparrow b$ such that $(t_m, \varphi^*(t_m)) \to (b, \xi) \in K \subseteq D$. Choose $r > 0$ such that $B((b, \xi), r) \subseteq D$ and set $\varepsilon = \frac{r}{3}, B \equiv B((b, \xi), 2\varepsilon) \subseteq D$. Let $M = \sup\{|f(t, x)| : (t, x) \in \overline{B}\} < \infty$. Since $\overline{B}$ is a compact set, from the above claim, $\{(t, \varphi^*(t)) : t \in [t_m, b)\}$ is not contained in $\overline{B}$. Hence there exists $\tau_m \in (t_m, b)$ such that $(\tau_m, \varphi^*(\tau_m)) \notin \overline{B}$. Without loss of generality, we assume $t_m < \tau_m < t_{m+1}$. Let m^* be sufficiently large such that for all $m \geq m^*$ we have $|t_m - b| + |\varphi^*(t_m) - \xi| < \varepsilon$. Obviously $|\tau_m - b| + |\varphi^*(\tau_m) - \xi| > 2\varepsilon$. Then there exists $\overline{t_m} \in (t_m, \tau_m)$ such that $|\overline{t_m} - b| + |\varphi^*(\overline{t_m}) - \xi| = 2\varepsilon$. Then

$$\varepsilon < |\overline{t_m} - b| + |\varphi^*(\overline{t_m}) - \xi| - |t_m - b| - |\varphi^*(t_m) - \xi|$$
$$\leq (b - \overline{t_m}) - (b - t_m) + |\varphi^*(\overline{t_m}) - \varphi^*(t_m)|$$
$$< \left| \int_{t_m}^{\overline{t_m}} f(s, \varphi^*(s)) ds \right| \leq M |\overline{t_m} - t_m|$$
$$\leq M |\tau_m - t_m| \to 0 \text{ as } m \to \infty.$$

This is a contradiction. Thus the proof is complete. $\qquad\Box$

Corollary 2.3.2 *If the solution $x(t)$ of (2.1) has an apriori bound M, i.e., $|x(t)| \leq M$ whenever $x(t)$ exists. Then $x(t)$ exists for all $t \in \mathbb{R}$.*

Proof. Let $T > t_0$ be arbitrary. Define the rectangle $D_1 = \{(t, x) : t_0 \leq t \leq T, |x| \leq M\}$. Then $f(t, x)$ is bounded on D_1. Then the solution $x(t)$ can be continued to the boundary of D_1. Hence $x(t)$ exists for $t_0 \leq t \leq T$ and T is arbitrary. Similarly $x(t)$ exists for $T_1 \leq t \leq t_0, T_1$ arbitrary. Thus we have global existence with $J^* = (-\infty, \infty)$. $\qquad\Box$

Remark 2.3.2 To get an apriori bound M for the solution $x(t)$ of (2.1), we apply differential inequalities (see §2.6) or Lyapunov method (see Chapter 5).

Example 2.3.2 There exists a unique solution $\varphi(t)$ of the linear system

$$\frac{dx}{dt} = A(t)x + h(t) = f(t, x),$$
$$x(t_0) = x_0,$$

for all $t \in \mathbb{R}$ where $A(t) \in \mathbb{R}^{n \times n}$ and $h(t) \in \mathbb{R}^n$ are continuous functions on $\mathbb{R}$.

Proof. Let $\varphi(t)$ be the solution for $t_0 \le t < t_0 + c$. Then

$$\varphi(t) - x_0 = \int_{t_0}^t f(s, \varphi(s)) ds$$

$$= \int_{t_0}^t \left[f(s, \varphi(s)) - f(s, x_0) \right] ds + \int_{t_0}^t f(s, x_0) ds,$$

and it follows that

$$|\varphi(t) - x_0| \le \int_{t_0}^t \| A(s) \| \, |\varphi(s) - x_0| ds + \int_{t_0}^t |A(s) x_0 + h(s)| ds$$

$$\le L \int_{t_0}^t |\varphi(s) - x_0| ds + \delta,$$

where

$$L = \sup_{t_0 \le t \le t_0 + c} \| A(t) \| \quad \text{and}$$

$$\delta = c \cdot \max\{|A(s)x_0 + h(s)| : t_0 \le s \le t_0 + c\}.$$

From Gronwall's inequality, we have

$$|\varphi(t) - x_0| \le \delta \exp(Lc).$$

Hence $\varphi(t)$ is bounded on $[t_0, t_0 + c)$. Then by Theorem 2.3.1, we can extend φ to go beyond $t_0 + c$. $\square$

2.4   Continuous Dependence Properties

In the theory of differential equations, well-posedness is an important property. We say a problem is well-posed if
(1) a solution exists locally.
(2) the solution is unique.
(3) The solution's behavior changes continuously with initial conditions and parameters.

Hardmard was the first one to give the examples of well-posedness in partial differential equation [E].

For the initial valued problem in ordinary differential equations, we have proved Theorem 2.2.6 which states if $f(t, x)$ is locally Lipschitz then the solution exists in a small interval $I = \{t : |t - t_0| \le \alpha\}$ for some $\alpha > 0$ and the solution is unique. In this section we shall prove in following Theorem 2.4.1 and Theorem 2.4.2 the solution's behavior continuously depends on the initial condition and parameters. Hence the initial-value problem of (2.1) is well-posed.

In the following we shall prove the continuous dependence properties of I.V.P. (2.1). First we state discrete version of continuous dependence properties.

Theorem 2.4.1 *Let* $\{f_n(t,x)\}$ *be a sequence of continuous functions on* D *and* $\lim_{n\to\infty} f_n = f$ *uniformly on any compact set in* D. *Let* $(\tau_n, \xi_n) \in D$, $(\tau_n, \xi_n) \to (\tau, \xi)$ *and* $\varphi_n(t)$ *be any noncontinuable solution of* $x' = f_n(t,x)$, $x(\tau_n) = \xi_n$. *If* $\varphi(t)$ *is the unique solution of* $x' = f(t,x)$, $x(\tau) = \xi$, *defined on* $[a,b]$. *Then* $\varphi_n(t)$ *is defined on* $[a,b]$ *for* n *large and* $\varphi_n \to \varphi$ *uniformly on* $[a,b]$.

Then we state the continuous version of continuous dependence properties.

Theorem 2.4.2 *Let* $f(t,x,\lambda)$ *be a continuous function of* (t,x,λ) *for all* (t,x) *in open set* D *and for all* λ *near* λ_0 *and let* $\varphi(t;\tau,\xi,\lambda)$ *be any noncontinuable solution of*

$$x' = f(t,x,\lambda),$$
$$x(\tau) = \xi.$$

If $\varphi(t;t_0,\xi_0,\lambda_0)$ *is defined on* $[a,b]$ *and is unique, then* $\varphi(t;\tau,\xi,\lambda)$ *is defined on* $[a,b]$ *for all* (τ,ξ,λ) *near* (t_0,ξ_0,λ_0) *and is a continuous function of* (t,τ,ξ,λ).

The proofs will be deferred to Appendix A.2.

2.5 Differentiability of I.C. and Parameters

For fixed τ,ξ, if $f(t,x,\lambda)$ is C^1 in x and λ then we shall show that the unique solution $\varphi(t,\lambda)$ of the I.V.P.

$$\frac{dx}{dt} = f(t,x,\lambda)$$
$$x(\tau,\lambda) = \xi$$

is differentiable with respect to λ. Furthermore from

$$\frac{d}{d\lambda}\left(\frac{d}{dt}\varphi(t,\lambda)\right) = \frac{d}{d\lambda}\left(f(t,\varphi(t,\lambda),\lambda)\right),$$

we have

$$\frac{d}{dt}\left(\frac{d}{d\lambda}\varphi(t,\lambda)\right) = f_x\left(t,\varphi(t,\lambda),\lambda\right)\left(\frac{d}{d\lambda}\varphi(t,\lambda)\right) + \frac{\partial f}{\partial\lambda}(t,\varphi(t,\lambda),\lambda).$$

Hence $\frac{d}{d\lambda}\varphi(t,\lambda) \equiv \psi(t,\lambda)$ satisfies the variational equation

$$\begin{cases} \frac{dy}{dt} = f_x(t,\varphi(t,\lambda),\lambda)y + f_\lambda(t,\varphi(t,\lambda),\lambda), \\ \\ y(\tau) = 0. \end{cases}$$

Similarly, for fixed τ, if $f(t,x)$ is C^1 in x then we shall show that the unique solution $\varphi(t,\xi)$ of the I.V.P.

$$\frac{dx}{dt} = f(t,x)$$
$$x(\tau) = \xi$$

is differentiable with respect to ξ. Furthermore from

$$\frac{d}{d\xi}\left(\frac{d}{dt}\varphi(t,\xi)\right) = \frac{d}{d\xi}\left(f(t,\varphi(t,\xi))\right),$$

we obtain

$$\frac{d}{dt}\left(\frac{d}{d\xi}\varphi(t,\xi)\right) = f_x(t,\varphi(t,\xi))\frac{d}{d\xi}\varphi(t,\xi),$$

$$\frac{d}{d\xi}\varphi(\tau,\xi) = I, \quad I \text{ is the } n \times n \text{ identity matrix.}$$

Then the $n \times n$ matrix $\frac{d}{d\xi}\varphi(t,\xi) \equiv \psi(t,\xi)$ satisfies the linear variational equation

$$\begin{cases} \frac{dY}{dt} = f_x(t,\varphi(t,\xi))Y \\ \\ Y(\tau) = I. \end{cases}$$

In this following we state Theorem 2.5.1 and defer its proof to Appendix A.3.

Theorem 2.5.1 [Cop].
Let $\varphi(t,\lambda_0)$ be the unique solution of

$$\frac{dx}{dt} = f(t,x,\lambda_0),$$
$$x(t_0) = \xi_0,$$

on compact interval $J = [a, b]$. *Assume* $f \in C^1$ *in* x *and* λ *at all points* $(t, \varphi(t, \lambda_0), \lambda_0)$, $t \in J$. *Then for* λ *sufficiently near* λ_0, *the system* E_λ : $\frac{dx}{dt} = f(t, x, \lambda)$, $x(t_0) = \xi_0$ *has a unique solution* $\varphi(t, \lambda)$ *defined on* J. *Moreover,* $\varphi_\lambda(t, \lambda_0)$ *exists and satisfies*

$$y' = f_x(t, \varphi(t, \lambda_0), \lambda_0)y + f_\lambda(t, \varphi(t, \lambda_0), \lambda_0),$$
$$y(t_0) = 0.$$

Theorem 2.5.2 (Peano) [Cop].
Let $\varphi(t, t_0, \xi_0)$ *be the unique solution of*

$$\frac{dx}{dt} = f(t, x)$$
$$x(t_0) = \xi_0$$

on $J = [a, b]$. *Assume* f_x *exists and is continuous at all points* $(t, \varphi(t, t_0, \xi_0))$, $t \in J$. *Then the system* $E_{\tau, \xi}$: $\frac{dx}{dt} = f(t, x)$, $x(\tau) = \xi$, *has a unique solution* $\varphi(t, \tau, \xi)$ *for* (τ, ξ) *near* (t_0, ξ_0). *Moreover* $\varphi_\xi(t, t_0, \xi_0)$ *exists (t* $\in J$) *and satisfies the linear homogeneous equation*

$$\frac{dY}{dt} = f_x(t, \varphi(t, t_0, \xi_0))Y$$
$$Y(t_0) = I.$$

Proof. Put $x = u + \xi$. Then the I.V.P.

$$\frac{dx}{dt} = f(t, x), \ x(t_0) = \xi$$

is transformed into the I.V.P.

$$\frac{du}{dt} = f(t, u + \xi), \ u(t_0) = 0$$

with ξ as a parameter. Then the first part of theorem follows from Theorem 2.5.1. From Theorem 2.5.1, we have

$$\frac{d}{dt}\left[\frac{du}{d\xi}(t, \xi_0)\right] = f_x(t, u(t, \xi_0) + \xi_0)\left(\frac{du}{d\xi}(t, \xi_0) + I\right),$$

or

$$\frac{d}{dt}\left[\frac{dx}{d\xi}(t, \xi_0)\right] = f_x(t, x(t, \xi_0))\frac{dx}{d\xi}(t, \xi_0).$$

$\square$

2.6   Differential Inequalities

We shall apply differential inequality to estimate bounds of a solution of I.V.P. It is usually in the scalar form. In the following, we shall apply the property of continuous dependence on initial conditions and parameters to obtain differential inequalities.

Theorem 2.6.1 Let $x(t)$ be a scalar, differentiable function. Let $\varphi(t)$ be the unique solution of the I.V.P. $x' = f(t, x)$, $x(t_0) = x_0$.

(i) If

$$x'(t) \geq f(t, x(t)), \quad t_0 \leq t \leq b$$
$$x(t_0) \geq x_0. \tag{2.8}$$

Then $\varphi(t) \leq x(t)$ for $t_0 \leq t \leq b$.

(ii) If

$$x'(t) \leq f(t, x(t)), \quad t_0 \leq t \leq b$$
$$x(t_0) \leq x_0. \tag{2.9}$$

Then $\varphi(t) \geq x(t)$ for $t_0 \leq t \leq b$.

Proof. Consider the following I.V.P.

$$x'_n = f(t, x_n) + \frac{1}{n},$$
$$x_n(t_0) = x_0 + \frac{1}{n}.$$

Then, from the continuously dependence on initial conditions and parameters, we have $x_n(t) \to \varphi(t)$ uniformly on $[t_0, b]$. We shall only consider the case (2.9). The other case (2.8) can be done by similar arguments. It suffices to show that $x(t) \leq x_n(t)$ on $[t_0, b]$ for n sufficiently large. If not, then there exists a large n such that there exist t_1, t_2, with $t_0 < t_1 < t_2 < b$ satisfying (see Fig. 2.1)

$$x(t) > x_n(t) \text{ on } (t_1, t_2)$$
$$x(t_1) = x_n(t_1).$$

Then for $t > t_1$, t near t_1, we have

$$\frac{x(t) - x(t_1)}{t - t_1} > \frac{x_n(t) - x_n(t_1)}{t - t_1}. \tag{2.10}$$

Fig. 2.1

Let $t \to t_1$ in (2.10), then it follows that

$$x'(t_1) \geq x_n'(t_1) = f(t_1, x_n(t_1)) + 1/n$$
$$= f(t_1, x(t_1)) + 1/n$$
$$> f(t_1, x(t_1)).$$

This contradicts to the assumption (2.9). Hence we have completed the proof of Theorem 2.6.1. $\square$

Corollary 2.6.1 *Let* $D_r x(t) = \lim\limits_{h \to 0^+} \dfrac{x(t+h) - x(t)}{h}$. *Then the conclusions of Theorem 2.6.1 also hold if we replace* $x'(t)$ *in (2.9) by* $D_r x(t)$.

Example 2.6.1 Consider the Lotka-Volterra two species competition model

$$\frac{dx_1}{dt} = r_1 x_1 \left(1 - \frac{x_1}{K_1}\right) - \alpha x_1 x_2,$$
$$\frac{dx_2}{dt} = r_2 x_2 \left(1 - \frac{x_2}{K_2}\right) - \beta x_1 x_2,$$
$$x_1(0) > 0, \quad x_2(0) > 0.$$

It is easy to verify that $x_1(t) > 0$, $x_2(t) > 0$ for all $t > 0$. Then we have

$$\frac{dx_i}{dt} \leq r_i x_i \left(1 - \frac{x_i}{K_i}\right),$$
$$x_i(0) > 0.$$

By Theorem 2.6.1, for any $\varepsilon > 0$, $x_i(t) \leq K_i + \varepsilon$ for $t \geq T$ for some T sufficiently large.

For system of differential equations, we have the following comparison the-orem Theorem 2.6.2. First we prove the following lemma.

Lemma 2.6.1 *Let* $x(t) = (x_1(t), \cdots, x_n(t)) \in \mathbb{R}^n$ *be differentiable. Then* $D_r|x(t)| \leq |x'(t)|$.

Proof. For $h > 0$, from the triangle inequality, we have

$$\frac{|x(t+h)| - |x(t)|}{h} \leq \left| \frac{x(t+h) - x(t)}{h} \right|.$$

Then the lemma follows directly by letting $h \to 0^+$ in the above inequality. □

Theorem 2.6.2 (Comparison theorem).
Let $F(t, v)$ *be continuous and* $|f(t, x)| \leq F(t, |x|)$. *Let* $\varphi(t)$ *be a solution of the system* $\frac{dx}{dt} = f(t, x)$ *and* $v(t)$ *be the unique solution of scalar equation* $\frac{dv}{dt} = F(t, v)$, $v(t_0) = \eta$ *with* $|\varphi(t_0)| \leq \eta$. *Then* $|\varphi(t)| \leq v(t)$ *for all* $t \geq t_0$.

Proof. Let $u(t) = |\varphi(t)|$. From Lemma 2.6.1, we have

$$D_r u(t) = D_r|\varphi(t)| \leq |\varphi'(t)| = |f(t, \varphi(t))|$$
$$\leq F(t, |\varphi(t)|) = F(t, u(t)),$$
$$u(t_0) = |\varphi(t_0)| \leq \eta.$$

Then, by Theorem 2.6.1, $|\varphi(t)| \leq v(t)$ for all $t \geq 0$. □

Corollary 2.6.2 *Let* $f(t, x) = A(t)x + h(t)$, *and* $A(t) \in \mathbb{R}^{n \times n}, h(t) \in \mathbb{R}^n$ *be continuous on* $\mathbb{R}$. *Then we have global existence for the solution of I.V.P.*, $\frac{dx}{dt} = f(t, x)$, $x(t_0) = x_0$.

Proof. For $t \geq t_0$,

$$|f(t, x)| \leq \|A(t)\| |x| + |h(t)|$$
$$\leq \max_{t_0 \leq \tau \leq t} \{\|A(\tau)\|, |h(\tau)|\}(|x| + 1)$$
$$= g(t)\psi(|x|),$$

where $\psi(u) = u + 1$.

Claim: The solution of

$$\frac{du}{dt} = g(t)\psi(u)$$
$$u(t_0) = |x_0| = u_0$$

exists globally on $\mathbb{R}$. If not, we assume that the maximal interval of existence is $[t_0, b)$. Then

$$\int_{u_0}^{+\infty} \frac{du}{\psi(u)} = \int_{t_0}^{b} g(t)dt.$$

In the above identity, the left-hand side is infinite while the right-hand side is finite. Thus we obtain a contradiction. Hence $u(t)$ exists on $\mathbb{R}$. From Theorem 2.6.2, we have

$$|x(t)| \leq u(t) \quad \text{for } t \geq t_0.$$

Thus $x(t)$ exists for all t by Theorem 2.3.1. For $t \leq t_0$, we reverse time by introducing new time scale $\tau = t_0 - t$, then the system becomes

$$\frac{dx}{d\tau} = -f(t_0 - \tau, x),$$
$$x(0) = x_0.$$

Then using the above argument, we can show that $x(\tau)$ exists for $\tau \geq 0$, i.e. $x(t)$ exists for $t \leq t_0$. $\qquad\qquad\qquad\qquad\qquad\qquad\qquad\qquad\quad\square$

Now we consider the differential inequalities for certain special type of systems.

Let $\mathbb{R}^n_+ = \{x = (x_1, \cdots, x_n) \in \mathbb{R}^n : x_i \geq 0, \ i = 1, \cdots, n\}$ be the nonnegative cone. Define the following partial orders:

$x \leq y$ if $y - x \in \mathbb{R}^n_+$, i.e., $x_i \leq y_i$ for all i.

$x < y$ if $x \leq y$ and $x \neq y$.

$x \ll y$ if $x_i < y_i$ for all i.

Definition 2.6.1 *We say $f = (f_1, \cdots, f_n) : D \subseteq \mathbb{R}^n \to \mathbb{R}^n$ is of type K on D if for each i, $f_i(a) \leq f_i(b)$ for any $a, b \in D$ satisfying $a \leq b$ and $a_i = b_i$.*

Theorem 2.6.3 (Kamke Theorem [Cop]).
Let $f(t, x)$ be of type K for each fixed t and let $x(t)$ be a solution of $\frac{dx}{dt} = f(t, x)$ on $[a, b]$. If $y(t)$ is continuous on $[a, b]$ and satisfies $D_r y(t) \geq f(t, y)$

and $y(a) \geq x(a)$, then $y(t) \geq x(t)$ for $a \leq t \leq b$. If $z(t)$ is continuous on $[a, b]$ and satisfies $D_\ell z(t) \leq f(t, z)$ and $z(a) \leq x(a)$ then $z(t) \leq x(t)$ for $a \leq t \leq b$.

Proof. We only prove the second case. First we prove that if $D_\ell z(t) \ll f(t, z(t))$, $z(a) \ll x(a)$ then $z(t) \ll x(t)$ for $a \leq t \leq b$. If not, we let c be the least number in $[a, b]$ such that $z(t) < x(t)$ for $a \leq t < c$ and $z(c) \leq x(c)$, $z_i(c) = x_i(c)$ for some $1 \leq i \leq n$. Then

$$D_\ell z_i(c) < f_i(c, z(c)) \leq f_i(c, x(c)) = x_i'(c).$$

Since $z_i(c) = x_i(c)$, it follows that $z_i(t) > x_i(t)$ for certain values of t less than and near c. This contradicts the definition of c. Now we are in a position to complete the proof of Theorem 2.6.3. Assume $D_\ell z(t) \leq f(t, z(t))$, $z(a) \leq x(a)$. We want to show that $z(t) \leq x(t)$ for $a \leq t \leq b$. Let c be the largest value of t such that $z(s) \leq x(s)$ for $a \leq s \leq t$. Suppose, contrary to the theorem, that $c < b$. Let vector $\epsilon = (1, ...1)^T$ and let $\varphi_n(t)$ be the solution of the following I.V.P.

$$\begin{aligned} \frac{dw}{dt} &= f(t, w) + \frac{\epsilon}{n} \\ w(c) &= x(c) + \frac{\epsilon}{n} \end{aligned} \quad , \ c \leq t \leq c + \delta.$$

Then $\varphi_n(t) \to x(t)$ on $[c, c + \delta]$ as $n \to \infty$ and $z(t) < \varphi_n(t)$ on $[c, c + \delta]$. Let $n \to \infty$, then $z(t) \leq x(t)$ on $[c, c + \delta]$. This contradicts the definition of c. Hence $c = b$ and $z(t) \leq x(t)$ on $[a, b]$. $\qquad \square$

Remark 2.6.1 Kamke's Theorem is an essential tool to study the behavior of the solutions of cooperative system, competitive system [H1] and the monotone flow [Str]. See Chapter 10.

2.7   Exercises

Exercise 2.1 Find a bounded sequence $\{f_m\}_{m=1}^\infty \subseteq C(I), I = [a, b]$ such that there is no convergent subsequence.

Exercise 2.2 Let $J = [a, b]$, $a, b < \infty$ and $\mathcal{F}$ be a subset of $C(J)$. Show that if each sequence in $\mathcal{F}$ contains a uniformly convergent subsequence, then $\mathcal{F}$ is both equicontinuous and uniformly bounded.

Exercise 2.3 Consider initial value problem

$$\frac{dy}{dt} = f(t, y) = \begin{cases} \frac{4t^3 y}{t^4 + y^2}, & (t, y) \neq (0, 0), \\ 0, & (t, y) = (0, 0), \end{cases}$$

and $y(0) = 0$. Verify that $f(t, y)$ is continuous at $(t, y) = (0, 0)$ but does not satisfy the Lipschitz condition. Show that the initial value problem has infinitely many solutions

$$y(t) = c^2 - \sqrt{t^4 + c^4} \text{ for } c \in \mathbb{R}.$$

Exercise 2.4 Show that if $g \in C^1(\mathbb{R})$ and $f \in C(\mathbb{R})$ then the solution of I.V.P.

$$y'' + f(y)y' + g(y) = 0, \ y(t_0) = A, \ y'(t_0) = B$$

exists locally, is unique and can be continued so long as y and y' remains bounded.
Hint: Let $z = y' + F(y)$ where F is any anti-derivative of f, and convert the equation to an equation of (y, z).

Exercise 2.5 If $f : W \subseteq \mathbb{R}^n \to \mathbb{R}$ is locally Lipschitz and $A \subseteq W$ is a compact set, then $f|_A$ is Lipschitz.

Exercise 2.6 Prove that any solution of the system $x' = f(t, x)$ can be extended indefinitely if $|f(t, x)| \leq k|x|$ for all t and all $|x| \geq r, r, k > 0$.

Exercise 2.7 Let $K = K(x, y) : [a, b] \times [a, b] \to \mathbb{R}$ be continuous with $0 \leq K(x, y) \leq d$ for all $x, y \in [a, b]$. Let $2(b - a)d \leq 1$ along with $u_0(x) \equiv 0$ and $v_0(x) \equiv 2$. Then the iterates

$$u_{n+1}(x) = \int_a^b K(x, y)u_n(y)dy + 1,$$

$$v_{n+1}(x) = \int_a^b K(x, y)v_n(y)dy + 1,$$

converge uniformly on $[a, b]$ to a unique solution $u \in X = C(a, b)$ of the integral equation

$$u(x) = \int_a^b K(x, y)u(y)dy + 1, \quad x \in [a, b]$$

where $u_0 \leq u_1(x) \leq \dots \leq v_1(x) \leq v_0(x)$.

Exercise 2.8 Consider the pendulum equation with constant torque $\ddot{\theta} = a - \sin\theta$, $\theta(0, a) = 0$, $\dot{\theta}(0, a) = 0$. Compute $\frac{d}{da}\theta(t, a)|_{a=0}$.

Exercise 2.9 Let $f : \mathbb{R} \to \mathbb{R}$ be C^k with $f(0) = 0$. Then $f(x) = xg(x)$, $g \in C^{k-1}$.

Exercise 2.10 Prove the second part of Theorem 2.6.1.

Exercise 2.11 Let the continuous function $f(t, x)$ be nondecreasing in x for each fixed value of t and let $x(t)$ be a ***unique solution*** of the differential equation $\frac{dx}{dt} = f(t, x)$ on an interval $[a, b]$. If the continuous function $y(t)$ satisfies the integral inequality

$$y(t) \le x(a) + \int_a^t f(s, y(s))ds,$$

for $a \le t \le b$, then $y(t) \le x(t)$ on $[a, b]$.

A solution $x(t)$ is said to be a ***unique solution*** *of the differential equation* $dx/dt = f(t, x)$ *on a interval* $[a, b]$ if for every solution y on $[a, b]$ of this equation with $y(a) = x(a)$, we have that $y(b) \le x(b)$.

Exercise 2.12 Consider $\dot{x} = f(x)$, $f : D \subseteq \mathbb{R}^n \to \mathbb{R}^n$ is of K-type on D and $x_0, y_0 \in D$. Let $<_r$ denote any of the relations $\le, <$, or $\ll$. If $x_0 <_r y_0$, $t > 0$ and the solutions $\varphi(t, x_0)$, $\varphi(t, y_0)$ are defined then $\varphi(t, x_0) <_r \varphi(t, y_0)$.

Exercise 2.13 Consider the system $\frac{dx}{dt} = f(x)$, where $f : D \to \mathbb{R}^n$, D is a convex domain in $\mathbb{R}^n$. We say that the system is a cooperative system if $\frac{\partial f_i}{\partial x_j}(x) \ge 0$, $i \ne j$, $x \in D$ and a competitive system if $\frac{\partial f_i}{\partial x_j}(x) \le 0$, $i \ne j$, $x \in D$. Show that

(1) if $\varphi(t)$ and $\psi(t)$ are solutions of a cooperative system $\frac{dx}{dt} = f(x)$ and $\varphi(0) \le \psi(0)$, then $\varphi(t) \le \psi(t)$ for all $t \ge 0$.
(2) if $n = 2$ and $\psi(t)$ are solutions of competitive system $\frac{dx}{dt} = f(x)$, $\varphi_1(0) \le \psi_1(0)$, $\varphi_2(0) \ge \psi_2(0)$, then $\varphi_1(t) \le \psi_1(t)$, $\varphi_2(t) \ge \psi_2(t)$ for all $t > 0$.

Exercise 2.14 Let $f \in C(D), D \subseteq \mathbb{R}^n$ and let f be smooth enough so that solutions $\psi(t, \tau, \xi)$ of

$$x' = f(x), \quad x(\tau) = \xi$$

are unique. Show that $\psi(t, \tau, \xi) = \psi(t - \tau, 0, \xi)$ for all $\xi \in D$, all $\tau \in \mathbb{R}$ and all t such that ψ is defined.

Exercise 2.15 Let $f \in C(D)$ and $f(t + T, x) = f(t, x)$, i.e. f is periodic in t with period T and let f be smooth. Show that for any integer m

$$\varphi(t, t_0, \xi) = \varphi(t + mT, \ t_0 + mT, \ \xi),$$

where $\varphi(t, t_0, \xi)$ regarded as a function in t is the unique solution of the I.V.P.

$$\begin{cases} \frac{dx}{dt} = f(t, x), \\ x(t_0) = \xi. \end{cases}$$

Exercise 2.16 Let $x : \mathbb{R}_+ \to [a, \infty)$, $y : \mathbb{R}_+ \to [b, \infty)$ and $f : [a, \infty) \times [b, \infty) \to \mathbb{R}$ be continuously differentiable and satisfy

$$x'(t) \leq f\left(x(t), y(t)\right), \, t \geq 0.$$

Suppose $\frac{\partial f}{\partial x}(x, y) < 0$, $\frac{\partial f}{\partial y}(x, y) > 0$ and for each $y \in [b, \infty)$, there exists a unique solution $x^* = x^*(y) \in [a, \infty)$ of $f(x, y) = 0$. Show that if $\limsup_{t \to \infty} y(t) \leq \alpha$ then $\limsup_{t \to \infty} x(t) \leq x^*(\alpha)$.

Exercise 2.17 Consider the following predator-prey system

$$\begin{cases} x' = \gamma x \left(1 - \frac{x}{K}\right) - \alpha xy, \\ y' = y(\beta x - d), \qquad \qquad \gamma, K, \alpha, \beta, d > 0 \\ x(0) > 0, \quad y(0) > 0. \end{cases}$$

Show that the solutions $x(t)$, $y(t)$ are defined for all $t > 0$ and the solutions are positive and bounded for all $t > 0$.

Exercise 2.18 Consider the equation $\dot{x} = f(t, x)$ where f is continuous on $\mathbb{R} \times \mathbb{R}^n$. Suppose $|f(t, x)| \leq \phi(t) |x|$ for all t, x, where ϕ is a non-negative continuous function defined on $\mathbb{R}$ satisfying $\int_c^\infty \phi(t) \, dt < \infty$ for every $c \in \mathbb{R}$.

(1) Prove that every solution approaches to a constant as $t \to \infty$.

(2) If, in addition,

$$|f(t,x) - f(t,y)| \leq \phi |x - y| \qquad \text{for all } x, y \in \mathbb{R}^n$$

prove that there is a one-to-one correspondence between the initial values and the limit value of the solution, that is, the map which sends the value of a solution x at a fixed t_0 to the limit of x at infinity is a bijection on $\mathbb{R}^n$.
(Hint: First, take the initial time sufficient large to obtain the desired correspondence.)

(3) Does the above result imply anything for the equation

$$\dot{x} = -x + a(t)x,$$

where a is a continuous function satisfying

$$\int_c^\infty |a(t)| \, dt < \infty \quad \text{for every} \quad c \in \mathbb{R}?$$

(Hint: Consider the transformation $x = e^{-t}y$.)

(4) Does this imply anything about the system

$$\dot{x}_1 = x_2,$$
$$\dot{x}_2 = -x_1 + a(t)x,$$

where x_1 and x_2 are scalar functions, and a is a continuous function satisfying

$$\int_c^\infty |a(t)| \, dt < \infty \quad \text{for every} \quad c \in \mathbb{R}?$$

Exercise 2.19 Consider the population model

$$\frac{dx}{dt} = rx \left(1 - \frac{x}{k}\right) - \frac{mx}{a+x}y,$$
$$\frac{dy}{dt} = sy \left(1 - \frac{y}{hx}\right),$$
$$x(0) > 0, \quad y(0) > 0, \quad a, h, k, m, r, s > 0.$$

Show that the solutions $x(t)$, $y(t)$ are positive and bounded.

Exercise 2.20 Consider the initial value problem

$$\ddot{z} + \alpha(z, \dot{z})\dot{z} + \beta(z) = u(t), \ z(0) = \xi, \ \dot{z}(0) = \eta,$$

where α, β together with all of their first partial derivatives are continuous, and u is a bounded, continuous function defined on $\mathbb{R}$. Suppose $\alpha \geq 0$

on $\mathbb{R}^2$ and $z\beta(z) \geq 0$ for all $z \in \mathbb{R}$. Show that there is one and only one solution to this problem and the solution can be defined on $[0, \infty)$. (Hint: Write the equations as a system by letting $z = x$, $\dot{z} = y$, define $V(x, y) = y^2/2 + \int_0^x \beta(s)ds$ and study the rate of change of $V(x(t), y(t))$ along the solutions of the two-dimensional system.)

Exercise 2.21 Let $g : R \to R$ be Lipschitz and $f : R \to R$ continuous. Show that the system

$$x' = g(x),$$
$$y' = f(x)y,$$

has at most one solution on any interval, for a given initial value. (Hint: Use Gronwall's inequality.)

Exercise 2.22 Consider the differential equation $x' = x^{2/3}$.

(1) There are infinitely many solutions satisfying $x(0) = 0$ on every interval $[0, \beta]$.
(2) For what values of α are there infinitely many solutions on $[0, \alpha]$ satisfying $x(0) = -1$.

Exercise 2.23 Let $f \in C^1$ on the (t, x, y) set given by $0 \leq t \leq 1$, and all x, y. Let φ be a solution of the second-order equation $x'' = f(t, x, x')$ on $[0, 1]$, and let $\varphi(0) = a$, $\varphi(1) = b$. Suppose $\partial f/\partial x > 0$ for $t \in [0, 1]$ and for all x, y. Prove that if β is near b then there exists a solution ψ of $x'' = f(t, x, x')$ such that $\psi(0) = a$, $\psi(1) = \beta$. Hint: Consider the solution θ (as a function of (t, α)) with initial values $\theta(0, \alpha) = a$, $\theta'(0, \alpha) = \alpha$. Let $\varphi'(0) = \alpha_0$. Then for $|\alpha - \alpha_0|$ small, θ exists for $t \in [0, 1]$. Let

$$u(t) = \frac{\partial \theta}{\partial \alpha}(t, \alpha_0).$$

Then

$$u'' - \frac{\partial f}{\partial y}(t, \varphi(t), \varphi'(t))u' - \frac{\partial f}{\partial x}(t, \varphi(t), \varphi'(t))u = 0,$$

where $u(0) = 0$, $u'(0) = 1$. Since $\partial f/\partial x > 0$, u is monotone nondecreasing and thus $u(1) = (\partial \theta/\partial \alpha) > 0$. Thus the equation $\theta(1, \alpha) - \beta = 0$ can be solved for α as a function of β for (α, β) in a neighborhood of (α_0, b).

Exercise 2.24 Let $\text{Lip}(\varphi)$ be defined as

$$\text{Lip}(\varphi) = \sup_{x,y \in E,\ x \neq y} \frac{\parallel \varphi(x) - \varphi(y) \parallel}{\parallel x - y \parallel},$$

where $\varphi : E \to E$ is Lipschitz continuous, E is a Banach space. Assume $L : E \to E$ to be a linear, invertible operator. Show that if $\text{Lip}(\varphi) < \frac{1}{\|L^{-1}\|}$ then $L + \varphi$ is invertible with

$$\text{Lip}\left((L + \varphi)^{-1}\right) < \frac{1}{(\parallel L^{-1} \parallel^{-1} - \text{Lip}(\varphi))}.$$

Hint:

(i) Show that $L + \varphi$ is 1-1 by

$$\parallel (L + \varphi)x - (L + \varphi)y \parallel \geq \left(\frac{1}{\parallel L^{-1} \parallel} - \text{Lip}(\varphi) \right) \parallel x - y \parallel .$$

(ii) Prove that $L + \varphi$ is onto by solving $(L + \varphi)(x) = y$, given $y \in E$. Apply the contraction mapping principle.

Exercise 2.25 Prove that the Initial Value Problem

$$\begin{cases} \frac{dx}{dt} = f(t, x), \\ x(t_0) = x_0, \end{cases}$$

has a unique solution defined on the interval $[t_0, t_1]$ if $f(t, x)$ is continuous in the strip $t_0 \leq t \leq t_1$, $|x| < \infty$ and satisfies

$$|f(t, x_1) - f(t, x_2)| \leq L|x_1 - x_2| \text{ for some } L > 0.$$

Hint: Let X be $C(I)$, $I = [t_0, t_1]$, $K > L$ and define the norm

$$\|x\|_K = \sup_{t_0 \leq t \leq t_1} \left\{ e^{-K(t - t_0)} |x(t)| \right\}.$$

Then $(X, \|\ \|_K)$ is a Banach space and apply the contraction mapping principle.

Chapter 3

LINEAR SYSTEMS

3.1 Introduction

In this chapter we first study the general properties of the linear homogeneous system

$$\frac{dx}{dt} = A(t)x, \qquad (LH)$$

and linear nonhomogeneous system

$$\frac{dx}{dt} = A(t)x + g(t), \qquad (LN)$$

where $A(t) = (a_{ij}(t)) \in \mathbb{R}^{n \times n}$ and $g(t) \in \mathbb{R}^n$ are continuous on $\mathbb{R}$. There are two important special cases, namely

(i) $A(t) \equiv A$ is an $n \times n$ constant matrix;
(ii) $A(t + \omega) = A(t)$, $A(t)$ is a periodic matrix.

Example 3.1.1 From elementary ordinary differential equations, we know that the solution of scalar equations

$$x' = a(t)x, \quad x(0) = x_0,$$
$$\text{and}$$
$$x' = ax, \quad x(0) = x_0, \quad a \in \mathbb{R}$$

are $x(t) = x_0 \exp\left(\int_0^t a(s)ds\right)$ and $x(t) = x_0 e^{at}$ respectively. For the systems of linear equations,

$$\frac{dx}{dt} = Ax \quad \text{and} \quad \frac{dx}{dt} = A(t)x$$
$$x(0) = x_0 \qquad\qquad x(0) = x_0.$$

Is it true that $x(t) = e^{At}x_0$ and $x(t) = \exp\left(\displaystyle\int_0^t A(s)ds\right) x_0$ (where e^{At} will be defined in Section 3.3)?

Remark 3.1.1 Theory of linear system has many applications in engineering, especially in linear control theory. It is also very important to understand linear systems in order to study nonlinear systems. For instance, if we do the linearization about an equilibrium x^* of a nonlinear system $\frac{dx}{dt} = f(x)$, then the behavior of the solutions of linear system $x' = Ax$, $A = Df(x^*)$ gives us the local behavior of the solutions of nonlinear system $\frac{dx}{dt} = f(x)$ in a neighborhood of x^*. To understand the local behavior of the solutions near a periodic orbit $\{p(t)\}_{0 \leq t \leq T}$, of the nonlinear system $\frac{dx}{dt} = f(x)$, the linearization yields the linear periodic system $\frac{dx}{dt} = A(t)x$, where $A(t) = Df(p(t))$. This is the orbital stability we shall study later in Chapter 4. Also there are some well-known equations arising from physics like Matheiu's equations and Hill's equations which are second order linear periodic equations.

3.2   Fundamental Matrices

In this section we shall study the structure of the solutions of (LH) and (LN).

Theorem 3.2.1 *The set V of all solutions of (LH) on $J = (-\infty, \infty)$ is an n-dimensional vector space.*

Proof. It is easy to verify that V is a vector space over $\mathbb{C}$. We shall show $\dim V = n$. Let $\varphi_i(t)$, $i = 1, \cdots, n$ be the unique solution of $\frac{dx}{dt} = A(t)x$, $x(0) = e_i = (0 \cdots 1 \cdots 0)^T$.
Claim: $\{\varphi_i\}_{i=1}^n$ are linearly independent. Let $\sum_{i=1}^n \alpha_i \varphi_i = 0$. It follows that $\sum_{i=1}^n \alpha_i \varphi_i(t) = 0$ for all t. Setting $t = 0$ yields $\sum_{i=1}^n \alpha_i e_i = 0$, i.e., $\alpha_i = 0$ for all $i = 1, 2, \cdots, n$. Next we show that any solution φ of (LH) can be generated by $\varphi_1, \cdots, \varphi_n$. Let $\varphi(0) = \xi = (\xi_1, \cdots, \xi_n)^T$. Then $y(t) = \sum_{i=1}^n \xi_i \varphi_i(t)$ is a solution of (LH) with $y(0) = \xi$. From uniqueness of solutions, we have $y(t) \equiv \varphi(t)$ for all t, i.e., $\varphi = \sum_{i=1}^n \xi_i \varphi_i$. $\qquad\square$

Next we introduce the fundamental matrices.

Definition 3.2.1 *Let* $\varphi_1, \cdots, \varphi_n$ *be* n *linearly independent solutions of (LH) on* $\mathbb{R}$. *We call the matrix* $\Phi = [\varphi_1, \cdots, \varphi_n] \in \mathbb{R}^{n \times n}$ *a fundamental matrix of (LH).*

There are infinitely many fundamental matrices. Consider the following matrix equation

$$X' = A(t)X, \tag{3.1}$$

where $X = (x_{ij}(t))$ and $X' = (x'_{ij}(t))$.

Theorem 3.2.2 *A fundamental matrix* $\Phi(t)$ *of (LH) satisfies (3.1).*

Proof.

$$
\begin{aligned}
\Phi'(t) &= [\varphi'_1(t), \cdots, \varphi'_n(t)] \\
&= [A(t)\varphi_1, \cdots, A(t)\varphi_n] \\
&= A(t)[\varphi_1, \cdots, \varphi_n] = A(t)\Phi(t).
\end{aligned}
$$

$\square$

Obviously for any $\xi \in \mathbb{R}^n$, $\Phi(t)\xi$ is a solution of (LH) since $(\Phi(t)\xi)' = \Phi'(t)\xi = A(t)(\Phi(t)\xi)$.

In the following we prove the Abel's formula in which $\det \Phi(t)$ is a generalization of Wronskian for n-th order scalar linear equations.

Theorem 3.2.3 *(Liouville's formula or Abel's formula).*
Let $\Phi(t)$ *be a solution of (3.1) on* $J = (-\infty, \infty)$ *then*

$$\det \Phi(t) = \det \Phi(\tau) \exp \left(\int_{\tau}^{t} tr A(s) ds \right). \tag{3.2}$$

Proof. Let $\Phi(t) = (\varphi_{ij}(t))$, $A(t) = (a_{ij}(t))$. From $\Phi' = A(t)\Phi$, we have $\varphi'_{ij} = \sum_{k=1}^{n} a_{ik}\varphi_{kj}$. By induction (Exercises!), we have

$$\frac{d}{dt}(\det \Phi(t)) = \frac{d}{dt} \begin{vmatrix} \varphi_{11} & \cdots\cdots & \varphi_{1n} \\ \varphi_{21} & & \varphi_{2n} \\ \vdots & & \vdots \\ \varphi_{n1} & & \varphi_{nn} \end{vmatrix}$$

$$= \begin{vmatrix} \varphi_{11}' & \varphi_{12}' & \cdots & \varphi_{1n}' \\ \varphi_{21} & \varphi_{22} & & \varphi_{2n} \\ \vdots & & & \\ \varphi_{n1} & \varphi_{n2} & & \varphi_{nn} \end{vmatrix} + \begin{vmatrix} \varphi_{11} & \varphi_{12} & \cdots & \varphi_{1n} \\ \varphi_{21}' & \varphi_{22}' & & \varphi_{2n}' \\ \vdots & & & \\ \varphi_{n1} & & & \varphi_{nn} \end{vmatrix}$$

$$+ \cdots + \begin{vmatrix} \varphi_{11} & \cdots & \varphi_{1n} \\ \varphi_{21} & & \varphi_{2n} \\ \vdots & & \\ \varphi_{n1}' & \cdots & \varphi_{nn}' \end{vmatrix}$$

$$= \begin{vmatrix} \sum_{k=1}^{n} a_{1k}\varphi_{k1}, & \cdots\cdots & , & \sum_{k=1}^{n} a_{1k}\varphi_{kn} \\ \varphi_{21} & \cdots\cdots & & \varphi_{2n} \\ & & & \\ \varphi_{n1} & \cdots\cdots & & \varphi_{nn} \end{vmatrix} + \cdots +$$

$$\begin{vmatrix} \varphi_{11} & \varphi_{12} & \cdots & & \varphi_{1n} \\ \vdots & & & & \\ \varphi_{n-1,1} & & \cdots\cdots & & \varphi_{n-1,n} \\ \sum_{k=1}^{n} a_{nk}\varphi_{k1}, & & & , & \sum_{k=1}^{n} a_{nk}\varphi_{kn} \end{vmatrix}$$

$$= a_{11}(t) \det \Phi(t) + \cdots + a_{nn}(t) \det \Phi(t)$$

$$= \text{trace}(A(t)) \det \Phi(t).$$

Hence we have

$$\det \Phi(t) = \det \Phi(\tau) \exp\left(\int_{\tau}^{t} tr(A(s))ds \right).$$

□

Remark 3.2.1 The Abel's formula can be interpreted geometrically as follows: From (3.1) and Theorem 3.2.2, we have $\Phi' = A(t)\Phi$. Obviously (3.1) describes the evolutions of the initial vector $(\varphi_1(\tau), \cdots, \varphi_n(\tau))$ according to the dynamics of (3.1). Then Liouville's formula describes the evolution of the volume of a parallelepiped generated by $\varphi_1(\tau), \cdots, \varphi_n(\tau)$.

Theorem 3.2.4 *A solution $\Phi(t)$ of (3.1) is a fundamental matrix of (LH) iff* $\det \Phi(t) \neq 0$ *for all t.*

Proof. From (3.2) we have $\det \Phi(t) \neq 0$ for all t iff $\det \Phi(t) \neq 0$ for some t. If $\Phi(t)$ is a fundamental matrix, then $\varphi_1, \cdots, \varphi_n$ are linearly independent and hence $\det \Phi(t) \neq 0$ for all t. Conversely, if $\Phi(t)$ satisfies (3.1) and $\det \Phi(t) \neq 0$ for all t, then $\varphi_1, \cdots, \varphi_n$ are linearly independent and $\Phi(t)$ is a fundamental matrix of (LH). $\qquad\qquad\square$

Example 3.2.1 Consider n-th order linear equation

$$x^{(n)}(t) + a_1(t)x^{(n-1)}(t) + \cdots + a_n(t)x(t) = 0. \tag{3.3}$$

It can be reduced to a first order linear system. Let $x_1 = x$, $x_2 = x', \cdots, x_n = x^{(n-1)}$. Then we have

$$
\begin{bmatrix} x_1 \\ x_2 \\ \vdots \\ x_n \end{bmatrix}' =
\begin{bmatrix} 0 & 1 & 0 & \cdots & 0 \\ 0 & 0 & 1 & & 0 \\ \vdots & & & & \vdots \\ -a_n(t) & \cdots & & & -a_1(t) \end{bmatrix}
\begin{bmatrix} x_1 \\ x_2 \\ \vdots \\ x_n \end{bmatrix}. \tag{3.4}
$$

If $\varphi_1(t), \varphi_2(t), \cdots, \varphi_n(t)$ are n linearly independent solutions of (3.3), then

$$
\begin{bmatrix} \varphi_1 \\ \varphi_1' \\ \vdots \\ \varphi_1^{(n-1)} \end{bmatrix}, \cdots,
\begin{bmatrix} \varphi_n \\ \varphi_n' \\ \vdots \\ \varphi_n^{(n-1)} \end{bmatrix}
$$

are n linearly independent solutions of (3.4).

Let

$$
\Phi(t) =
\begin{bmatrix} \varphi_1 & \cdots & \varphi_n \\ \varphi_1' & & \varphi_n' \\ \vdots & & \vdots \\ \varphi_1^{(n-1)} & & \varphi_n^{(n-1)} \end{bmatrix}
$$

and

$$
W(\varphi_1, \cdots, \varphi_n) = \det \Phi(t)
$$

is the Wronskian of (3.3).

From (3.2) we have

$$W(\varphi_1, \cdots, \varphi_n)(t) = W(\varphi_1, \cdots, \varphi_n)(t_0) \cdot \exp\left(-\int_{t_0}^t a_1(s)ds\right).$$

Remark 3.2.2 It is very difficult to compute explicitly a fundamental matrix $\Phi(t)$ for general $A(t)$. In the case of constant coefficients, $A(t) \equiv A$, we shall show that the exponential matrix e^{At} is a fundamental matrix in the following section.

Theorem 3.2.5 *Let $\Phi(t)$ be a fundamental matrix of (LH) and $C \in \mathbb{R}^{n \times n}$ be nonsingular. Then $\Phi(t)C$ is also a fundamental matrix. Conversely if $\Psi(t)$ is also a fundamental matrix, then there exists a nonsingular matrix P such that $\Psi(t) = \Phi(t)P$ for all t.*

Proof. Since

$$(\Phi(t)C)' = \Phi'(t)C = A(t)(\Phi(t)C)$$

and

$$\det(\Phi(t)C) = \det \Phi(t) \cdot \det C \neq 0,$$

then $\Phi(t)C$ is a fundamental matrix. Now we consider $\Phi^{-1}(t)\Psi(t)$ and we want to show $\frac{d}{dt}(\Phi^{-1}(t)\Psi(t)) \equiv 0$. Then we complete the proof by setting $P \equiv \Phi^{-1}(t)\Psi(t)$. From

$$\frac{d}{dt}(\Phi^{-1}\Psi) = (\Phi^{-1})'\Psi + \Phi^{-1}\Psi'.$$

Since $\Phi\Phi^{-1} = I$, we have $\Phi'(\Phi^{-1}) + \Phi(\Phi^{-1})' = 0$ and $(\Phi^{-1})' = -\Phi^{-1}\Phi'\Phi^{-1}$. Then,

$$\frac{d}{dt}\left(\Phi^{-1}\Psi\right) = -\Phi^{-1}\Phi'\Phi^{-1}\Psi + \Phi^{-1}\Psi'$$
$$= -\Phi^{-1}A\Phi\Phi^{-1}\Psi + \Phi^{-1}A\Psi = 0.$$

$\square$

Example 3.2.2 The solution $\varphi(t)$ of the scalar equation

$$x'(t) = ax(t) + g(t)$$
$$x(\tau) = \xi$$

can be found by method of integrating factor. Then

$$\varphi(t) = e^{a(t-\tau)}\xi + \int_\tau^t e^{a(t-\eta)}g(\eta)d\eta.$$

In the following, we shall discuss the "variation of constant formula" which is very important in the linearized stability theory.

Theorem 3.2.6 *(Variation of Constant formula).*
Let $\Phi(t)$ be a fundamental matrix of (LH) and $\varphi(t, \tau, \xi)$ be the solution of (LN)

$$\begin{cases} \frac{dx}{dt} = A(t)x + g(t), \\ x(\tau) = \xi. \end{cases} \qquad (LN)$$

Then

$$\varphi(t, \tau, \xi) = \Phi(t)\Phi^{-1}(\tau)\xi + \int_{\tau}^{t} \Phi(t)\Phi^{-1}(\eta)g(\eta)d\eta. \qquad (3.5)$$

Proof. Let $\psi(t)$ be the RHS of (3.5). Then

$$\begin{aligned} \psi'(t) &= \Phi'(t)\Phi^{-1}(\tau)\xi + g(t) + \int_{\tau}^{t} \Phi'(t)\Phi^{-1}(\eta)g(\eta)d\eta \\ &= A(t)\left[\Phi(t)\Phi^{-1}(\tau)\xi + \int_{\tau}^{t} \Phi(t)\Phi^{-1}(\eta)g(\eta)d\eta\right] + g(t) \\ &= A(t)\psi(t) + g(t). \end{aligned}$$

Since $\psi(\tau) = \xi$, then by the uniqueness of the solutions, it follows that $\psi(t) \equiv \varphi(t, \tau, \xi)$ for all t. Thus we complete the proof. $\square$

Remark 3.2.3 Set $\xi = 0$ then $\varphi_\rho(t) = \int_{\tau}^{t} \Phi(t)\Phi^{-1}(\eta)g(\eta)d\eta$ is a particular solution of (LN). Obviously $\Phi(t)\Phi^{-1}(\tau)\xi$ is a general solution of (LH). Thus we have

general solution of (LN)

= general solution of (LH) + particular solution of (LN).

Remark 3.2.4 If $A(t) \equiv A$, then $\Phi(t) = e^{At}$ and $\varphi(t, \tau, \xi) = e^{A(t-\tau)}\xi + \int_{\tau}^{t} e^{A(t-\eta)}g(\eta)d\eta$ which will be used in the theory of linearization.

3.3 Linear Systems with Constant Coefficients

In this section we shall study the linear system with constant coefficients,

$$\begin{cases} x' = Ax, \quad A = (a_{ij}) \in \mathbb{R}^{n \times n}, \\ x(0) = x_0. \end{cases} \qquad (LC)$$

It is easy to guess that the solution of the above should be $x(t) = e^{At}x_0$. We need to define the exponential matrix e^A.

Definition 3.3.1 *For $A \in \mathbb{R}^{n \times n}$,*

$$e^A = I + A + \frac{A^2}{2!} + \cdots + \frac{A^n}{n!} + \cdots$$
$$= \sum_{n=0}^{\infty} \frac{A^n}{n!}.$$

In order to show that the above series of matrices is well-defined, we need to define the norms of matrix $A \in \mathbb{R}^{m \times n}$ for any m, $n \in \mathbb{Z}^+$.

Definition 3.3.2 *Let $A \in \mathbb{R}^{m \times n}, A = (a_{ij})$. Define*

$$\| A \| = \sup_{x \neq 0} \frac{\| Ax \|}{\| x \|} .$$

Remark 3.3.1 Obviously we have

$$\| A \| = \sup\{ \| Ax \| : \| x \| = 1 \}$$
$$= \sup\{ \| Ax \| : \| x \| \leq 1 \}.$$

The following facts are well known from Linear Algebra ([IK], p. 9). Let $x = (x_1, \cdots, x_n)$ and $A = (a_{ij})$. If $\| x \|_\infty = \sup |x_i|$, then $\| A \|_\infty = \sup_i \sum_k |a_{ik}|$; if $\| x \|_1 = \sum_i |x_i|$, then $\| A \|_1 = \sup_k \sum_i |a_{ik}|$; if $\| x \|_2 = \left(\sum_i |x_i|^2 \right)^{1/2}$, then $\| A \|_2 = \sqrt{\rho(A^T A)}$, where

$$\rho(A^T A) = \text{Spectral radius of } A^T A$$
$$= \text{Max}\{ |\lambda| : \lambda \in \sigma(A^T A) \},$$

and

$$\sigma(A^T A) = \text{Spectrum of } A^T A$$
$$= \text{set of eigenvalues of } A^T A.$$

From the definition of $\| A \|$, it follows that

$\| Ax \| \leq \| A \| \| x \|$, $A \in \mathbb{R}^{m \times n}$, $x \in \mathbb{R}^n$;
$\| A + B \| \leq \| A \| + \| B \|$, $A, B \in \mathbb{R}^{m \times n}$;
$\| cA \| = \| c \| \| A \|$, $c \in \mathbb{R}$;
$\| AB \| \leq \| A \| \| B \|$, $A \in \mathbb{R}^{m \times n}$, $B \in \mathbb{R}^{n \times p}$;
$\| A^n \| \leq \| A \|^n$, $A \in \mathbb{R}^{m \times m}$.

Theorem 3.3.1 e^A *is well-defined and* $\| e^A \| \leq e^{\| A \|}$.

Proof. Since for any $a \in \mathbb{R}$, $e^a = \sum_{n=0}^{\infty} \frac{a^n}{n!}$ converges, it follows that

$$\| \sum_{m}^{m+p} \frac{A^n}{n!} \| \le \sum_{m}^{m+p} \frac{\| A \|^n}{n!},$$

we complete the proof. □

Next we state and prove the properties of e^A.

Theorem 3.3.2
(i) $e^O = I$;
(ii) If $AB = BA$ then $e^{A+B} = e^A e^B$;
(iii) e^A is invertible and $(e^A)^{-1} = e^{-A}$;
(iv) If $B = PAP^{-1}$ then $e^B = Pe^A P^{-1}$.

Proof. (i) follows directly from the definition of e^A.

(ii) Since $AB = BA$, then $(A + B)^n = \sum_{m=0}^{n} \binom{n}{m} A^m B^{n-m}$ and

$$e^{A+B} = \sum_{n=0}^{\infty} \frac{(A + B)^n}{n!} = \sum_{n=0}^{\infty} \frac{1}{n!} \sum_{m=0}^{n} \binom{n}{m} A^m B^{n-m}$$

$$= \sum_{n=0}^{\infty} \frac{1}{n!} \sum_{m=0}^{n} \frac{n!}{m!(n-m)!} A^m B^{n-m}$$

$$= \sum_{n=0}^{\infty} \sum_{j+k=n} \frac{A^j}{j!} \frac{B^k}{k!}$$

$$= \left(\sum_{j=0}^{\infty} \frac{A^j}{j!} \right) \left(\sum_{j=0}^{\infty} \frac{B^k}{k!} \right) = e^A e^B.$$

The 2nd last equality holds because the two series converge absolutely.
(iii) Since $A(-A) = (-A)(A)$, then $I = e^O = e^{A+(-A)} = e^A e^{-A}$ and (iii)
follows.
(iv) Since $B = PAP^{-1}$ then $B^k = PA^k P^{-1}$ and $P \left(\sum_{k=0}^{n} \frac{A^k}{k!} \right) P^{-1} = \sum_{k=0}^{n} \frac{B^k}{k!}$.
Let $n \to \infty$ and we complete the proof of (iv).

□

Theorem 3.3.3 e^{At} is a fundamental matrix of $x' = Ax$.

Proof. Let $Y(t) = e^{At}$. Then

$$Y(t+h) - Y(t) = e^{A(t+h)} - e^{At}$$
$$= \left(e^{Ah} - I\right) e^{At} = \left(hA + O(h^2)\right) e^{At}.$$

Thus we have

$$Y'(t) = \lim_{h \to 0} \frac{Y(t+h) - Y(t)}{h} = Ae^{At} = AY(t).$$

$\square$

Consider the linear nonhomogeneous system

$$\begin{aligned} x' &= Ax + g(t), \\ x(\tau) &= \xi. \end{aligned} \qquad (3.6)$$

From (3.5) and Theorem 3.3.3, we have the following important variation of constant formula: the solution of (3.6) is

$$\varphi(t, \tau, \xi) = e^{A(t-\tau)}\xi + \int_\tau^t e^{A(t-\eta)} g(\eta) d\eta. \qquad (3.7)$$

Equation (3.7) is very important in showing linearized stability implies the nonlinear stability for the equilibrium solution of the nonlinear system $x' = f(x)$.

Since the solution of I.V.P. of (LC) is

$$x(t) = e^{At} x_0,$$

it is important to estimate $|x(t)|$ by

$$|x(t)| \le \| e^{At} \| \, |x_0|.$$

Thus we need to estimate $\| e^{At} \|$ and compute e^{At}.

In the following we compute e^{At}.

Example 3.3.1 If $A = D = \text{diag}\,(\lambda_1, \cdots, \lambda_n)$ then by the definition of e^{At}, we have

$$e^{At} = I + At + \frac{A^2 t^2}{2!} + \cdots = \text{diag}\,(e^{\lambda_1 t}, \cdots, e^{\lambda_n t}).$$

Example 3.3.2 If A is diagonalizable, then we have $D = P^{-1}AP = \text{diag}\,(\lambda_1, \cdots, \lambda_n)$ for some nonsingular P. Since $AP = PD$, if $P = [x_1, \cdots, x_n]$, then we have $Ax_k = \lambda_k x_k$, $k = 1, \cdots, n$ i.e. x_k is an eigenvector of the corresponding eigenvalue λ_k of A. From Theorem 3.3.2(iv), it follows that

$$e^{Dt} = P^{-1}e^{At}P \ \ \text{or} \ \ e^{At} = Pe^{Dt}P^{-1}.$$

Example 3.3.3 Let J be the Jordan block of A and $J = P^{-1}AP$, where

$$J = \begin{bmatrix} J_0 & & & \\ & J_1 & & \\ & & \ddots & \\ & & & J_s \end{bmatrix},$$

$$J_0 = \text{diag} \ (\lambda_1, \cdots, \lambda_k),$$

and

$$J_i = \begin{bmatrix} \lambda_i & 1 & & O \\ & \ddots & \ddots & \\ & & \ddots & 1 \\ O & & & \lambda_i \end{bmatrix}.$$

Then, $e^{Jt} = P^{-1}e^{At}P$.

In order to evaluate $\| e^{Jt} \|$ we need to review the Jordan forms.

Review of Jordan forms:

Let $A \in \mathbb{R}^{n \times n}$ and the characteristic polynomial of A be

$$f(\lambda) = \det(\lambda I - A) = (\lambda - \lambda_1)^{n_1} \cdots (\lambda - \lambda_s)^{n_s},$$

where $\lambda_1, \cdots, \lambda_s$ are distinct eigenvalues of A. We call n_i the algebraic multiplicity of λ_i. Let $V_i = \{v \in \mathbb{C}^n : Av = \lambda_i v\}$. We call V_i the eigenspace of λ_i and $m_i = \dim V_i$, the geometric multiplicity of λ_i. It is a well-known fact that $m_i \leq n_i$. Obviously $V_i = $ Null space of $(A - \lambda_i I) = N(A - \lambda_i I)$ and $N(A - \lambda_i I) \subseteq N(A - \lambda_i I)^2 \subseteq N(A - \lambda_i I)^3 \subseteq \cdots$. From linear algebra, it is a well-known result that

$$N(A - \lambda_i I)^{n_i} = N(A - \lambda_i I)^{n_i + 1} \subseteq \mathbb{C}^n.$$

Let $M(\lambda_i, A) = N(A - \lambda_i I)^{n_i}$. We call $M(\lambda_i, A)$ the generalized eigenspace of λ_i. We say that an eigenvalue λ has simple elementary divisor if $M(\lambda, A) = N(A - \lambda I)$, i.e., algebraic multiplicity of λ is equal to the geometric multiplicity. The following is a theorem from linear algebra.

Theorem 3.3.4 ([HS1]).
Let $\mathbb{C}^n = M(\lambda_1, A) \oplus \cdots \oplus M(\lambda_s, A)$.

To understand the structure of Jordan blocks, for simplicity, we may assume that the $n \times n$ matrix A has only one eigenvalue λ. Let $v_1, \cdots, v_k$ be a basis of eigenspace of λ. Let $v_i^{(1)}$ be solution of $(A - \lambda I)v = v_i$ and $v_i^{(j)}$ be a solution of

$$(A - \lambda I)v = v_i^{(j-1)}, \quad j = 2, \cdots, \ell_i.$$

We call $\{v_i^{(j)}\}_{j=1}^{\ell_i}$ the generalized eigenvectors of eigenvector v_i. It is easy to show that $\{v_i, v_i^{(1)}, \cdots, v_i^{(\ell_i)}\}_{i=1}^{k}$ is a basis of $\mathbb{C}^n$, $\sum_{i=1}^{k}(\ell_i + 1) = n$. Let

$$P = \left[v_1, v_1^{(1)}, \cdots, v_1^{(\ell_1)}, v_2, v_2^{(1)}, \cdots, v_2^{(\ell_2)}, \cdots, v_k, v_k^{(1)}, \cdots, v_k^{(\ell_k)} \right],$$

and

$$P^{-1} = \left[w_1, w_1^{(1)}, \cdots, w_1^{(\ell_1)}, \cdots, w_k, w_k^{(1)}, \cdots, w_k^{(\ell_k)} \right]^T.$$

Then it follows that

$$AP = \left[Av_1, Av_1^{(1)}, \cdots, Av_1^{(\ell_1)}, \cdots, Av_k, Av_k^{(1)}, \cdots, Av_k^{(\ell_k)} \right]$$

$$= \left[\lambda v_1, \lambda v_1^{(1)} + v_1, \lambda v_1^{(2)} + v_1^{(1)}, \cdots, \lambda v_1^{(\ell_1)} \right.$$

$$\left. + v_1^{(\ell_1-1)}, \cdots, \lambda v_k, \lambda v_k^{(1)} + v_k, \cdots, \lambda v_k^{(\ell_k)} + v_k^{(\ell_k-1)} \right]$$

$$= \lambda \left[v_1, v_1^{(1)}, \cdots, v_1^{(\ell_1)}, \cdots, v_k, v_k^{(1)}, \cdots, v_k^{(\ell_k)} \right]$$

$$+ \left[0, v_1, v_1^{(1)}, \cdots, v_1^{(\ell_1-1)}, 0, v_2, \cdots, v_2^{(\ell_2-1)}, \cdots, 0, v_k, \cdots, v_k^{(\ell_k-1)} \right].$$

Then,

$$P^{-1}AP = \lambda I + \begin{bmatrix} w_1 \\ w_1^{(1)} \\ \vdots \\ w_1^{(\ell_1)} \\ \vdots \\ w_k \\ \vdots \\ w_k^{(1)} \\ \vdots \\ w_k^{(\ell_k)} \end{bmatrix} \left[0, v_1, v_1^{(1)}, \cdots, v_1^{(\ell_1-1)}, \cdots, 0, v_k, \cdots, v_k^{(\ell_k-1)} \right]$$

$$= \lambda I + \begin{bmatrix} \begin{matrix} 0\ 1\ 0\ \cdots\cdot\ 0 \\ 0\ 0\ 1\ \cdots\cdot\ 0 \\ \ddots\quad 1 \\ \ddots \\ 0\ \cdots\cdots\ 0 \end{matrix} & & \\ & \begin{matrix} 0\ 1 \\ \ddots\ \ddots \\ \ddots\ 1 \\ 0 \end{matrix} & \\ & & \begin{matrix} 0\ 1 \\ \ddots\ \ddots \\ \ddots\ 1 \\ 0 \end{matrix} \end{bmatrix}$$

$$= \begin{bmatrix} \begin{matrix} \lambda\ 1 \\ \ddots\ \ddots \\ \ddots\ 1 \\ \lambda \end{matrix} & & \\ & \begin{matrix} \lambda\ 1 \\ \ddots\ \ddots \\ \ddots\ 1 \\ \lambda \end{matrix} & \\ & & \begin{matrix} \lambda\ 1 \\ \ddots\ \ddots \\ \ddots\ 1 \\ \lambda \end{matrix} \end{bmatrix}.$$

Thus, for general case we have

$$P^{-1}AP = \begin{bmatrix} J_0 & & & \\ & J_1 & & \\ & & \ddots & \\ & & & J_s \end{bmatrix},$$

where

$$J_0 = \text{diag } (\lambda_1 \cdots \lambda_I),$$

$$I = \text{cardinal number of } \lambda \text{ with } n_\lambda = m_\lambda,$$

$$J_i = \begin{bmatrix} J_{i1} & & \\ & \ddots & \\ & & J_{i\ell} \end{bmatrix}, \quad J_{ij} = \begin{bmatrix} \lambda_i & 1 & & O \\ & \ddots & \ddots & \\ & & \ddots & 1 \\ O & & & \lambda_i \end{bmatrix}.$$

Lemma 3.3.1 *Let* $J = \begin{bmatrix} \lambda & 1 & & O \\ & \ddots & \ddots & \\ & & \ddots & 1 \\ O & & & \lambda \end{bmatrix}$ *be an* $n_p \times n_p$ *matrix.*

Then

$$e^{Jt} = e^{\lambda t} \begin{bmatrix} 1, \, t, \, \frac{t^2}{2!}, \, \cdots & \frac{t^{n_p-1}}{(n_p-1)!} \\ 0, 1, \, t \, , & , \frac{t^{n_p-2}}{(n_p-2)!} \\ & \ddots \\ & \ddots \\ & 1 \end{bmatrix}. \qquad (3.8)$$

Proof. $e^{Jt} = e^{(\lambda I + N)t} = e^{\lambda It + Nt} = e^{\lambda t} e^{Nt}$.
Since $N^{n_p} = 0$, we have $e^{Jt} = e^{\lambda t} \sum_{k=0}^{n_p-1} \frac{t^k N^k}{k!}$. Then Lemma 3.3.1 follows
directly from routine computations.   □

Let $Re \; \lambda(A) = \max\{\text{Re}\lambda : \lambda \text{ is an eigenvalue of } A\}$.

Theorem 3.3.5 *If* $Re \; \lambda(A) < 0$ *then there exists* $\alpha > 0$ *and* $K > 0$ *such that* $\left\| e^{At} \right\| \le K e^{-\alpha t}, t \ge 0$.

If $Re \; \lambda(A) \le 0$ *and assume that those eigenvalues with zero real parts have simple elementary divisors, then* $\left\| e^{At} \right\| \le M$ *for* $t \ge 0$ *and some* $M > 0$.

Proof. Let $Re\ \lambda(A) < 0$. From (3.8) and Theorem 3.3.4, it follows that

$$e^{At} = Pe^{Jt}P^{-1},$$

and

$$\left\|e^{At}\right\| \leq \|P\| \left\|e^{Jt}\right\| \left\|P^{-1}\right\| \leq Ke^{-\alpha t},$$

where $0 < \alpha < -Re\ \lambda(A)$ and $K > 0$.

When $Re\ \lambda(A) \leq 0$ and those eigenvalues with zero real parts have simple elementary divisors, then

$$\left\|e^{At}\right\| \leq \|P\| \left\|e^{Jt}\right\| \left\|P^{-1}\right\| \leq M.$$

$\square$

Theorem 3.3.6 *Let* $\lambda_1, \cdots, \lambda_s$ *be distinct eigenvalues of* A *and* $M(\lambda_1, A)$, $\cdots, M(\lambda_s, A)$ *be the corresponding generalized eigenspaces. Let* $x_0 = \sum_{j=1}^{s} x^{0,j}$ *where* $x^{0,j} \in M(\lambda_j, A)$. *Then, the solution of (LC) can be written as:*

$$x(t) = e^{At}x_0 = \sum_{j=1}^{s} \left[\sum_{k=1}^{n_j-1} (A - \lambda_j I)^k \frac{t^k}{k!} \right] x^{0,j} e^{\lambda_j t}. \tag{3.9}$$

Proof. Then,

$$e^{At}x^0 = e^{(A-\lambda I)t}e^{\lambda t}x^0 = \left(e^{(A-\lambda I)t}x^0 \right) e^{\lambda t}$$

$$= \left[\sum_{k=0}^{\infty} (A - \lambda I)^k \frac{t^k}{k!} \right] x^0 e^{\lambda t}.$$

If $x^0 \in M(\lambda, A)$ then

$$e^{(A-\lambda I)t}x^0 = \sum_{k=0}^{n_\lambda-1} \left[(A - \lambda I)^k \frac{t^k}{k!} \right] x^0.$$

Hence for any $x^0 \in \mathbb{C}^n$,

$$x(t) = \sum_{j=1}^{s} \left[\sum_{k=1}^{n_j-1} (A - \lambda_j I)^k \frac{t^k}{k!} \right] x^{0,j} e^{\lambda_j t}.$$

$\square$

Remark 3.3.2 Given $A \in \mathbb{R}^{n \times n}$. How do we verify analytically that A is a stable matrix, i.e. $Re\ \lambda(A) < 0$? Suppose the characteristic polynomial of A is

$$g(z) = \det(zI - A) = a_0 z^n + a_1 z^{n-1} + \cdots + a_n, \quad (a_0 > 0).$$

The Routh-Hurwitz Criterion provides a necessary and sufficient condition for a real polynomial $g(z) = a_0 z^n + \cdots + a_n$, $a_0 > 0$ to have all roots with negative real parts (see [Cop] p. 158). In the following we list the conditions for $n = 2, 3, 4$ which are frequently used in applications.

n	$f(z)$	R–H criterion
2	$a_0 z^2 + a_1 z + a_2$	$a_2 > 0, \ a_1 > 0$
3	$a_0 z^3 + a_1 z^2 + a_2 z + a_3$	$a_3 > 0, \ a_1 > 0$
		$a_1 a_2 > a_0 a_3$
4	$a_0 z^4 + a_1 z^3 + a_2 z^2 + a_3 z + a_4$	$a_4 > 0, \ a_2 > 0,$
		$a_1 > 0,$
		$a_3(a_1 a_2 - a_0 a_3) > a_1^2 a_4$

3.4   Two-Dimensional Linear Autonomous Systems

In this section we shall apply Theorem 3.3.6 to classify the behavior of the solutions of two-dimensional linear Systems [H1]

$$\dot{x} = Ax, \ A = \begin{bmatrix} a & b \\ c & d \end{bmatrix}, \ \det A \neq 0 \qquad (3.10)$$

where a, b, c, d are real constants. Then $(0,0)$ is the unique rest point of (3.10). Let λ_1, λ_2 be the eigenvalues of A, consider the following cases:

Case 1: λ_1, λ_2 are real and $\lambda_2 < \lambda_1$.
    Let v^1, v^2 be unit eigenvectors of A associated with λ_1, λ_2 respectively. Then from (3.9), the general real solution of (3.10) is

$$x(t) = c_1 e^{\lambda_1 t} v^1 + c_2 e^{\lambda_2 t} v^2.$$

Case 1a (Stable node) $\lambda_2 < \lambda_1 < 0$.
    Let L_1, L_2 be the lines generated by v^1, v^2 respectively. Since $\lambda_2 < \lambda_1 < 0$, $x(t) \approx c_1 e^{\lambda_1 t} v^1$ as $t \to \infty$ and the trajectories are tangent to L_1. The origin is a stable node. (See Fig. 3.1.)

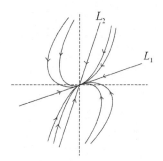

Fig. 3.1

Case 1b (Unstable node) $0 < \lambda_2 < \lambda_1$.

Then $x(t) \approx c_1 e^{\lambda_1 t} v^1$ as $t \to \infty$. The origin is an unstable node. (See Fig. 3.2.)

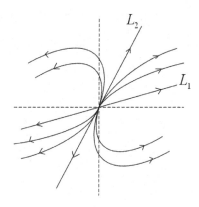

Fig. 3.2

Case 1c (Saddle point) $\lambda_2 < 0 < \lambda_1$. In this case, the origin is called a saddle point and L_1, L_2 are called unstable manifold and stable manifold of the rest point $(0,0)$ respectively. (See Fig. 3.3.)

Case 2: λ_1, λ_2 are complex.

Let $\lambda_1 = \alpha + i\beta$, $\lambda_2 = \alpha - i\beta$ and $v^1 = u + iv$ and $v^2 = u - iv$ be complex eigenvectors. Then

$$x(t) = ce^{(\alpha+i\beta)t} v^1 + \bar{c} e^{(\alpha-i\beta)t} \overline{v^1} = 2Re\left(ce^{(\alpha+i\beta)t} v^1\right).$$

Let $c = ae^{i\delta}$. Then

$$x(t) = 2ae^{\alpha t}\left(u\cos(\beta t + \delta) - v\sin(\beta t + \delta)\right).$$

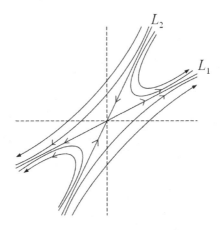

Fig. 3.3

Let U and V be the lines generated by u, v respectively.

Case 2a (Center) $\alpha = 0$, $\beta \neq 0$. The origin is called a center. (See Fig. 3.4.)

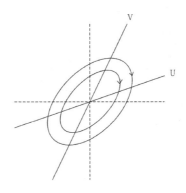

Fig. 3.4

Case 2b (Stable focus, spiral) $\alpha < 0$, $\beta \neq 0$. The origin is called a stable focus or stable spiral. (See Fig. 3.5.)

Case 2c (Unstable focus, spiral) $\alpha > 0$, $\beta \neq 0$. The origin is called an unstable focus or unstable spiral. (See Fig. 3.6.)

Case 3: (Improper nodes) $\lambda_1 = \lambda_2 = \lambda$.

Case 3a There are two linearly independent eigenvectors v^1 and v^2 of

Fig. 3.5

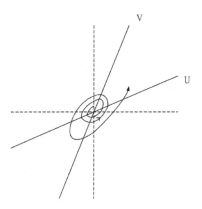

Fig. 3.6

the eigenvalue λ. Then,

$$x(t) = \left(c_1 v^1 + c_2 v^2\right) e^{\lambda t}.$$

If $\lambda > 0$ ($\lambda < 0$) then the origin 0 is called an unstable (stable) improper node. (See Fig. 3.7.)

Case 3b There is only one eigenvector v^1 associated with eigenvalue λ. Then from (3.9)

$$x(t) = (c_1 + c_2 t)\, e^{\lambda t} v^1 + c_2 e^{\lambda t} v^2$$

where v^2 is any vector independent of v^1. (See Fig. 3.8.)

Fig. 3.7

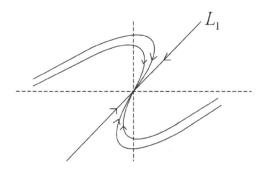

Fig. 3.8

3.5   Linear Systems with Periodic Coefficients

In this section, we shall study the linear periodic systems

$$x' = A(t)x, \quad A(t) = (a_{ij}(t)) \in \mathbb{R}^{n \times n}, \qquad (LP)$$

where $A(t)$ is continuous on $\mathbb{R}$ and is periodic with period T, i.e., $A(t) = A(t + T)$ for all t. We shall analyze the structure of the solutions $x(t)$ of (LP). Before we prove the main results we need the following theorem concerning the logarithm of a nonsingular matrix.

Theorem 3.5.1 *Let $B \in \mathbb{R}^{n \times n}$ be nonsingular. Then there exists $A \in \mathbb{C}^{n \times n}$, called logarithm of B, satisfying $e^A = B$.*

Proof. Let $B = PJP^{-1}$ where J is a Jordan form of B, $J = \text{diag}$ $(J_0, J_1, \cdots, J_s)$ with

$$J_0 = \begin{bmatrix} \lambda_1 & & \\ & \ddots & 0 \\ 0 & & \ddots \\ & & & \lambda_k \end{bmatrix} \quad \text{and} \quad J_i = \begin{bmatrix} \lambda_i & 1 & & 0 \\ & \ddots & \ddots & \\ 0 & & \ddots & 1 \\ & & & \lambda_i \end{bmatrix} \in \mathbb{C}^{n_i \times n_i},$$

$i = 1, \cdots, s$.

Since B is nonsingular, $\lambda_i \neq 0$ for all i. If $J = e^{\tilde{A}}$ for some $\tilde{A} \in \mathbb{C}^{n \times n}$ then it follows that $B = Pe^{\tilde{A}}P^{-1} = e^{P\tilde{A}P^{-1}} \stackrel{\text{def}}{=} e^A$. Hence it suffices to show that the theorem is true for Jordan blocks J_i, $i = 1, \cdots, s$. Write

$$J_i = \lambda_i \left(I + \frac{1}{\lambda_i} N_i \right), \quad N_i = \begin{pmatrix} 0 & 1 & & 0 \\ & \ddots & \ddots & \\ & & \ddots & 1 \\ 0 & & & 0 \end{pmatrix}.$$

Then $N_i^{n_i} = O$. From the identity

$$\log(1+x) = \sum_{p=1}^{\infty} \frac{(-1)^{p+1}}{p} x^p, \quad |x| < 1$$

and

$$e^{\log(1+x)} = 1 + x, \tag{3.11}$$

we formally write

$$\log J_i = (\log \lambda_i) I + \log \left(I + \tfrac{1}{\lambda_i} N_i \right)$$

$$= (\log \lambda_i) I + \sum_{p=1}^{\infty} \frac{(-1)^{p+1}}{p} \left(\frac{N_i}{\lambda_i} \right)^p. \tag{3.12}$$

From (3.12) we define

$$A_i = (\log \lambda_i) I + \sum_{p=1}^{n_i-1} \frac{(-1)^{p+1}}{p} \left(\frac{N_i}{\lambda_i} \right)^p.$$

Then from (3.11) we have

$$e^{A_i} = \exp((\log \lambda_i) I) \exp \left(\sum_{p=1}^{n_i-1} \frac{(-1)^{p+1}}{p} \left(\frac{N_i}{\lambda_i} \right)^p \right) = \lambda_i \left(I + \frac{N_i}{\lambda_i} \right) = J_i.$$

$\square$

Now we state our main results about the structure of the solutions of (LP).

Theorem 3.5.2 (Floquet Theorem).
If $\Phi(t)$ is a fundamental matrix of (LP), then so is $\Phi(t+T)$. Moreover there exists $P(t) \in \mathbb{C}^{n \times n}$ which is nonsingular and satisfies $P(t) = P(t+T)$ and there exists $R \in \mathbb{C}^{n \times n}$ such that

$$\Phi(t) = P(t)e^{tR}.$$

Proof. Let $\Psi(t) = \Phi(t+T)$. Then

$$\Psi'(t) = \Phi'(t+T) = A(t+T)\Phi(t+T) = A(t)\Psi(t).$$

From Theorem 3.2.5, there exists $C \in \mathbb{R}^{n \times n}$ nonsingular such that $\Phi(t+T) = \Phi(t)C$. By Theorem 3.5.1, $C = e^{TR}$ for some $R \in \mathbb{C}^{n \times n}$. Then we have

$$\Phi(t+T) = \Phi(t)e^{TR}. \tag{3.13}$$

Define

$$P(t) = \Phi(t)e^{-tR}.$$

Then $\Phi(t) = P(t)e^{tR}$ and we have

$$P(t+T) = \Phi(t+T)e^{-(t+T)R} = \Phi(t)e^{TR}e^{-(t+T)R}$$
$$= \Phi(t)e^{-tR} = P(t).$$

Thus we complete the proof of the theorem. $\qquad\square$

Definition 3.5.1 *The eigenvalues of $C = e^{TR}$ are called the characteristic multipliers (Floquet multipliers) of (LP). The eigenvalues of R are called the characteristic exponents (Floquet exponents) of (LP).*

Next we shall establish the relationship between Floquet multipliers and Floquet exponents. We also show that the Floquet multipliers are uniquely determined by the system (LP).

Theorem 3.5.3 *Let $\rho_1, \cdots, \rho_n$ be the characteristic exponents of (LP). Then the characteristic multipliers are $e^{T\rho_1}, \cdots, e^{T\rho_n}$. If ρ_i appears in the Jordan block of R, then $e^{T\rho_i}$ appears in the Jordan block of $C = e^{TR}$ with the same size.*

Proof. Since $\rho_1 \cdots \rho_n$ are eigenvalues of R, $T\rho_1 \cdots T\rho_n$ are eigenvalues of TR. Let $P^{-1}RP = J = \text{diag}\,(J_0, J_1, \cdots, J_s)$. Then $C = e^{TR} = Pe^{TJ}P^{-1}$ where $e^{TJ} = \text{diag}\left(e^{TJ_0},\ e^{TJ_1},\ \cdots, e^{TJ_s}\right)$. $\square$

Theorem 3.5.4 *The characteristic multipliers $\lambda_1, \cdots, \lambda_n$ are uniquely determined by $A(t)$ and all characteristic multipliers are nonzero.*

Proof. Let $\Phi_1(t)$ be another fundamental matrix of $x' = A(t)x$. Then there exists a nonsingular matrix C_1 such that $\Phi(t) = \Phi_1(t)C_1$. Then it follows that

$$\Phi_1(t + T)C_1 = \Phi(t + T) = \Phi(t)C = \Phi(t)e^{TR} = \Phi_1(t)C_1 e^{TR},$$

or

$$\Phi_1(t + T) = \Phi_1(t)C_1 e^{TR} C_1^{-1} = \Phi_1(t)e^{TR'}.$$

Then $e^{TR'} = C_1 e^{TR} C_1^{-1}$ and $e^{TR'}$, e^{TR} have the same eigenvalues. Hence the characteristic multipliers $\lambda_1 \cdots \lambda_n$ are uniquely determined by the system (LP). The fact that λ_i are nonzero follows directly from $\lambda_1 \cdots \lambda_n = \det e^{TR} \neq 0$. $\square$

Now we are in a position to study the asymptotic behavior of the solution $x(t)$ of (LP).

Theorem 3.5.5

(i) *Each solution $x(t)$ of (LP) approaches zero as $t \to \infty$ iff $|\lambda_i| < 1$ for each characteristic multiplier λ_i or, equivalently, $\text{Re}\ \rho_i < 0$ for each characteristic exponent ρ_i.*

(ii) *Each solution of (LP) is bounded iff*
  (1) $|\lambda_i| \leq 1$
  and
  (2) If $|\lambda_i| = 1$ then the corresponding Jordan block e^{TJ_i} is $[\lambda_i]$.

Proof. Consider the change of variable $x = P(t)y$. Then we have

$$x' = P'(t)y + P(t)y' = A(t)x = A(t)P(t)y,$$

or

$$P'(t)y + P(t)y' = A(t)P(t)y. \tag{3.14}$$

To compute $P'(t)$, we differentiate both sides of $\Phi(t) = P(t)e^{tR}$. Then we have

$$A(t)\Phi(t) = P'(t)e^{tR} + P(t)Re^{tR},$$

or

$$P'(t) = A(t)P(t) - P(t)R. \tag{3.15}$$

Then from (3.14) and (3.15) it follows that

$$y' = Ry. \tag{3.16}$$

Since $x(t) = P(t)y(t)$ and $P(t)$ is periodic and continuous, the theorem follows directly from Theorem 3.3.5. $\square$

Remark 3.5.1 If the fundamental matrix $\Phi(t)$ satisfies $\Phi(0) = I$, then from (3.13) the characteristic multipliers are eigenvalues of $\Phi(T)$. Hence we may compute the Floquet multipliers by solving the numerical solutions of the I.V.P.

$$x' = A(t)x$$
$$x(0) = e_i$$

and denote the solution by $y_i(t)$, $i = 1, 2, ..., n$. Then

$$\Phi(T) = [y_1(T), \cdots, y_n(T)].$$

Example 3.5.1 Hill's and Mathieu's equations ([JS] p. 237):
Consider a pendulum of length a with a bob mass m is suspended from a support which is constrained to move vertically with displacement $\xi(t)$. (See Fig. 3.9.)
    The kinetic energy T and potential energy V are given by

$$T = \frac{1}{2}m\left[\left(\dot{\xi} - a\dot{\theta}\sin\theta\right)^2 + a^2\dot{\theta}^2\cos^2\theta\right],$$
$$V = -mg(\xi + a\cos\theta).$$

The Lagrange's equation for $L = T - V$ is

$$\frac{d}{dt}\left(\frac{\partial L}{\partial \dot{\theta}}\right) = \frac{\partial L}{\partial \theta}.$$

From routine computations, it follows that

$$a\ddot{\theta} + (g - \ddot{\xi})\sin\theta = 0.$$

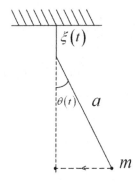

Fig. 3.9

For small oscillation, $\sin\theta \approx \theta$, we have

$$a\ddot{\theta} + \left(g - \ddot{\xi}\right)\theta = 0.$$

As a standard form for the above equation, we consider

$$x'' + (\alpha + p(t))x = 0.$$

When $p(t)$ is periodic, the above equation is known as Hill's equation. If $p(t) = \beta \cos t$, then the equation

$$\ddot{x} + (\alpha + \beta \cos t)\, x = 0$$

is called Mathieu's equation.

Now we consider the Mathieu's equation. Write it as a linear periodic system

$$\begin{pmatrix} x \\ y \end{pmatrix}' = \begin{pmatrix} 0 & 1 \\ -\alpha - \beta\cos t & 0 \end{pmatrix} \begin{pmatrix} x \\ y \end{pmatrix} = A(t)\begin{pmatrix} x \\ y \end{pmatrix}.$$

Let $\Phi(t)$ be the fundamental matrix with $\Phi(0) = I$. From $trace(A(t)) = 0$ and Abel's formula, the characteristic multipliers μ_1, μ_2 satisfy

$$\mu_1\mu_2 = \det \Phi(T) = \det \Phi(0)\exp\left(\int_0^T tr(A(t))dt\right) = 1,$$

and μ_1, μ_2 are the solutions of

$$\mu^2 - \mu\left(Trace(\Phi(T))\right) + \det \Phi(T) = 0.$$

Let $\varphi = \varphi(\alpha, \beta) = Trace(\Phi(T))$. Then

$$\mu_{1,2} = \frac{1}{2}\left[\varphi \pm \sqrt{\varphi^2 - 4}\right].$$

From Floquet Theorem, (3.16) and (3.9), we have five cases:

(i) $\varphi > 2$.

Let μ_1, $\mu_2 > 0$ and $\mu_1 > 1$, $\mu_2 = \frac{1}{\mu_1} < 1$. Then the characteristic exponents satisfying $\rho_1 > 0$ and $\rho_2 < 0$ and

$$x(t) = c_1 e^{\rho_1 t} p_1(t) + c_2 e^{\rho_2 t} p_2(t),$$

where $p_i(t)$ is periodic with period 2π, $i = 1, 2$.
Hence $\{(\alpha, \beta) : \varphi(\alpha, \beta) > 2\}$ is an unstable region.

(ii) $\varphi = 2$.

Then, $\mu_1 = \mu_2 = 1$, $\rho_1 = \rho_2 = 0$ and one solution is of period 2π, the other is unbounded.

(iii) $-2 < \varphi < 2$.

The characteristic multipliers are complex. In fact $|\mu_1| = |\mu_2| = 1$, $\rho_1 = iv$, $\rho_2 = -iv$ and $x(t) = c_1 e^{ivt} p_1(t) + c_2 e^{-ivt} p_2(t)$. Obviously the solutions are bounded and $\{(\alpha, \beta) : -2 < \varphi(\alpha, \beta) < 2\}$ is a stable region. We note that the solutions are oscillatory but not periodic for $v \neq 2\pi$ in general.

(iv) $\varphi = -2$.

Then $\mu_1 = \mu_2 = -1$. Set $T = 2\pi$ and let x_0 be an eigenvector of $\Phi(T)$ corresponding to the eigenvalue $\mu_1 = -1$. Consider $x(t) = \Phi(t)x_0$. Then,

$$x(t + T) = \Phi(t + T)x_0 = \Phi(t)\Phi(T)x_0 \text{ (Exercises!)}$$
$$= \Phi(t)(-x_0) = -\Phi(t)x_0.$$

Hence

$$x(t + 2T) = -\Phi(t + T)x_0 = -\Phi(t)\Phi(T)x_0,$$
$$= \Phi(t)x_0 = x(t)$$

and there is one periodic solution with period $2T = 4\pi$.

(v) $\varphi < -2$.

Then μ_1, μ_2 are real and negative and $\mu_1\mu_2 = 1$. Let $-1 < \mu_1 < 0$ and $\mu_2 < -1$. Write

$$\mu_1 = e^{T\rho_1} = e^{2\pi(-\sigma + \frac{i}{2})} = -e^{-2\pi\sigma}$$
$$\mu_2 = e^{2\pi(\sigma + \frac{i}{2})} = -e^{2\pi\sigma}$$

The general solutions take the form

$$x(t) = c_1 e^{(\sigma + \frac{i}{2})t} p_1(t) + c_2 e^{(-\sigma + \frac{i}{2})t} p_2(t),$$

where p_1, p_2 are periodic with period 2π or equivalently

$$x(t) = c_1 e^{\sigma t} q_1(t) + c_2 e^{-\sigma t} q_2(t),$$

where $q_1(t)$, $q_2(t)$ are periodic with period 4π.

The following theorem of Lyapunov gives a sufficient condition for stable solutions.

Theorem 3.5.6 (Lyapunov [H1] p. 130).
Let $p(t + \pi) = p(t) \not\equiv 0$ for all t, $\int_0^\pi p(t)dt \geq 0$ and $\int_0^\pi |p(t)|\, dt \leq \frac{4}{\pi}$, $p(t)$ is continuous on $\mathbb{R}$. Then, all solutions of the equation $u'' + p(t)u = 0$ are bounded.

Proof. It suffices to show that no characteristic multiplier is real. If μ_1, μ_2 are complex numbers, then $\mu_1\mu_2 = 1$ implies $\mu_1 = e^{i\nu}$ and $\mu_2 = e^{-i\nu}$ and $u(t) = c_1 e^{i\nu t} p_1(t) + c_2 e^{-i\nu t} p_2(t)$ are bounded. If μ_1 is real, then there exists a real solution $u(t)$ with $u(t + \pi) = \rho u(t)$ for all t for some $\rho \neq 0$. Then, either $u(t) \neq 0$ for all t or $u(t)$ has infinitely many zeros with two consecutive zeros a and b, $0 \leq b - a \leq \pi$.

Case I: $u(t) \neq 0$, for all t.
Then, we have $u(\pi) = \rho u(0)$, $u'(\pi) = \rho u'(0)$ and $u'(\pi)/u(\pi) = u'(0)/u(0)$.
Since $\frac{u''(t)}{u(t)} + p(t) = 0$, it follows that

$$\int_0^\pi \frac{u''(t)}{u(t)} dt + \int_0^\pi p(t)dt = 0.$$

Write

$$\int_0^\pi \frac{u''(t)}{u(t)} dt = \frac{u'}{u}\bigg|_0^\pi + \int_0^\pi \frac{(u'(t))^2}{u^2(t)} dt = \int_0^\pi \left(\frac{u'(t)}{u(t)}\right)^2 dt > 0.$$

Thus, we obtain a contradiction to the assumption $\int_0^\pi p(t)dt \geq 0$.
Case II: Assume $u(t) > 0$ for $a < t < b$, $u(a) = u(b) = 0$.
Let $u(c) = \max_{a \leq t \leq b} |u(t)|$. Then, we have

$$\frac{4}{\pi} \geq \int_0^\pi |p(t)|dt \geq \int_a^b |p(t)|dt = \int_a^b \left|\frac{u''(t)}{u(t)}\right| dt$$

$$\geq \frac{1}{u(c)} \int_a^b |u''(t)|dt \geq \frac{1}{u(c)}|u'(\alpha) - u'(\beta)|$$

for any $\alpha < \beta$ in (a, b). We note that

$$|u'(\alpha) - u'(\beta)| = \left|\int_\alpha^\beta u''(t)dt\right| \leq \int_\alpha^\beta |u''(t)|dt \leq \int_a^b |u''(t)|dt.$$

By Mean Value Theorem we have

$$\frac{u(c) - u(a)}{c - a} = u'(\alpha), \quad \frac{u(c) - u(b)}{b - c} = -u'(\beta),$$

for some $\alpha \in (a, c)$, $\beta \in (c, b)$. Hence it follows that

$$\frac{4}{\pi} > \frac{1}{u(c)} \left| u'(\alpha) - u'(\beta) \right| = \frac{1}{u(c)} \left| \frac{u(c)}{c - a} + \frac{u(c)}{b - c} \right|$$

$$= \frac{1}{c - a} + \frac{1}{b - c} = \frac{b - a}{(c - a)(b - c)} > \frac{4}{b - a} \geq \frac{4}{\pi}$$

here we apply the inequality $4xy \leq (x + y)^2$ with $x = c - a$, $y = b - c$. Thus we obtain a contradiction.

$\square$

3.6   Adjoint Systems

Let $\Phi(t)$ be a fundamental matrix of $x' = A(t)x$. Then we have

$$\left(\Phi^{-1} \right)' = -\Phi^{-1} \Phi' \Phi^{-1} = -\Phi^{-1} A(t).$$

Taking conjugate transpose yields

$$\left(\Phi^{*^{-1}} \right)' = -A^*(t) \left(\Phi^* \right)^{-1}.$$

Hence $(\Phi^*)^{-1}$ is a fundamental matrix of

$$y' = -A^*(t)y$$

and we call $Y' = -A^*(t)Y$ the adjoint of $Y' = A(t)Y$.

Theorem 3.6.1 *Let $\Phi(t)$ be a fundamental matrix of $x' = A(t)x$. Then $\Psi(t)$ is a fundamental matrix of $Y' = -A^*(t)Y$ iff $\Psi^* \Phi = C$ for some non-signular matrix C.*

Proof. If $\Psi(t)$ is a fundamental matrix of $Y' = -A^*(t)Y$, then $\Psi = (\Phi^*)^{-1}P$ for some nonsingular matrix P and $\Psi^* \Phi = P^* \Phi^{-1} \Phi = P^* \stackrel{\text{def}}{=} C$. On the other hand if $\Psi^* \Phi = C$, then $\Psi = (\Phi^*)^{-1}C^*$ and hence Ψ is a fundamental matrix of the adjoint system. $\square$

Example 3.6.1 Consider n-th order scalar linear equation

$$D^n y + a_1(t)D^{n-1}y + \cdots + a_n(t)y = 0.$$

Then we rewrite it in the form of first order system with $x_1 = y$, $x_2 = y', \cdots, x_n = y^{(n-1)}$ and we have

$$\begin{pmatrix} x_1 \\ \vdots \\ \vdots \\ x_n \end{pmatrix}' = \begin{pmatrix} 0 & 1 & \cdots\cdots & 0 \\ 0 & 0 & 1 & \cdots & 0 \\ & & & & \\ -a_n & -a_{n-1} & \cdots\cdots & \cdots\cdots & -a_1 \end{pmatrix} \begin{pmatrix} x_1 \\ \vdots \\ \vdots \\ x_n \end{pmatrix}$$

or

$$x' = A(t)x.$$

Then the adjoint equation is

$$y' = -A(t)^T y,$$

or

$$\begin{pmatrix} y_1 \\ \vdots \\ \vdots \\ y_n \end{pmatrix}' = \begin{pmatrix} 0, & \cdots\cdots & 0 & a_n \\ -1\,0 & \cdots & 0 & a_{n-1} \\ \vdots & \ddots & & \vdots \\ 0 & \cdots\cdots & -1 & a_1 \end{pmatrix} \begin{pmatrix} y_1 \\ \vdots \\ \vdots \\ y_n \end{pmatrix}.$$

Hence we have

$$y_1' = a_n y_n,$$
$$y_2' = -y_1 + a_{n-1}y_n,$$
$$y_3' = -y_2 + a_{n-2}y_n,$$
$$\vdots$$
$$y_n' = -y_{n-1} + a_1 y_n.$$

Let $z = y_n$ then we obtain the adjoint equation

$$D^n z - D^{n-1}(a_1 z) + D^{n-2}(a_2 z) + \ldots + (-1)^n a_n z = 0.$$

Fredholm Alternatives

We recall that in Linear Algebra the solvability of the linear system $Ax = b$, $A \in \mathbb{R}^{m \times n}$, $b \in \mathbb{R}^m$ is stated as follows:

(Uniqueness): The solution of $Ax = b$ is unique iff $Ax = 0$ has only trivial solution $x = 0$.
(Existence): The equation $Ax = b$ has a solution iff $\langle b, v \rangle = 0$ for every vector v satisfying $A^*v = 0$, where A^* is the complex conjugate of A, i.e., the adjoint of A.

Theorem 3.6.2 ([H1] p. 145).
Let $A(t) \in \mathbb{R}^{n \times n}$ be a continuous periodic matrix with period T and $f \in P_T$, where $P_T = \{f | f : \mathbb{R} \to \mathbb{R}^n$ is continuous with period $T\}$. Then,

$$x' = A(t)x + f(t) \tag{3.17}$$

has a solution in P_T iff

$$\langle y, f \rangle \overset{\text{def}}{=} \int_0^T y^*(t)f(t)dt = 0,$$

for all $y \in P_T$ satisfying $y' = -A^*(t)y$.

Proof. Let $x(t)$ be a periodic solution of (3.17) in P_T. Then $x(0) = x(T)$. Let $\Phi(t)$ be the fundamental matrix of $x' = A(t)x$ with $\Phi(0) = I$. From the variation of constant formula (3.5), we have

$$x(0) = x(T) = \Phi(T)x(0) + \int_0^T \Phi(T)\Phi^{-1}(s)f(s)ds.$$

Then, it follows that

$$\Phi^{-1}(T)x(0) = x(0) + \int_0^T \Phi^{-1}(s)f(s)ds$$

or

$$\left[\Phi^{-1}(T) - I\right]x(0) = \int_0^T \Phi^{-1}(s)f(s)ds$$

or

$$Bx(0) = b.$$

From Fredholm's Alternative of linear algebra, the necessary and sufficient condition of the solvability of $Bx(0) = b$ is

$$\langle v, b \rangle = 0 \text{ for all } v \text{ satisfying } B^*v = 0.$$

We note that

$$B^*v = 0 \text{ iff } \left(\Phi^{-1}(T)\right)^* v = v,$$

and $y(t) = \left(\Phi^{-1}(t)\right)^* v$ is a solution of the adjoint equation $y' = -A^* y$, where v is an initial value of a T-periodic solution of $y' = -A^* y$. Then,

$$\langle v, b \rangle = 0 \text{ iff } \left\langle v, \int_0^T \Phi^{-1}(s) f(s) ds \right\rangle = 0.$$

For $y \in P_T$, $y(t) = (\Phi^{-1}(t))^* v$,

$$\langle y(t), f(t) \rangle = 0 \text{ iff } \int_0^T f^*(t) \left(\Phi^{-1}(t)\right)^* v \, dt = 0,$$

$$\text{iff } \left\langle v, \int_0^T \Phi^{-1}(s) f(s) ds \right\rangle = 0.$$

Thus we complete the proof of Theorem 3.6.2. $\qquad\square$

Example 3.6.2 $u'' + u = \cos wt$.
Rewrite the equation in the form

$$\begin{pmatrix} x_1 \\ x_2 \end{pmatrix}' = \begin{pmatrix} 0 & 1 \\ -1 & 0 \end{pmatrix} \begin{pmatrix} x_1 \\ x_2 \end{pmatrix} + \begin{pmatrix} 0 \\ \cos wt \end{pmatrix}. \tag{3.18}$$

The adjoint equation is

$$\begin{pmatrix} y_1 \\ y_2 \end{pmatrix}' = \begin{pmatrix} 0 & 1 \\ -1 & 0 \end{pmatrix} \begin{pmatrix} y_1 \\ y_2 \end{pmatrix}, \tag{3.19}$$

or

$$y_1' = y_2, \ y_2' = -y_1.$$

Then,

$$\begin{aligned} y_1(t) &= a \cos t + b \sin t, \\ y_2(t) &= -a \sin t + b \cos t. \end{aligned} \tag{3.20}$$

If $w \neq 1$, there are two cases:

(i) $w = \frac{1}{n}$, $n \in \mathbb{N}, n \geq 2$. Then the adjoint system (3.19) has nontrivial periodic solutions of period $\frac{2\pi}{w} = 2n\pi$. For all $y = \begin{pmatrix} y_1 \\ y_2 \end{pmatrix} \in P_{2n\pi}$

$$\left\langle \begin{pmatrix} y_1(t) \\ y_2(t) \end{pmatrix}, \begin{pmatrix} 0 \\ \cos \frac{t}{n} \end{pmatrix} \right\rangle = \int_0^{2n\pi} (-a \sin t + b \cos t) \cos \frac{t}{n} dt$$

$$= -a \left(\frac{2(\sin^2 n\pi)}{1 - (\frac{1}{n})^2} \right) + b \left(\frac{n \sin n\pi}{1 - (\frac{1}{n})^2} \right) = 0.$$

(ii) $w \neq \frac{1}{n}$. Then (3.19) has no "nontrivial" periodic solution of period $\frac{2\pi}{w}$.

Then for all $y = \begin{pmatrix} y_1 \\ y_2 \end{pmatrix} \in P_{\frac{2\pi}{w}}, y = \begin{pmatrix} 0 \\ 0 \end{pmatrix} \left\langle \begin{pmatrix} y_1(t) \\ y_2(t) \end{pmatrix}, \begin{pmatrix} 0 \\ \cos wt \end{pmatrix} \right\rangle =$

$\left\langle \begin{pmatrix} 0 \\ 0 \end{pmatrix}, \begin{pmatrix} 0 \\ \cos wt \end{pmatrix} \right\rangle = 0.$

Hence (3.18) has a unique periodic solution of period $2\pi/w$.

If $w = 1$ then every solution of the adjoint system is of period 2π. From (3.19), we have

$$\int_0^T \begin{pmatrix} y_1(t) \\ y_2(t) \end{pmatrix}^T \begin{pmatrix} 0 \\ \cos t \end{pmatrix} dt = \int_0^T \left(-a \sin t \cos t + b \cos^2 t \right) dt = \pi b \neq 0,$$

for some solutions of (3.18). In fact, from elementary differential equations, $u(t) = a \sin t + b \cos t + \frac{t \sin t}{2}$ and we have the "resonance" phenomena. The reader can find the generalization of "resonance" in Exercise 3.25.

3.7   Exercises

Exercise 3.1

(i) Show that $\det e^A = e^{tr A}$.

(ii) If A is skew symmetric, i.e., $A^T = -A$, then e^A is orthogonal.

(iii) Is it true that the solution of I.V.P. $\frac{dx}{dt} = A(t)x$. $x(0) = x_0$, is $x(t) = \exp \left(\int_0^t A(s)ds \right) x_0$? If not, give a counterexample. Show that if

$A(t)A(s) = A(s)A(t)$ for all $s, t > 0$, then $x(t) = \exp \left(\int_0^t A(s)ds \right) x_0$.

(iv) $e^A = \lim_{m \to \infty} \left(I + \frac{A}{m} \right)^m$, $m \in \mathbb{Z}^+$.

Exercise 3.2 Find the general solution of $\dot{x} = Ax$ with

$$A = \begin{bmatrix} 2 & 1 & 3 \\ 0 & 2 & -1 \\ 0 & 0 & 2 \end{bmatrix}, \quad x(0) = \begin{bmatrix} 1 \\ 2 \\ 1 \end{bmatrix}.$$

Exercise 3.3 Prove that $Be^{At} = e^{At}B$ for all t if and only if $BA = AB$.

Exercise 3.4 Let A be an invertible matrix. Show that the only invariant lines for the linear system $dx/dt = Ax$, $\mathbf{x} \in \mathbb{R}^2$, are the lines $ax_1 + bx_2 = 0$, where $\mathbf{v} = (-b, a)^T$ is an eigenvector of A.

A line $L \subset \mathbb{R}^2$ is called an **invariant line** for a linear system $x' = f(t, x)$, $x \in \mathbb{R}^2$ if for every solution x of this system, we have $x(t) \in L$ for all $t \in \mathbb{R}$ if $x(t_0) \in L$ for some $t_0 \in \mathbb{R}$.

Exercise 3.5 Consider $x' = Ax$ where $A \in \mathbb{R}^{n \times n}$ with $\operatorname{Re} \lambda(A) \leq 0$.

(1) Show that all the solutions are bounded in positive time, when all the eigenvalues are simple.
(2) Give two examples demonstrating that when the eigenvalues are not all simple, the solutions may or may not all be bounded in positive time.
(3) Prove that all the solutions are bounded in positive time when the matrix A is symmetric.

Exercise 3.6 Consider the differential equation $\dot{x} = Ax$, where

$$A = \begin{pmatrix} a & -b & 0 & 0 \\ b & a & 0 & 0 \\ 0 & 0 & 2 & -5 \\ 1 & 0 & 1 & -2 \end{pmatrix}.$$

Find all real (a, b) such that every solution is bounded on $[0, \infty)$.

Exercise 3.7 Consider the two-dimensional system:

$$X' = (A + B(t)) X,$$

where

$$A = \begin{pmatrix} -2 & 1 \\ 1 & -2 \end{pmatrix},$$

$$B(t) = \begin{pmatrix} e^{-t^2} & \frac{1}{1+t^2} \\ \frac{1}{1+t^{2002}} & e^{-t} \end{pmatrix}.$$

Prove that $\lim_{t \to \infty} X(t) = 0$ for every solution $X(t)$.

Exercise 3.8 Find a fundamental matrix solution of the system

$$\dot{x} = \begin{pmatrix} 1 & -1/t \\ 1+t & -1 \end{pmatrix} x, \quad t > 0.$$

Hint: $x(t) = \begin{pmatrix} 1 \\ t \end{pmatrix}$ is a solution.

Exercise 3.9 Suppose that A is a continuous matrix and every solution of $\dot{x} = A(t)x$ is bounded for $t \geq 0$ and let $\Phi(t)$ be a fundamental matrix solution. Prove that $\Phi^{-1}(t)$ is bounded for $t \geq 0$ if and only if the function $t \mapsto \int_0^t \operatorname{tr} A(s)ds$ is bounded below. Hint: The inverse of a matrix is the adjugate of the matrix divided by its determinant.

Exercise 3.10 Suppose that A is continuous, the linear system $\dot{x} = A(t)x$ is defined on an open interval containing the origin whose right-hand end point is $w \leq \infty$, and the norm of every solution has a finite limit as $t \to w$. Show that there is a non-trivial solution converging to zero as $t \to w$ if and only if $\int_0^w \operatorname{tr} A(s)ds = -\infty$. Hint: A matrix has a nontrivial kernel if and only if its determinant is zero.

Exercise 3.11 Let A be a continuous $n \times n$ matrix such that the system $\dot{x} = A(t)x$ has a bounded fundamental matrix $\Phi(t)$ over $0 \leq t < \infty$.

(1) Show that all fundamental matrices are bounded on $[0, \infty)$.
(2) Show that if

$$\liminf_{t \to \infty} Re \left[\int_0^t tr A(s)ds \right] > -\infty,$$

then $\Phi^{-1}(t)$ is also bounded on $[0, \infty)$.

Exercise 3.12 Assume $a(t)$ is a bounded continuous function, on $[0, \infty)$ and $\phi(t)$ is a nontrivial solution of

$$y'' + a(t)y = 0, \qquad (*)$$

satisfying $\phi(t) \to 0$ as $t \to \infty$. Show that $(*)$ has a solution which is not bounded over $[0, \infty)$.

Exercise 3.13 Let g be a bounded continuous function on $(-\infty, \infty)$ and let B and $-C$ be matrices of dimensions k and $n - k$. All eigenvalues of B and $-C$ have negative real parts. Let

$$A = \begin{bmatrix} B & 0 \\ 0 & C \end{bmatrix} \quad \text{and} \quad g(t) = \begin{bmatrix} g_1(t) \\ g_2(t) \end{bmatrix},$$

so that the system

$$x' = Ax + g(t), \qquad (**)$$

is equivalent to

$$x'_1 = Bx_1 + g_1(t), \quad x'_2 = Cx_2 + g_2(t).$$

Show that the functions

$$\phi_1(t) = \int_{-\infty}^{t} e^{B(t-s)} g_1(s)ds, \quad \phi_2(t) = -\int_{t}^{\infty} e^{C(t-s)} g_2(s)ds,$$

are defined for all $t \in R$ and determine a solution of (**).

Exercise 3.14 Show that if g is a bounded continuous function on R^1 and if A has no eigenvalues with zero real part, then (**) has at least one bounded solution. (Hint: Use Exercise 3.13.)

Exercise 3.15 Let $A = \lambda I + N$ consist of a single Jordan block. Show that for any $\alpha > 0$, A is similar to a matrix $B = \lambda I + \alpha N$. (Hint: Let $P = [\alpha^{i-1}\delta_{ij}]$ and compute $P^{-1}AP$.)

Exercise 3.16 Let A be a real $n \times n$ matrix. Show that there exists a real nonsingular matrix P such that $P^{-1}AP = B$ has the real Jordan canonical form $J = \text{diag}(J_0, J_1, \cdots, J_s)$ where J_k is given as before for real eigenvalues λ_j while for complex eigenvalue $\lambda = \alpha \pm i\beta$ the corresponding J_k has the form

$$J_k = \begin{bmatrix} \Lambda & I_2 & \cdots & 0_2 & 0_2 \\ 0_2 & \Lambda & I_2 & 0_2 & 0_2 \\ \vdots & \vdots & \ddots & \vdots & \vdots \\ 0_2 & 0_2 & \cdots & 0_2 & \Lambda \end{bmatrix}.$$

Here 0_2 is the 2×2 zero matrix, I_2 the 2×2 identity matrix, and

$$\Lambda = \begin{bmatrix} \alpha & -\beta \\ \beta & \alpha \end{bmatrix}.$$

Exercise 3.17 Use the Jordan form to prove that all eigenvalues of A^2 have the form λ^2, where λ is an eigenvalue of A.

Exercise 3.18 If $A = C^2$, where C is a real nonsingular $n \times n$ matrix, show that there is a real matrix L such that $e^L = A$. Hint: Use Exercise 3.16 and the fact that if $\lambda = \alpha + i\beta = re^{i\theta}$, then

$$\Lambda = \exp \begin{bmatrix} \log r & -\theta \\ \theta & \log r \end{bmatrix}.$$

Exercise 3.19 Let $A(t)$ be a 2×2 continuous, periodic matrix of period T and let $Y(t)$ be the fundamental matrix of $y' = A(t)y$ satisfying $Y(0) = I$. Let $B(t)$ be a 3×3 continuous periodic matrix of period T of the form

$$B(t) = \begin{bmatrix} A(t), \, b_1(t) \\ b_2(t) \\ 0, 0, \, b_3(t) \end{bmatrix}.$$

Show that the fundamental matrix $\Phi(t)$ of

$$z' = B(t)z$$

is given by

$$\Phi(t) = \begin{bmatrix} Y(t), \, z_1 \\ z_2 \\ 0, 0, \, z_3 \end{bmatrix}$$

where

$$z_3(t) = e^{\int_0^t b_3(s)ds}$$

and $z = (z_1, z_2)$ is given by

$$z(t) = \int_0^t Y(t)Y^{-1}(s)b(s)z_3(s)ds$$

where $b(s) = col(b_1(s), b_2(s))$.

In particular, show that if ρ_1, ρ_2 are the Floquet multipliers associated with the 2×2 system eigenvalues of $Y(t)$ then the Floquet multipliers of the 3×3 system are ρ_1, ρ_2 and $\rho_3 = z_3(T)$.

Exercise 3.20 Let $a_0(t)$ and $a_1(t)$ be continuous and T-periodic functions and let ϕ_1 and ϕ_2 be solutions of $y'' + a_1(t)y' + a_0(t)y = 0$ such that

$$\Phi(0) = \begin{bmatrix} \phi_1(0) & \phi_2(0) \\ \phi_1'(0) & \phi_2'(0) \end{bmatrix} = \begin{bmatrix} 1 & 0 \\ 0 & 1 \end{bmatrix} = I_2.$$

Show that the Floquet multipliers λ satisfy $\lambda^2 + \alpha\lambda + \beta = 0$, where

$$\alpha = -\left[\phi_1(T) + \phi_2'(T)\right], \quad \beta = \exp\left[-\int_0^T a_1(t)dt\right].$$

Exercise 3.21 In Exercise 3.20, we assume $a_1(t) = 0$.

(1) Show that if $-2 < \alpha < 2$, then all solutions y are bounded over $-\infty < t < \infty$.

(2) Show that if $\alpha > 2$ or $\alpha < -2$, then for every non-trivial solution y, $y(t)^2 + y'(t)^2$ must be unbounded over R.

(3) Show that if $\alpha = -2$, then there is at least one solution y of period T.

(4) Show that if $\alpha = 2$, then there is at least one solution y of period $2T$.

Exercise 3.22 Show that $\Phi(t + T) = \Phi(t)\Phi(T)$ where $\Phi(t)$ is the fundamental matrix of linear T-periodic system $X' = A(t)X$ with $\Phi(0) = I$.

Exercise 3.23 Let A be a continuous T-periodic $n \times n$ matrix.

(1) Show that the linear system $x' = A(t)x$ has at least one nontrivial solution $x = x(t)$ such that $x(t + T) = \mu x(t)$, μ is a characteristic multiplier.

(2) Suppose $\mu_1, ..., \mu_k$ are k distinct characteristic multipliers of $x' = A(t)x$. Show that $x' = A(t)x$ has k linearly independent solutions $x_i(t) = p_i(t)e^{\rho_i t}$, $p_i(t)$ is periodic with period T, $i = 1, 2, ..., k$.

Exercise 3.24 (Markus and Yamabe [MY]).

If $A(t) = \begin{bmatrix} -1 + \frac{3}{2}\cos^2 t & 1 - \frac{3}{2}\cos t \sin t \\ -1 - \frac{3}{2}\sin t \cos t & -1 + \frac{3}{2}\sin^2 t \end{bmatrix}$ then the eigenvalue $\lambda_1(t)$, $\lambda_2(t)$ of $A(t)$ are $\lambda_1(t) = [-1 + i\sqrt{7}]/4$, $\lambda_2(t) = \overline{\lambda_1(t)}$ and, in particular, the real parts of the eigenvalues have negative real parts. On the other hand, one can verify directly that the vector $(-\cos t, \sin t)\exp(\frac{t}{2})$ is a solution of $x' = A(t)x$ and this solution is unbounded as $t \to \infty$. Show that one of the characteristic multipliers is $-e^{\frac{\pi}{2}}$ and the other multiplier is $-e^{-\pi}$.

Exercise 3.25

(1) Consider linear inhomogeneous system
$$x' = Ax + f(t) \qquad (***)$$
where $f(t)$ is a continuous 2π-periodic function. If there is a 2π-periodic solution $y(t)$ of adjoint equation
$$y' = -A^T y$$
such that
$$\int_0^{2\pi} y^T(t)f(t)dt \neq 0,$$
show that every solution $x(t)$ of $(***)$ is unbounded.
Hint: Compute $\frac{d}{dt}\left(y^T(t)x(t)\right)$ and integrate from 0 to ∞.

(2) Show that the resonance occurs for the second order linear equation

$$x'' + w_0^2 x = F \cos wt,$$

when $w = w_0$.

Exercise 3.26 In $x'' + a_1(t)x' + a_2(t)x = 0$, make the change of variable $s = F(t)$, where $F'(t) = exp[-\int^t a_1(s)ds]$ and let $t = G(s)$. Show that this loads to

$$\frac{d^2x}{ds^2} + g(s)x = 0 \quad where \quad g(s) \quad is \quad \frac{a_2(t)}{(F'(t))^2}$$

evaluated at $t = G(s)$.

Exercise 3.27

(1) Consider the system $\dot{x} = Ax$ where

$$A = \begin{pmatrix} 1 & 0 & 0 \\ 1 & 0 & -1 \\ 1 & 0 & 0 \end{pmatrix}.$$

Let $x(t)$ be a solution of the above system. We are saying that $x(t)$ is *growing linearly* if there exists $\lim_{x \to \infty} \frac{|x(t)|}{t} = c > 0$, and $x(t)$ is *growing exponentially* if $\lim_{x \to \infty} \frac{\log|x(t)|}{t} = \chi > 0$.
Find all initial conditions $x(0)$ such that the respective solution $x(t)$ are a) bounded; b) growing linearly; c) growing exponentially. In case c) find the respective constant χ.

(2) Let $f : \mathbb{R} \to \mathbb{R}$ be an odd continuous function such that $f(0) = 0, f(x) > 0$ for $x > 0$ and $\int_0^{2012} \frac{1}{f(x)}dx = 1$. Determine all values of a for which there is a solution $x(t)$ of the equation $\dot{x} = f(x)$ with $x(0) = 0, x(1) = a$.

Chapter 4

STABILITY OF NONLINEAR SYSTEMS

4.1 Definitions

Let $x(t)$ be a solution of the system

$$\frac{dx}{dt} = f(t, x). \tag{4.1}$$

In this section, we shall introduce three types of stability for a solution $x(t)$ of (4.1), namely stability, instability and asymptotic stability in the sense of Lyapunov.

Definition 4.1.1 *We say that a solution $x(t)$ of (4.1) is stable, more precisely, stable over the interval $[t_0, \infty)$ if*

(i) *for each $\epsilon > 0$ there exists $\delta = \delta(\varepsilon) > 0$ such that for any solution $\bar{x}(t)$ of (4.1) with $|\bar{x}(t_0) - x(t_0)| < \delta$ the inequality $|\bar{x}(t) - x(t)| < \epsilon$ holds for all $t \geq t_0$.*

$x(t)$ is said to be asymptotically stable if it is stable and

(ii) *$|\bar{x}(t) - x(t)| \to 0$ as $t \to \infty$ whenever $|\bar{x}(t_0) - x(t_0)|$ is sufficiently small.*

A solution $x(t)$ is said to be unstable if it is not stable, i.e., there exist $\varepsilon > 0$ and sequences $\{t_m\}$ and $\{\xi_m\}$, $\xi_m \to \xi_0$, as $m \to \infty$ where $\xi_0 = x(t_0)$, such that $|\varphi(t_0 + t_m, t_0, \xi_m) - x(t_0 + t_m)| \geq \varepsilon$.

Example 4.1.1 The constant solution $x(t) \equiv \xi$ of the equation $x' = 0$ is stable, but it is not asymptotically stable.

Example 4.1.2 The solutions of the equation $x' = a(t)x$ are

$$x(t) = \xi_1 \exp\left(\int_0^t a(s)ds\right).$$

The solution $x(t) \equiv 0$ is asymptotically stable if and only if

$$\int_0^t a(s)ds \to -\infty \text{ as } t \to \infty.$$

Example 4.1.3 $\begin{pmatrix} x_1 \\ x_2 \end{pmatrix}' = \begin{pmatrix} 0 & 1 \\ 1 & 0 \end{pmatrix}\begin{pmatrix} x_1 \\ x_2 \end{pmatrix}.$

Then, $(0,0)$ is a saddle point. Hence $\begin{pmatrix} x_1(t) \\ x_2(t) \end{pmatrix} \equiv \begin{pmatrix} 0 \\ 0 \end{pmatrix}$ is unstable, even though there exist points, namely, the points (ξ_1, ξ_2) on stable manifold such that $\varphi(t, (\xi_1, \xi_2)) \to (0,0)$ as $t \to \infty$.

Example 4.1.4 Consider the equation $v''(t) + t \sin v(t) = 0$. Then, $\begin{pmatrix} v(t) \\ v'(t) \end{pmatrix}$ $\equiv \begin{pmatrix} \pi \\ 0 \end{pmatrix}$ is unstable and $\begin{pmatrix} v(t) \\ v'(t) \end{pmatrix} \equiv \begin{pmatrix} 0 \\ 0 \end{pmatrix}$ is asymptotically stable. (Exercise 4.1.)

Example 4.1.5 For the equation $x'' + x = 0$, the null solution $\begin{pmatrix} x(t) \\ x'(t) \end{pmatrix} \equiv$ $\begin{pmatrix} 0 \\ 0 \end{pmatrix}$ is stable, but not asymptotically stable.

Remark 4.1.1 For applications, we mostly deal with autonomous systems $x' = f(x)$ or periodic systems $x' = f(t, x)$ where f is periodic in t. For autonomous systems $x' = f(x)$, $f : \Omega \subseteq \mathbb{R}^n \to \mathbb{R}^n$, we consider the stability of the following specific solutions, namely

(i) Equilibrium solutions $x(t) \equiv x^*$, $f(x^*) = 0$.

(ii) Periodic solutions whose existence usually follows from Poincaré-Bendixson theorem for $n = 2$ (see Chapter 6) or Brouwer fixed point Theorem for $n \geq 3$ (see Chapter 8). We shall discuss the "orbital stability" for a periodic orbit of $x' = f(x)$ in Section 4.4.

4.2   Linearization

Stability of linear systems:

The stability of a solution $x(t)$ of the linear system $x' = A(t)x$ is equivalent to the stability of the null solution $x(t) \equiv 0$. Consider the linear system $x' = Ax$, where A is an $n \times n$ constant matrix. From Chapter 3, we have the following:

(i) If all eigenvalues of A have negative real parts, then $x(t) \equiv 0$ is asymptotically stable. Moreover, there are positive constants K and α such that
$$\| e^{At}x_0 \| \leq Ke^{-\alpha t} \| x_0 \| \text{ for all } t \geq 0, \ x_0 \in \mathbb{R}^n.$$

(ii) $x(t) \equiv 0$ is stable iff all eigenvalues of A have nonpositive real parts and those with zero real parts have simple elementary divisors.

(iii) $x(t) \equiv 0$ is unstable iff there exists an eigenvalue with positive real part or those with zero real part have no simple elementary divisors.

There are two ways to justify the stability of equilibrium solutions of nonlinear autonomous system $x' = f(x)$, namely, linearization method and Lyapunov method. In this chapter we discuss the method of linearization and we shall discuss the Lyapunov method in the next chapter.

Given a nonlinear autonomous system $x' = f(x)$, with initial condition $x(0) = x_0$, it is very difficult to study the global asymptotic behavior of the solution $\varphi(t, x_0)$. As a first step, we find all equilibria of the system, i.e., solving the nonlinear system $f(x) = 0$. Then, we check the stability property of each equilibrium x^*. With these stability information, we may predict the behavior of the solution $\varphi(t, x_0)$.

Let x^* be an equilibrium of autonomous system
$$x' = f(x). \tag{4.2}$$
The Taylor series expansion of $f(x)$ around $x = x^*$ has the form
$$f(x) = f(x^*) + A(x - x^*) + g(x - x^*)$$
$$= A(x - x^*) + g(x - x^*),$$

where

$$A = D_x f(x^*) = \begin{bmatrix} \frac{\partial f_1}{\partial x_1}, \cdots, \frac{\partial f_1}{\partial x_n} \\ \vdots \\ \vdots \\ \frac{\partial f_n}{\partial x_1}, \cdots, \frac{\partial f_n}{\partial x_n} \end{bmatrix}_{x=x^*},$$

and

$$g(y) = o(|y|) \text{ as } y \to 0,$$

i.e.

$$\lim_{y \to 0} \frac{|g(y)|}{|y|} = 0.$$

The Jacobian $D_x f(x^*)$ is called the variational matrix of (4.2) evaluated at $x = x^*$. Let $y = x - x^*$, the linear system

$$y' = D_x f(x^*) y \qquad (4.3)$$

is said to be the linearized system of (4.2) about the equilibrium x^*. The stability property of the linear system (4.3) is called the linearized stability of the equilibrium x^*. The stability property of the equilibrium x^* of (4.2) is called the nonlinear stability of x^*. Does the linearized stability of x^* imply the nonlinear stability of x^*? From the following examples, in general, the answer is "No".

Example 4.2.1 Consider the system

$$\begin{cases} \frac{dx_1}{dt} = x_2 - x_1 \left(x_1^2 + x_2^2 \right), \\ \frac{dx_2}{dt} = -x_1 - x_2 \left(x_1^2 + x_2^2 \right). \end{cases}$$

$(0,0)$ is an equilibrium. The variational matrix A evaluated at $(0,0)$ is

$$\begin{bmatrix} -\left(x_1^2 + x_2^2 \right) - 2x_1^2 & 1 - 2x_1 x_2 \\ -1 - 2x_1 x_2 & -\left(x_1^2 + x_2^2 \right) - 2x_2^2 \end{bmatrix}_{\substack{x_1 = 0 \\ x_2 = 0}} = \begin{bmatrix} 0 & 1 \\ -1 & 0 \end{bmatrix}$$

whose eigenvalues are $\pm i$. Hence $(0,0)$ is a center for the linearized system $x' = Ax$. However

$$x_1 \frac{dx_1}{dt} + x_2 \frac{dx_2}{dt} = -\left(x_1^2 + x_2^2 \right)^2$$

or

$$\frac{1}{2} \frac{d}{dt} \left(r^2(t) \right) = -\left(r^2(t) \right)^2, \quad r^2(t) = x_1^2(t) + x_2^2(t).$$

Then

$$r^2(t) = \frac{c}{1 + 2ct}, \quad c = x_1^2(0) + x_2^2(0)$$

and $r(t) \to 0$ as $t \to \infty$. It follows that $(0,0)$ is asymptotically stable.

Example 4.2.2 For the system

$$\begin{cases} \frac{dx_1}{dt} = x_2 + \left(x_1^2 + x_2^2\right) x_1, \\ \frac{dx_2}{dt} = -x_1 + \left(x_1^2 + x_2^2\right) x_2, \end{cases}$$

and the equilibrium $(0,0)$, the linearized system is

$$x' = \begin{pmatrix} 0 & 1 \\ -1 & 0 \end{pmatrix} x, \text{with } (0,0) \text{ being a center.}$$

However

$$\frac{d}{dt}\left(r^2(t)\right) = 2\left(r^2(t)\right)^2,$$

or

$$r^2(t) = \frac{r^2(0)}{1 - 2r^2(0)t}.$$

The equilibrium $(0,0)$ is unstable and the solution blows up in finite time. In the following Theorem 4.2.1 we show that if all eigenvalues of $D_x f(x^*)$ have negative real parts, then linearized stability implies nonlinear stability.

Theorem 4.2.1 *Let x^* be an equilibrium of the autonomous system (4.2). If the variational matrix $D_x f(x^*)$ has all eigenvalues with negative real parts, then the equilibrium x^* is asymptotically stable.*

Proof. Let $A = D_x f(x^*)$ and $y = x - x^*$, then we have

$$y' = f(x) = f(y + x^*) = Ay + g(y),$$

where

$$g(y) = o(|y|) \text{ as } y \to 0, \ g(0) = 0. \tag{4.4}$$

Since $\text{Re}\lambda(A) < 0$, there exists $M > 0, \sigma > 0$ such that

$$|e^{At}| \le Me^{-\sigma t} \text{ as } t \ge 0. \tag{4.5}$$

From (4.5) and the variation of constant formula

$$y(t) = e^{A(t-t_0)}y(t_0) + \int_{t_0}^{t} e^{A(t-s)}g(y(s))ds,$$

it follows that

$$|y(t)| \le M|y(t_0)|e^{-\sigma(t-t_0)} + M\int_{t_0}^{t} e^{-\sigma(t-s)}|g(y(s))|ds. \tag{4.6}$$

From (4.4), given ε, $0 < \varepsilon < \sigma$, there exists $\delta > 0$ such that

$$|g(y)| < \frac{\varepsilon}{M}|y| \text{ for } |y| < \delta. \tag{4.7}$$

Let $0 < \gamma < \min\left\{\frac{\delta}{2M}, \frac{\delta}{2}\right\}$. We claim that if $|y(t_0)| < \gamma$ then $|y(t)| < \delta$ and $|y(t)| \leq \frac{\delta}{2}e^{-(\sigma-\epsilon)(t-t_0)}$ for all $t \geq t_0$. If $|y(t)| < \frac{\delta}{2}$ for $t_0 \leq t < t^*$ and $|y(t^*)| = \frac{\delta}{2}$, then from (4.7), for $t_0 \leq t < t^*$ we have

$$|y(t)| \leq M|y(t_0)|e^{-\sigma(t-t_0)} + \epsilon \int_{t_0}^{t} e^{-\sigma(t-s)}|y(s)|ds.$$

Applying Gronwall's inequality to $e^{\sigma t}|y(t)|$ yields

$$|y(t)| \leq M|y(t_0)|e^{-(\sigma-\epsilon)(t-t_0)} < \frac{\delta}{2}e^{-(\sigma-\epsilon)(t-t_0)}.$$

This is a desired contradiction to $|y(t^*)| = \frac{\delta}{2}$. Hence for $t \geq t_0$ we have $|y(t)| < \frac{\delta}{2}$ and $|y(t)| \leq \frac{\delta}{2}e^{-(\sigma-\epsilon)(t-t_0)}$. Hence the equilibrium $y = 0$ is asymptotically stable. In fact, it is exponentially stable. □

Definition 4.2.1 *An equilibrium x^* of the autonomous system (4.2) is hyperbolic if the variational matrix $D_x f(x^*)$ has no eigenvalues with zero real parts.*

Definition 4.2.2 *We say two flows φ^t and ψ^t on $U \subseteq \mathbb{R}^n$ are topologically conjugate if there is a homeomorphism $h : U \to h(U) \subseteq U \subseteq \mathbb{R}^n$ such that $h \circ \varphi^t(x) = \psi^t \circ h(x)$ for all $x \in U$ (see Fig. 4.1).*

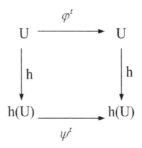

Fig. 4.1

If two flows φ^t and ψ^t are topologically conjugate, then obviously the homeomorphism h takes the trajectories of φ^t into the trajectories of ψ^t. In the following we state without proof the Hartman-Grobman Theorem which says that if an equilibrium x^* is hyperbolic, then linearized stability of x^* implies nonlinear stability of x^* in the sense of topological conjugacy.

Hartman-Grobman Theorem [R]

Let x^* be a hyperbolic equilibrium of the system $\frac{dx}{dt} = f(x)$. Then the flow φ^t of f is topologically conjugate in a neighborhood U of x^* to the affine flow $x^* + e^{At}(x - x^*)$ where $A = D_x f(x^*)$. More precisely, there is a neighborhood U of x^* and a homeomorphism $h : U \to U$ such that $\varphi^t (h(x)) = h (x^* + e^{At}(x - x^*))$ as long as $x^* + e^{At}(x - x^*) \in U$.

Example 4.2.3 Predator-Prey system with Holling's type II functional responses ([HHW]). Let $x(t)$, $y(t)$ be population densities of prey and predator species respectively. Assume the prey grows logistically in the absence of predation with intrinsic growth rate γ and carrying capacity K. Assume the per capital growth rate of predator, $\frac{y'}{y}$ is $\frac{mx}{a+x}$ where a is the half saturation constant or Michaelis-Menten constant and m is the maximum birth rate of predator. Let c be the conversion rate of predator, i.e., c is the ratio of the consumption of prey to the reproduction of predators and d is the death rate of predator. Then the equations take the form

$$\frac{dx}{dt} = \gamma x \left(1 - \frac{x}{K}\right) - c\frac{mx}{a+x}y,$$
$$\frac{dy}{dt} = \left(\frac{mx}{a+x} - d\right)y,$$
$$x(0) > 0, \quad y(0) > 0.$$

We may assume $c = 1$ if we do the scaling $y \to cy$.

Let $m > d$. Then we have three equilibria: $(0,0), (K,0)$ and (x^*,y^*) where $x^* = \frac{a}{\left(\frac{m}{d}\right)-1} > 0, y^* > 0$. The interior equilibrium (x^*,y^*) exists if $x^* < K$.

Assume $0 < x^* < K$. Then the variational matrix is

$$A(x,y) = \begin{bmatrix} \gamma\left(1 - \frac{2x}{K}\right) - \frac{ma}{(a+x)^2}y & -\frac{mx}{a+x} \\ \frac{ma}{(a+x)^2}y & \left(\frac{mx}{a+x} - d\right) \end{bmatrix}.$$

At $E_0 = (0,0)$

$$A(0,0) = \begin{bmatrix} \gamma & 0 \\ 0 & -d \end{bmatrix}.$$

E_0 is a saddle point for $\gamma > 0, -d < 0$.

At $E_1 = (K, 0)$,

$$A(K, 0) = \begin{bmatrix} -\gamma & -\frac{mK}{a+K} \\ 0 & \frac{mK}{a+K} - d \end{bmatrix}.$$

E_1 is a saddle point since $x^* < K$.

At $E^* = (x^*, y^*)$,

$$A(x^*, y^*) = \begin{bmatrix} \gamma \left(1 - \frac{2x^*}{K}\right) - \frac{ma}{(a+x^*)^2} y^* & -\frac{mx^*}{a+x^*} \\ \frac{ma}{(a+x^*)^2} y^* & 0 \end{bmatrix}.$$

The characteristic equation of $A(x^*, y^*)$ is

$$\lambda^2 - \lambda \left[\gamma \left(1 - \frac{2x^*}{K}\right) - \frac{ma}{(a+x^*)^2} y^* \right] + \frac{mx^*}{a+x^*} \cdot \frac{ma}{(a+x^*)^2} y^* = 0.$$

E^* is asymptotically stable iff

$$\gamma \left(1 - \frac{2x^*}{K}\right) - \frac{ma}{(a+x^*)^2} y^* < 0,$$

or

$$\gamma \left(1 - \frac{2x^*}{K}\right) - \frac{a}{a+x^*} \gamma \left(1 - \frac{x^*}{K}\right) < 0,$$

or

$$\frac{K-a}{2} < x^*.$$

If $\frac{K-a}{2} > x^*$, then E^* is an unstable spiral.

If $\frac{K-a}{2} = x^*$, then the eigenvalues of $A(x^*, y^*)$ are $\pm \omega i$ for some $\omega \neq 0$. This is "Hopf Bifurcation" which will be discussed in Section 6.3.

For $\frac{K-a}{2} < x^* < K$, the interior equilibrium E^* is a stable spiral, E_0, E_1 are saddles. From isocline analysis, we predict $(x(t), y(t)) \to E^*$ as $t \to \infty$ (see Fig. 4.2). For $x^* < \frac{K-a}{2}$, E^* is an unstable spiral, E_0 and E_1 are saddles. It can be shown that the trajectory $(x(t), y(t))$ approaches a unique limit cycle [Cop] (see Fig. 4.3).

Example 4.2.4 Lotka-Volterra two species competition model [Wal]:

$$\begin{cases} \frac{dx_1}{dt} = \gamma_1 x_1 \left(1 - \frac{x_1}{K_1}\right) - \alpha_1 x_1 x_2, \\ \frac{dx_2}{dt} = \gamma_2 x_2 \left(1 - \frac{x_2}{K_2}\right) - \alpha_2 x_1 x_2, \\ x_1(0) > 0, \ x_2(0) > 0, \end{cases}$$

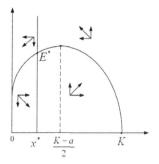

Fig. 4.2 $\frac{K-a}{2} < x^* < K$, E^* is a stable spiral.

Fig. 4.3 $0 < x^* < \frac{K-a}{2}$, E^* is an unstable spiral.

where γ_1, γ_2, K_1, K_2, α_1, $\alpha_2 > 0$. Three equilibria, $E_0 = (0,0)$, $E_1 = (K_1, 0)$ and $E_2 = (0, K_2)$ always exist. The interior equilibrium $E^* = (x^*, y^*)$ exists in the following cases (iii) and (iv). The variational matrix at $E(x, y)$ is

$$A(x, y) = \begin{bmatrix} \gamma_1 \left(1 - \frac{x_1}{K_1}\right) - \alpha_1 x_2 - \frac{\gamma_1}{K_1} x_1 & -\alpha_1 x_1 \\ -\alpha_2 x_2 & \gamma_2 \left(1 - \frac{x_2}{K_2}\right) - \alpha_2 x_1 - \frac{\gamma_2}{K_2} x_2 \end{bmatrix}.$$

At E_0,

$$A(0,0) = \begin{bmatrix} \gamma_1 & 0 \\ 0 & \gamma_2 \end{bmatrix}.$$

E_0 is an unstable node or a repeller for $\gamma_1 > 0$, $\gamma_2 > 0$.

At $E_1 = (K_1, 0)$

$$A(K_1, 0) = \begin{bmatrix} -\gamma_1 & -\alpha_1 K_1 \\ 0 & \gamma_2 - \alpha_2 K_1 \end{bmatrix}.$$

At $E_2 = (0, K_2)$

$$A(0, K_2) = \begin{bmatrix} \gamma_1 - \alpha_1 K_2 & 0 \\ -\alpha_2 K_2 & -\gamma_2 \end{bmatrix}.$$

There are four cases according to the positions of isoclines L_1 : $\gamma_1 \left(1 - \frac{x_1}{K_1}\right) - \alpha_1 x_2 = 0$ and $L_2 : \gamma_2 \left(1 - \frac{x_2}{K_2}\right) - \alpha_2 x_1 = 0$.

(i) $\frac{\gamma_2}{\alpha_2} > K_1$, $\frac{\gamma_1}{\alpha_1} < K_2$ (see Fig. 4.4).
In this case, $E_2 = (0, K_2)$ is a stable node, $E_1 = (K_1, 0)$ is a saddle point and $E_0 = (0,0)$ is a unstable node. We may predict that $(x_1(t), x_2(t)) \to (0, K_2)$ as $t \to \infty$.

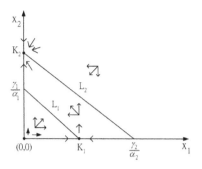

Fig. 4.4

(ii) $\frac{\gamma_2}{\alpha_2} < K_1$, $\frac{\gamma_1}{\alpha_1} > K_2$ (see Fig. 4.5).

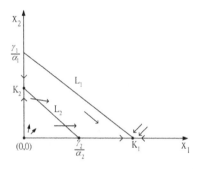

Fig. 4.5

In this case $E_1 = (K_1, 0)$ is a stable node, $E_2 = (0, K_2)$ is a saddle point, and $E_0 = (0,0)$ is an unstable node. We may predict that $(x_1(t), x_2(t)) \to (K_1, 0)$ as $t \to \infty$.

(iii) $\frac{\gamma_1}{\alpha_1} > K_2$, $\frac{\gamma_2}{\alpha_2} > K_1$ (see Fig. 4.6).
In this case, $E_1 = (K_1, 0)$ and $E_2 = (0, K_2)$ are saddle points, and $E_0 = (0,0)$ is an unstable node.

The variational matrix at E^* is

$$A(x^*, y^*) = \begin{bmatrix} -\frac{\gamma_1}{K_1} x_1^* & -\alpha_1 x_1^* \\ -\alpha_2 x_2^* & -\frac{\gamma_2}{K_2} x_2^* \end{bmatrix}.$$

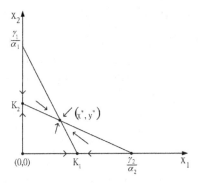

Fig. 4.6

The characteristic polynomial of $A(x^*, y^*)$ is

$$\lambda^2 + \left(\frac{\gamma_1}{K_1}x^* + \frac{\gamma_2}{K_2}x_2^*\right)\lambda + x_1^* x_2^* \left(\frac{\gamma_1 \gamma_2}{K_1 K_2} - \alpha_1 \alpha_2\right) = 0.$$

Since $\frac{\gamma_2}{\alpha_2} > K_1$, $\frac{\gamma_1}{\alpha_1} > K_2$, it follows that $E^* = (x^*, y^*)$ is a stable node. We may predict the solution $(x_1(t), x_2(t)) \to (x_1^*, x_2^*)$ as $t \to \infty$.

(iv) $\frac{\gamma_1}{\alpha_1} < K_2$, $\frac{\gamma_2}{\alpha_2} < K_1$ (see Fig. 4.7).

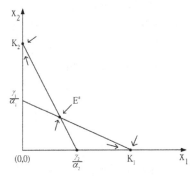

Fig. 4.7

In this case $E_1 = (K_1, 0)$ and $E_2 = (0, K_2)$ are stable nodes. $E_0 = (0, 0)$ is an unstable node. From $K_1 > \frac{\gamma_2}{\alpha_2}$, $K_2 > \frac{\gamma_1}{\alpha_1}$, it follows that $E^* = (x^*, y^*)$ is a saddle point. This is a well-known example of "bistability".

4.3   Saddle Point Property

In this section, we discuss the existence of stable and unstable manifolds for a hyperbolic equilibrium of an autonomous system and prove that they are submanifolds with the same smoothness properties as the vector field. Consider (4.2) with x^* as a hyperbolic equilibrium and $A = D_x f(x^*)$. Let $y = x - x^*$ then we have

$$y' = Ay + g(y),$$

where

$$g(0) = 0, \ Dg(0) = 0, \ \ g(y) = o(|y|) \ \text{ as } \ y \to 0.$$

Consider the differential equation

$$x' = Ax + g(x) \tag{4.8}$$

where the equilibrium 0 is hyperbolic and $g : \mathbb{R}^n \to \mathbb{R}^n$ is a Lipschitz continuous function satisfying

$$\begin{aligned} g(0) &= 0 \\ |g(x) - g(y)| &\leq \rho(\delta)|x - y| \text{ if } |x|, |y| \leq \delta \end{aligned} \tag{4.9}$$

where $\rho : [0, \infty] \to [0, \infty)$ is continuous with $\rho(0) = 0$. We note that (4.9) implies $g(x) = o(|x|)$ as $x \to 0$.

For any $x \in \mathbb{R}^n$, let $\varphi^t(x)$ be the solution of (4.8) through x. The unstable set $W^u(0)$ and the stable set $W^s(0)$ of 0 are defined as

$$W^u(0) = \{x \in \mathbb{R}^n : \varphi^t(x) \text{ is defined for } t \leq 0 \text{ and } \varphi^t(x) \to 0 \text{ as } t \to -\infty\},$$

$$W^s(0) = \{x \in \mathbb{R}^n : \varphi^t(x) \text{ is defined for } t \geq 0 \text{ and } \varphi^t(x) \to 0 \text{ as } t \to \infty\}.$$

The local unstable set $W^u_{loc}(0)$ and the local stable set $W^s_{loc}(0)$ of 0 corresponding to a neighborhood U of 0 are defined by

$$W^u_{loc}(0) \equiv W^u(0, U) = \{x \in W^u(0) : \varphi^t(x) \in U, \ t \leq 0\},$$

$$W^s_{loc}(0) \equiv W^s(0, U) = \{x \in W^s(0) : \varphi^t(x) \in U, \ t \geq 0\}.$$

Example 4.3.1 Consider case (iv) of Lotka-Volterra two species competition model in Example 4.2.4

$$x'_1 = \gamma_1 x_1 \left(1 - \frac{x_1}{K_1}\right) - \alpha_1 x_1 x_2,$$

$$x'_2 = \gamma_2 x_2 \left(1 - \frac{x_2}{K_2}\right) - \alpha_2 x_1 x_2.$$

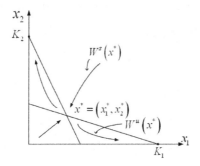

Fig. 4.8

If $\frac{\gamma_1}{\alpha_1} < K_2$, $\frac{\gamma_2}{\alpha_2} < K_1$ then $x^* = (x_1^*, x_2^*)$ is saddle point with one-dimensional stable set $W^s(x^*)$ (see Fig. 4.8) and one-dimensional unstable set $W^u(x^*)$.

Example 4.3.2 Consider the following nonlinear equation

$$\begin{cases} x_1' = -x_1, \\ x_2' = x_2 + x_1^2. \end{cases}$$

The equilibrium $(0,0)$ is a saddle since the linearized equation is

$$\begin{cases} x_1' = -x_1, \\ x_2' = x_2. \end{cases}$$

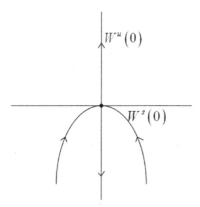

Fig. 4.9

Integrating the above nonlinear equations, we obtain

$$x_1(t) = e^{-t}x_1^0,$$

$$x_2(t) = e^t \left(x_2^0 + \frac{1}{3}(x_1^0)^2 \right) - \frac{1}{3}e^{-2t} \left(x_1^0 \right)^2.$$

From these formulas, we see that

$$W^u(0) = \left\{ x^0 = (x_1^0, x_2^0) : x_1^0 = 0 \right\},$$

$$W^s(0) = \left\{ x^0 = (x_1^0, x_2^0) : x_2^0 = -\frac{1}{3} \left(x_1^0 \right)^2 \right\},$$

(see Fig. 4.9).

Consider the linear system

$$x' = Ax \tag{4.10}$$

where $A \in \mathbb{R}^{n \times n}$ has k eigenvalues with negative real parts and $n - k$ eigenvalues with positive real parts. From the real Jordan form (see Exercise 3.16), there exists a nonsingular matrix $U = [u_1 \cdots u_k, u_{k+1} \cdots u_n] \in \mathbb{R}^{n \times n}$ such that

$$U^{-1}AU = \left[\begin{array}{c|c} A_- & 0 \\ \hline 0 & A_+ \end{array} \right], \tag{4.11}$$

where

$$A_- \in \mathbb{R}^{k \times k} \quad \text{with} \quad Re\lambda(A_-) < 0$$

and

$$A_+ \in \mathbb{R}^{(n-k) \times (n-k)} \quad \text{with} \quad Re\lambda(-A_+) < 0.$$

Then, there are constants $K_1 > 0$, $\alpha > 0$ such that

$$\begin{aligned} |e^{A_+t}| &\leq K_1 e^{\alpha t}, \quad t \leq 0, \\ |e^{A_-t}| &\leq K_1 e^{-\alpha t}, \quad t \geq 0. \end{aligned} \tag{4.12}$$

From (4.11), it follows that

$$A[u_1 \cdots u_k, u_{k+1} \cdots u_n] = U \left[\begin{array}{c|c} A_- & 0 \\ \hline 0 & A_+ \end{array} \right].$$

Hence, for the linear system (4.10), 0 is a saddle point with

$$E^s = \text{span}\langle u_1 \cdots u_k \rangle,$$

$$E^u = \text{span}\langle u_{k+1} \cdots u_n \rangle,$$

where $u_1 \cdots u_k$ (or $u_{k+1} \cdots u_n$) are eigenvectors or generalized eigenvectors associated with eigenvalues with negative (or positive) real parts. The stable set E^s and the unstable set E^u are invariant under A.

Define the projections

$$P : \mathbb{R}^n \to E^u \quad \text{and} \quad Q : \mathbb{R}^n \to E^s$$

with

$$P \left(\sum_{i=1}^{n} \alpha_i u_i \right) = \sum_{i=k+1}^{n} \alpha_i u_i,$$

$$Q \left(\sum_{i=1}^{n} \alpha_i u_i \right) = \sum_{i=1}^{k} \alpha_i u_i.$$

Then $P\mathbb{R}^n = E^u$, $Q\mathbb{R}^n = E^s$ and

$$PAx = PA \left(\sum_{i=1}^{n} \alpha_i u_i \right) = P \left(\sum_{i=1}^{n} \alpha_i A u_i \right)$$

$$= \sum_{i=k+1}^{n} \alpha_i A u_i = A \left(\sum_{k+1}^{n} \alpha_i u_i \right) = APx$$

i.e. $PA = AP$.

Similarly we have $QA = AQ$. Obviously $\mathbb{R}^n = P\mathbb{R}^n \bigoplus Q\mathbb{R}^n$. For any $x \in \mathbb{R}^n$, $Px = Uy$ for some $y = \begin{pmatrix} 0 \\ v \end{pmatrix}$, $v \in \mathbb{R}^{n-k}$. From (4.11), we have

$$e^{At} Px = e^{At} Uy = U \left[\begin{array}{c|c} e^{A_- t} & 0 \\ \hline 0 & e^{A_+ t} \end{array} \right] y = U \begin{pmatrix} 0 \\ e^{A_+ t} v \end{pmatrix}.$$

Then, from (4.12) it follows that

$$|e^{At} Px| \leq \| U \| \ |e^{A_+ t}| |v|$$
$$= \| U \| \ |e^{A_+ t}| |U^{-1} Px|$$
$$\leq \| U \| \ K_1 e^{\alpha t} \| U^{-1} \| \ |Px|, \quad t \leq 0.$$

Hence we have

$$|e^{At} P| \leq K e^{\alpha t}, \quad t \leq 0. \tag{4.13}$$

Similarly we have

$$|e^{At} Q| \leq K e^{-\alpha t}, \quad t \geq 0. \tag{4.14}$$

Q.E.D.

The following is a basic lemma.

Lemma 4.3.1 *If $x(t), t \leq 0$ is a bounded solution of (4.8), then $x(t)$ satisfies the integral equation*

$$y(t) = e^{At}Py(0) + \int_0^t e^{A(t-s)}Pg(y(s))ds + \int_{-\infty}^t e^{A(t-s)}Qg(y(s))ds.$$
(4.15)

If $x(t), t \geq 0$ is a bounded solution of (4.8) then $x(t)$ satisfies the integral equation

$$y(t) = e^{At}Qy(0) + \int_0^t e^{A(t-s)}Qg(y(s))ds - \int_t^\infty e^{A(t-s)}Pg(y(s))ds. \quad (4.16)$$

Conversely, if $y(t), t \leq 0$ (or $t \geq 0$) is a bounded solution of (4.15) (or (4.16)), then $y(t)$ satisfies (4.8).

Proof. Let $y(t) = x(t), t \leq 0$, be a bounded solution of (4.8). Since $PA = AP$, $QA = AQ$, from the variation of constant formula, for any $\tau \in (-\infty, 0]$, we have

$$Qy(t) = e^{A(t-\tau)}Qy(\tau) + \int_\tau^t e^{A(t-s)}Qg(y(s))ds.$$

From (4.14) and the assumption that $y(s)$ is bounded for $s \leq 0$, let $\tau \to -\infty$, we obtain

$$Qy(t) = \int_{-\infty}^t e^{A(t-s)}Qg(y(s))ds.$$

Since

$$Py(t) = e^{At}Py(0) + \int_0^t e^{A(t-s)}Pg(y(s))ds$$

then from $y(t) = Py(t) + Qy(t)$, we obtain (4.15). The proof for the case that $x(t) \geq 0$, $t \geq 0$, is bounded is similar.

The converse statement is proved by direct computation (Exercise 4.2). $\square$

Theorem 4.3.1 *(Stable Manifold Theorem).*
If g satisfies (4.9) and $\operatorname{Re}(\sigma(A)) \neq 0$, then there is a neighborhood U of 0 in $\mathbb{R}^n$ such that $W^u(0, U)$ (or $W^s(0, U)$) is a Lipschitz graph over $E^u := P\mathbb{R}^n$ (or $E^s := Q\mathbb{R}^n$) which is tangent to E^u (or E^s) at 0. Also, there are positive constants K_1, α_1 such that $x \in W^u(0, U)$ (resp. $W^s(0, U)$) then the solution $\varphi^t(x)$ of (4.8) satisfies

$$|\varphi^t(x)| \leq K_1 e^{-\alpha_1|t|}|x|, \quad t \leq 0 \ (resp. \ t \geq 0).$$

Furthermore if g is a C^k function (or an analytic function) in a neighborhood U of 0, then so are $W^u(0, U)$ and $W^s(0, U)$.

Then local stable and unstable manifold can be extended to global stable and unstable manifolds. Define

$$W^s(0) = \text{the global stable manifold of } 0$$
$$= \bigcup_{t \leq 0} \{\varphi_t(x), x \in W^s_{loc}(0)\}$$

and

$$W^u(0) = \text{the global unstable manifold of } 0$$
$$= \bigcup_{t \geq 0} \{\varphi_t(x), x \in W^u_{loc}(0)\}.$$

We defer the proof of stable manifold Theorem to Appendix B.

We note that $E^u := P\mathbb{R}^n$ and $E^s := Q\mathbb{R}^n$ are unstable and stable manifold of equilibrium 0 of linear system $x' = Ax$.

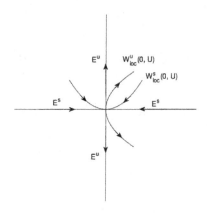

Fig. 4.10

4.4   Orbital Stability

Let $p(t)$ be a periodic solution of period T of the autonomous system

$$x' = f(x), \quad f : \Omega \subseteq \mathbb{R}^n \to \mathbb{R}^n. \tag{4.17}$$

It is inappropriate to consider the notion of asymptotic stability for the periodic solution $p(t)$. For example, consider $\varphi(t) = p(t + \tau)$, with $\tau > 0$ small, then $\varphi(t)$ is also a solution of (4.17), but it is impossible that $|\varphi(t) - p(t)| \to 0$ as $t \to \infty$ no matter how small τ is. Hence instead we consider the periodic orbit $\gamma = \{p(t) \in \mathbb{R}^n : 0 \le t < T\}$ and the concept of "orbital stability" of the periodic orbit γ.

Definition 4.4.1 *We say that a periodic orbit γ is orbitally stable if given $\varepsilon > 0$ there exists $\delta > 0$ such that*

$$dist\left(\varphi^t(x_0), \gamma\right) < \varepsilon \ \text{ for } \ t \ge 0,$$

provided $dist(x_0, \gamma) < \delta$.

If in addition, $dist(\varphi^t(x_0), \gamma) \to 0$ as $t \to \infty$ for $dist(x_0, \gamma)$ sufficiently small, then we say that γ is orbitally asymptotically stable. γ is orbitally unstable if γ is not orbitally stable.

Now the next question is that how do we verify the orbital stability of γ besides the definition.

Linearization about periodic orbit:
Let

$$y = x - p(t).$$

Then

$$\begin{aligned} y'(t) &= x'(t) - p'(t) = f(x(t)) - f(p(t)) \\ &= f(y + p(t)) - f(p(t)) \\ &= D_x f(p(t))y + h(t, y) \end{aligned} \tag{4.18}$$

where $h(t, y) = 0(|y|)$ as $y \to 0$ for all t. The linearized equation of (4.17) about $p(t)$ is

$$\begin{cases} y' = D_x f(p(t))y = A(t)y \\ A(t + T) = A(t). \end{cases} \tag{4.19}$$

We shall relate the characteristic multipliers of the periodic linear system (4.19) to the orbital stability of γ. Let $\Phi(t)$ be the fundamental matrix of (4.19) with $\Phi(0) = I$. Since $p(t)$ is a solution of (4.17), we have

$$p'(t) = f(p(t))$$

and

$$\frac{d}{dt}p'(t) = \frac{d}{dt}f(p(t)) = D_x f(p(t))p'(t) = A(t)p'(t).$$

Hence $p'(t)$ is a solution of (4.19) and it follows that

$$p'(t) = \Phi(t)p'(0).$$

Setting $t = T$ yields

$$p'(0) = p'(T) = \Phi(T)p'(0).$$

Thus, 1 is an eigenvalue of $\Phi(T)$ or equivalently 1 is a characteristic multiplier of (4.19). Assume μ_1, $\mu_2, \cdots, \mu_n$ are characteristic multipliers of T-periodic system (4.19) with $\mu_1 = 1$. Then from Liouville's formula

$$\mu_1 \cdots \mu_n = \mu_2 \cdots \mu_n = \det \Phi(T)$$

$$= \det \Phi(0) \exp\left(\int_0^T \operatorname{tr}(A(s))ds\right)$$

$$= \exp\left(\int_0^T \operatorname{tr}(A(s))ds\right).$$

We note that

$$A(t) = D_x f(p(t)) = \left(\frac{\partial f_i}{\partial x_j}\right)\bigg|_{x=p(t)}.$$

Hence

$$\mu_2 \cdots \mu_n = \exp\left(\int_0^T \frac{\partial f_1}{\partial x_1}(p(t)) + \cdots + \frac{\partial f_n}{\partial x_n}(p(t))dt\right)$$

$$= \exp\left(\int_0^T \operatorname{div}(f(p(t)))dt\right).$$

Definition 4.4.2 *We say that γ is orbitally stable with asymptotic phase if for any nearby solution $\varphi(t)$ we have*

$$|\varphi(t) - p(t+t_0)| \to 0 \quad as \quad t \to \infty$$

for some $0 \leq t_0 < T$.

Now we state the main result of this section.

Theorem 4.4.1 *If the* $(n-1)$ *characteristic multipliers* $\mu_2 \cdots \mu_n$ *satisfy* $|\mu_i| < 1$, $i = 2, \cdots, n$, *then* γ *is orbitally asymptotically stable and nearby solution* $\varphi(t)$ *of (4.17) possesses an asymptotic phase with*

$$|\varphi(t) - p(t + t_0)| \leq Le^{-\alpha t}, \ L, \alpha > 0$$

for some t_0, $0 \leq t_0 < T$.

Corollary 4.4.1 *For* $n = 2$. *If the periodic solution* $p(t)$ *satisfies*

$$\int_0^T \frac{\partial f_1}{\partial x_1}(p(t)) + \frac{\partial f_2}{\partial x_2}(p(t))dt < 0, \tag{4.20}$$

then γ *is orbitally stable with asymptotic phase.*

Remark 4.4.1 Even for the case of $n = 2$, it is difficult to verify (4.20) since the integral depends on how much we know about the location of the periodic orbit. Before we prove Theorem 4.4.1 where Poincaré map will be introduced, we study several properties of nonlinear map $f : \mathbb{R}^N \to \mathbb{R}^N$ and consider the difference equation

$$x_{n+1} = f(x_n). \tag{4.21}$$

Definition 4.4.3 *We say that* $\bar{x}$ *is an equilibrium or a fixed point of (4.21) if* $\bar{x} = f(\bar{x})$.

Definition 4.4.4 *We say that* $\bar{x}$ *is stable if for any* $\varepsilon > 0$ *there exists* $\delta > 0$ *such that for any* x_0 *with* $|x_0 - \bar{x}| < \delta$ *we have* $|x_n - \bar{x}| < \varepsilon$ *for all* $n \geq 0$. *If in addition,* $x_n \to \bar{x}$ *as* $n \to \infty$ *for* x_0 *sufficiently close to* $\bar{x}$ *then we say that* $\bar{x}$ *is asymptotically stable. We say that* $\bar{x}$ *is unstable if* $\bar{x}$ *is not stable.*

Linearization of the map:

$$\begin{aligned} x_{n+1} - \bar{x} &= f(x_n) - f(\bar{x}) \\ &= D_x f(\bar{x})(x_n - \bar{x}) + g(x_n - \bar{x}) \end{aligned} \tag{4.22}$$

where $g(y)$ satisfies $\dfrac{|g(y)|}{|y|} \to 0$ as $y \to 0$. $\tag{4.23}$

Let $y_n = x_n - \bar{x}$. The linearized equation of (4.21) about $\bar{x}$ is

$$y_{n+1} = Ay_n, \quad A = D_x f(\bar{x}). \tag{4.24}$$

From (4.24) we have

$$y_n = A^n y_0.$$

If $\| A \| < 1$ for some norm $\|'\|$, then $|y_n| \leq \| A^n \| |y_0| \leq (\| A \|)^n |y_0| \to 0$ or equivalently $|\lambda| < 1$ for all $\lambda \in \sigma(A)$, then $y_n \to 0$ as $n \to \infty$.

Lemma 4.4.1 *If all eigenvalues λ of $D_x f(\bar{x})$ satisfy $|\lambda| < 1$, then the fixed point $\bar{x}$ is asymptotically stable. Furthermore if $\| x_0 - \bar{x} \|$ is sufficiently small, then*

$$\| x_n - \bar{x} \| \leq ce^{-\alpha n} \| x_0 - \bar{x} \| \quad \text{for some} \quad c, \alpha > 0, \quad \text{for all} \quad n \geq 0.$$

Proof. (Exercise 4.14.) $\qquad\square$

Now let's consider (4.17) and its periodic solution $p(t)$ with period T and the periodic orbit γ. Take a local cross section $\sum \subseteq \mathbb{R}^n$ of dimension $n - 1$ intersecting γ. The hypersurface $\sum$ need not be planar, but it must be chosen in the way that the flow of (4.17) is transversal to $\sum$, i.e. $f(x){\cdot}n(x) \neq 0$ for all $x \in \Sigma$ where $n(x)$ is the normal vector to $\sum$ at x or equivalently the vector field $f(x)$ is not tangent to $\sum$ at x. Without loss of generality, we assume the periodic orbit γ intersect Σ at point p^* where $p^* = p(0)$. Let $U \subseteq \Sigma$ be a small neighborhood of p^*. Let's define the first return map (also called Poincaré map) $P : U \to \Sigma$

$$P(\xi) = \varphi^\tau(\xi) \in \Sigma$$

where $\tau = \tau(\xi)$ is the first time the trajectory $\{\varphi^t(\xi)\}_{t \geq 0}$ returns to Σ.

Obviously, p^* is a fixed point of P. The stability of the fixed point p^* will reflect the orbital stability of the periodic orbit γ. We shall show that $D_x P(p*)$ has eigenvalues $\lambda_1 \cdots \lambda_{n-1}$ which are exactly the characteristic multipliers $\mu_2, \cdots, \mu_n$ of (4.19). Without loss of generality, we assume $p^* = 0$. Then $p(0) = 0$ and $p'(0) = f(0)$. Let Π be the hyperplane $\{x \in \mathbb{R}^n : x \cdot f(0) = 0\}$ and $\varphi(t, \xi)$ be the solution of (4.17) with $\varphi(0, \xi) = \xi$.

Lemma 4.4.2 *For $\xi \in \mathbb{R}^n, |\xi|$ sufficiently small, there exists a unique real-valued C^1 function $t = \tau(\xi)$, the first return time, with $\tau(0) = T$, $\varphi(\tau(\xi), \xi) \in \Pi$.*

Proof. Let

$$F(t,\xi) = \varphi(t,\xi) \cdot f(0)$$

and consider the equation

$$F(t,\xi) = 0.$$

Since $F(T,0) = \varphi(T,0) \cdot f(0) = 0$,

$$\left. \frac{\partial F}{\partial t} \right|_{t=T,\ \xi=0} = f(\varphi(T,0)) \cdot f(0) = |f(0)|^2 > 0.$$

From Implicit Function Theorem, there exists a neighborhood of $\xi = 0$, t can be expressed as a C^1 function of ξ with $t = \tau(\xi)$, $T = \tau(0)$ and $F(\tau(\xi),\xi) = 0$. Then $\varphi(\tau(\xi),\xi) \cdot f(0) = 0$ and it follows that $\varphi(\tau(\xi),\xi) \in \Pi$. □

Lemma 4.4.3 *The eigenvalues $\lambda_1, \ldots, \lambda_{n-1}$ of $DP(p^*)$ are exactly the characteristic multipliers $\mu_2, \ldots, \mu_n$ of (4.19).*

Proof. The first return map $P : U \subseteq \Pi \to \Pi$ satisfies

$$P(\xi) = \varphi(\tau(\xi),\xi), \ P(0) = \varphi(T,0) = p(T) = 0. \tag{4.25}$$

Without loss of generality we may assume $f(0) = (0,\ldots,0,1)^T$ (why?) and hence $\Pi = \{x = (x_1,\ldots,x_n) : x_n = 0\}$. Since

$$\frac{d}{dt}\varphi(t,\xi) = f(\varphi(t,\xi)), \tag{4.26}$$
$$\varphi(0,\xi) = \xi,$$

then differentiating (4.26) with respect to ξ yields

$$\Phi'(t,\xi) = D_x f(\varphi(t,\xi))\Phi(t,\xi), \tag{4.27}$$
$$\Phi(0,\xi) = I,$$

where $\Phi(t,\xi) = \varphi_\xi(t,\xi)$.

Setting $\xi = 0$ in (4.27) yields

$$\Phi'(t,0) = D_x f(p(t))\Phi(t,0).$$
$$\Phi(0,0) = I.$$

Hence $\Phi(t,0)$ is the fundamental matrix of (4.19) with $\Phi(0,0) = I$ and the characteristic multipliers $\mu_1 = 1$, $\mu_2,\ldots,\mu_n$ of (4.19) are eigenvalues of $\Phi(T,0)$. Since $p'(0) = f(0)$ and $p'(0) = \Phi(T,0)p'(0)$, it follows that $f(0) = \Phi(T,0)f(0)$. From $f(0) = (0,\ldots,0,1)^T$, it follows that the last column of $\Phi(T,0)$ is $(0,\dots,0,1)^T$. Now we consider the Poincaré map P in (4.25), and compute $DP(0)$. Differentiating (4.25) with respect to ξ, we obtain

$$DP(\xi) = \varphi'(\tau(\xi),\xi)\frac{d}{d\xi}\tau(\xi) + \varphi_\xi(\tau(\xi),\xi). \qquad (4.28)$$

Setting $\xi = 0$ in (4.28), we have

$$DP(0) = \varphi'(T,0)\frac{d\tau(\xi)}{d\xi}\bigg|_{\xi=0} + \Phi(T,0)$$

$$= f(0)\cdot\frac{d\tau(\xi)}{d\xi}\bigg|_{\xi=0} + \Phi(T,0)$$

$$= (0,\dots,0,1)^T\left(\frac{\partial\tau}{\partial\xi_1}(0),\dots,\frac{\partial\tau}{\partial\xi_n}(0)\right) + \Phi(T,0)$$

$$= \begin{bmatrix} 0 & \cdots\cdots & 0 \\ 0 & \cdots\cdots & 0 \\ \frac{\partial\tau}{\partial\xi_1}(0) & \cdots & \frac{\partial\tau}{\partial\xi_n}(0) \end{bmatrix} + \begin{bmatrix} & & & & 0 \\ & & & & 0 \\ & & & & \vdots \\ & & & & 0 \\ x & x & x & x & 1 \end{bmatrix}$$

$$= \begin{bmatrix} & & & & 0 \\ & & & & \vdots \\ & & & & \vdots \\ & & & & 0 \\ x & x & x & x \end{bmatrix}.$$

Hence the eigenvalues of $DP(0)$ are exactly the characteristic multipliers of (4.19). $\qquad\square$

Proof of Theorem 4.4.1: [Ha]
Let $p(0) = 0$ and $P : \Pi \cap B(0,\delta) \to \Pi$ be the Poincaré map. From the assumption $|\mu_i| < 1$, $i = 2,\cdots,n$ and Lemma 4.4.3, it follows that 0 is an asymptotically stable fixed point for the first return map P on $U \subseteq \Pi$. Thus from Lemma 4.4.1, if $\|\xi_0\|$ is sufficiently small, then $\|\xi_n\| \leq e^{-n\alpha}$ $\|\xi_0\| \to 0$ as $n \to \infty$ for some $\alpha > 0$ where $\xi_n = P^n\xi_0$. From the continuous dependence on initial data, given $\varepsilon > 0$ there exists $\delta = \delta(\varepsilon) > 0$ such that if $\text{dist}(x_0,\gamma) < \delta$ then there exists $\tau^0 = \tau^0(x_0)$, where $\varphi(t,x_0)$ exists for $0 \leq t \leq \tau^0$, $\varphi(\tau^0,x_0) \in \Pi$ and $|\varphi(\tau^0,x_0)| < \varepsilon$.

Fig. 4.11

Let $\xi_0 = \varphi(\tau^0, x_0) \in \Pi$ and

$$\tau^1 = \tau^0 + \tau(\xi_0) \quad \text{such that} \quad \xi_1 = \varphi\left(\tau(\xi_0), \xi_0\right) \in \Pi,$$

$$\vdots$$

$$\tau^n = \tau^{n-1} + \tau(\xi_{n-1}) \quad \text{such that} \quad \xi_n = \varphi\left(\tau(\xi_{n-1}), \xi_{n-1}\right) \in \Pi,$$

i.e., τ^k is the travelling time from x_0 to ξ_k (see Fig. 4.11). We claim that

$\lim_{n \to \infty} \frac{\tau^n}{nT} = 1$ and there exists $t_0 \in \mathbb{R}$ such that $t_0 = \lim_{n \to \infty} (\tau^n - nT)$ and $|\tau^n - (nT + t_0)| \leq L_1 e^{-\alpha n} \parallel \xi_0 \parallel$.

$$(4.29)$$

We first show that $\{\tau^n - nT\}$ is a Cauchy sequence. Consider

$$
\begin{aligned}
|(\tau^n - nT) - (\tau^{n-1} - (n-1)T)| &= |\tau^n - \tau^{n-1} - T| \\
&= |\tau(\xi_{n-1}) - T| = |\tau(\xi_{n-1}) - \tau(0)| \\
\text{(Note that } \tau \text{ is } C^1) \quad &\leq \sup_{\theta \in B(0,\delta)} \left| \frac{d\tau}{d\xi}(\theta) \right| \parallel \xi_{n-1} \parallel = L_0 \parallel \xi_{n-1} \parallel \\
&\leq L_0 e^{-\alpha(n-1)} \parallel \xi_0 \parallel .
\end{aligned}
$$

Hence for $m > n$ we have

$$
\begin{aligned}
|(\tau^m - mT) - (\tau^n - nT)| &\leq |(\tau^m - mT) - (\tau^{m-1} - (m-1)T| \\
&\quad + \cdots + |(\tau^{n+1} - (n+1)T) - (\tau^n - nT)| \\
&\leq L_0 \parallel \xi_0 \parallel \left(e^{-\alpha(m-1)} + \cdots + e^{-\alpha n} \right) \\
&\leq L_0 \parallel \xi_0 \parallel \frac{e^{-\alpha n}}{1 - e^{-\alpha}} < \varepsilon \quad \text{if} \quad n \geq N.
\end{aligned}
$$

Then $\lim\limits_{n\to\infty} (\tau^n - nT)$ exists and equals to a number, say, t_0 and

$$
\begin{aligned}
|\tau^n - (nT + t_0)| &= |t_0 - (\tau^n - nT)| \\
&\leq |t_0 - (\tau^m - mT)| + |(\tau^m - mT) - (\tau^{m-1} - (m-1)T)| \\
&\quad + \cdots + |(\tau^{n+1} - (n+1)T) - (\tau^n - nT)| \\
&\leq \sum_{k=n}^{\infty} |(\tau^{(k+1)} - (k+1)T) - (\tau^k - kT)| \\
&\leq \sum_{k=n}^{\infty} L_0 e^{-\alpha(k-1)} \|\xi_0\| \leq \frac{L_0 e^{-\alpha n}}{1 - e^{-\alpha}} \| \xi_0 \| .
\end{aligned}
$$

Thus we prove the claim (4.29).

Next we want to show that
$$
|\varphi(t + t_0, x_0) - p(t)| \leq L e^{\frac{-\alpha}{T} t} |\xi_0|.
$$
Let $0 \leq t \leq T$,
$$
\begin{aligned}
|\varphi(t + \tau^n, x_0) - p(t)| &= |\varphi(t, \xi_n) - \varphi(t, 0)| \\
&\leq \sup_{0 \leq t \leq T, \ \theta \in B(0,\delta)} \| \varphi_\xi(t, \theta) \| \ |\xi_n| \leq L_2 |\xi_n| \\
&\leq L_2 e^{-n\alpha} |\xi_0|,
\end{aligned}
$$
and
$$
\begin{aligned}
&|\varphi(t + \tau^n, x_0) - \varphi(t + nT + t_0, x_0)| \\
&\leq |\varphi'(\tilde{t}, x_0)| \ |(t + \tau^n) - (t + nT + t_0)| \\
&\leq L_3 |\tau^n - (nT + t_0)| \leq L_3 L_1 e^{-\alpha n} |\xi_0|.
\end{aligned}
$$
Hence for $0 \leq t \leq T$, we have
$$
\begin{aligned}
&|\varphi(t + nT + t_0, x_0) - p(t)| \hspace{2cm} (4.30) \\
&\leq |\varphi(t + nT + t_0, x_0) - \varphi(t + \tau^n, x_0)| \\
&\quad + |\varphi(t + \tau^n, x_0) - p(t)| \\
&\leq (L_2 + L_3 L_1) e^{-\alpha n} \| \xi_0 \| .
\end{aligned}
$$
For any t, we assume $nT \leq t \leq (n+1)T$ for some n. Then $0 \leq t - nT \leq T$ and $t/T - 1 \leq n \leq t/T$. Substituting t in (4.30) by $t - nT$, we have
$$
\begin{aligned}
&|\varphi(t + t_0, x_0) - p(t - nT)| \\
&= |\varphi(t + t_0, x_0) - p(t)| \\
&\leq (L_2 + L_3) e^{-\alpha n} \| \xi_0 \| \ \leq (L_2 + L_3) e^{-\alpha(\frac{t}{T} - 1)} \| \xi_0 \| .
\end{aligned}
$$
Hence
$$
|\varphi(t + t_0, x_0) - p(t)| \leq L e^{\frac{-\alpha}{T} t} \| \xi_0 \| \quad \text{for some} \quad L > 0.
$$

Q.E.D.

4.5   Traveling Wave Solutions

In this section we shall apply stable manifold theorem to obtain the existence of traveling wave solutions for the scalar diffusion-reaction partial differential equation

$$u_t = Ku_{xx} + f(u), \quad K > 0. \tag{4.31}$$

A traveling wave solution $u(x,t)$ of PDE (4.31) is a special solution of the form

$$u(x,t) = U(x + ct), \ c \in \mathbb{R}, \tag{4.32}$$

where c is called the traveling speed. Let $\xi = x + ct$ and substitute (4.32) into (4.31), we obtain

$$K\frac{d^2U}{d\xi^2} - c\frac{dU}{d\xi} + f(U) = 0. \tag{4.33}$$

In (4.33) there are two unknowns, namely, the wave speed c and $U(\xi)$. We shall study two well-known nonlinear diffusion-reaction equations. One is the Fisher's equation, and the other is the "bistable" equation.

Fisher's equations (or KPP equation)

In 1936 famous geneticist and statistician R. A. Fisher proposed a model to describe the spatial distribution of an advantageous gene. Consider a one locus, two allele genetic trait in a large randomly mating population. Let $u(x,t)$ denote the frequency of one gene, say, A, in the gene pool at position x. If A is dominant and different sites are connected by random migration of the organism, then the gene pool can be described by a diffusion model (the details of the derivation of the model see [AW]):

$$u_t = Ku_{xx} + su(1 - u), \ s > 0. \qquad \text{(Fisher's equation)}$$

We shall consider the following general model

$$\begin{aligned}
&u_t = Ku_{xx} + f(u), \\
&f(u) > 0, 0 < u < 1, f(0) = f(1) = 0, \\
&f'(0) > \frac{f(u)}{u} \text{ for } 0 < u < 1.
\end{aligned} \tag{4.34}$$

In 1936, Kolmogoroff, Petrovsky and Piscounov analyzed Fisher's equation by looking for steady progressing waves (4.32) with boundary conditions $u(-\infty, t) = 0$, $u(+\infty, t) = 1$. We shall look for a solution $U(\xi)$ of (4.33) with some speed c with boundary conditions $U(-\infty) = 0$, $U(+\infty) = 1$ (see

Fig. 4.12). Let $W = \frac{dU}{d\xi}$, then (4.33) can be rewritten as the following first order system:

$$\frac{dU}{d\xi} = W,$$

$$\frac{dW}{d\xi} = \frac{c}{K}W - \frac{f(U)}{K}, \tag{4.35}$$

$$(U(-\infty), W(-\infty)) = (0,0), \quad (U(+\infty), W(+\infty)) = (1,0).$$

In (4.35), the system has two equilibria $(0,0)$ and $(1,0)$. The variational matrix of (4.35) is

$$M(U,W) = \begin{bmatrix} 0 & 1 \\ -\frac{f'(U)}{K} & \frac{c}{K} \end{bmatrix}.$$

For equilibrium $(0,0)$,

$$M(0,0) = \begin{bmatrix} 0 & 1 \\ -\frac{f'(0)}{K} & \frac{c}{K} \end{bmatrix}.$$

The eigenvalues λ of $M(0,0)$ satisfy

$$\lambda^2 - \frac{c}{K}\lambda + \frac{f'(0)}{K} = 0.$$

Hence there are two eigenvalues

$$\lambda_\pm = \frac{\frac{c}{K} \pm \sqrt{\left(\frac{c}{K}\right)^2 - 4\frac{f'(0)}{K}}}{2}.$$

Multiplying $\frac{dU}{d\xi}$ on both sides of (4.33) and integrating from $-\infty$ to $+\infty$ yields

$$c = \frac{\int_0^1 f(u)du}{\int_{-\infty}^{+\infty} (U'(\xi))^2 \, d\xi} > 0.$$

Hence if $c \geq \sqrt{4Kf'(0)}$ then $(0,0)$ is an unstable node; if $0 < c < \sqrt{4Kf'(0)}$ then $(0,0)$ is an unstable spiral. For the equilibrium $(1,0)$,

$$M(1,0) = \begin{pmatrix} 0 & 1 \\ -\frac{f'(1)}{K} & \frac{c}{K} \end{pmatrix}.$$

The eigenvalues λ satisfy

$$\lambda^2 - \frac{c}{K}\lambda + \frac{f'(1)}{K} = 0.$$

Hence $(1,0)$ is a saddle point for $f'(1) < 0$.

Let $c_0 = \sqrt{4Kf'(0)}$. If $0 < c < c_0$ then there are no heteroclinic orbits connecting $(0,0)$ and $(1,0)$ because $(0,0)$ is an unstable spiral and the solution $U(\xi)$ of (4.35) is positive (biological restriction!). We shall show for each $c \geq c_0$, there is a heteroclinic orbit connecting unstable node $(0,0)$ and saddle point $(1,0)$ (see Fig. 4.13). First we reverse the time $\xi \to -\xi$, then (4.35) becomes

$$\frac{dU}{d\xi} = -W, \tag{4.36}$$

$$\frac{dW}{d\xi} = -\frac{c}{K}W + \frac{f(U)}{K},$$

$$(U(+\infty), W(+\infty)) = (0,0), (U(-\infty), W(-\infty)) = (1,0).$$

Choose $m, \rho > 0$ such that

$$m > \frac{f(U)}{c} \quad \text{for all} \quad 0 \leq U \leq 1, \tag{4.37}$$

and

$$0 < \rho_1 < \rho < \rho_2,$$

where ρ_1, ρ_2 are the positive roots of $K\rho^2 - c\rho + A = 0$, $A = \sup_{0 < u < 1} \frac{f(u)}{u}$. Let

$$\Omega = \left\{ (U, W) : 0 \leq W \leq m, \frac{W}{\rho} \leq U \leq 1 \right\}.$$

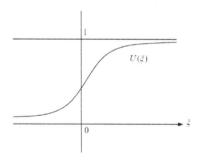

Fig. 4.12

We claim that with the choices of m and ρ in (4.37), the region Ω is positively invariant under the system (4.36). Let $\overrightarrow{F}$ be the vector field of (4.36) and $\overrightarrow{n}$ be the outward normal vector of $\partial\Omega$. (See Fig. 4.14.)

(i) For the line $W = \rho U$, $0 \leq W \leq m$, $\vec{n} = (-\rho, 1)$, from (4.37)

$$\vec{F} \cdot \vec{n} = \rho W + \left(-\frac{c}{K} W + \frac{f(U)}{K} \right)$$

$$= W \left(\rho - \frac{c}{K} + \frac{1}{K\rho} \frac{f(U)}{U} \right)$$

$$< W \left(\frac{K\rho^2 - c\rho + A}{K\rho} \right),$$

where $A = \sup_{0 < u < 1} \frac{f(u)}{u} < f'(0)$.

Since $c^2 - 4Kf'(0) \geq 0$ then $c^2 - 4KA > 0$. Choose ρ, $\rho_1 < \rho < \rho_2$ where ρ_1, ρ_2 are roots of $K\rho^2 - c\rho + A = 0$. Then $\vec{F} \cdot \vec{n} < 0$.

(ii) For $U = 1$, $0 < W < m$, $\vec{n} = (1, 0)$,

$$\vec{F} \cdot \vec{n} = -W < 0.$$

(iii) For $W = m$, $\frac{m}{\rho} \leq U \leq 1$, $\vec{n} = (0, 1)$, from (4.37) we have

$$\vec{F} \cdot \vec{n} = -\frac{c}{K} W + \frac{f(U)}{K} < 0.$$

(iv) For $W = 0$, $0 \leq U \leq 1$, $\vec{n} = (0, -1)$,

$$\vec{F} \cdot \vec{n} = -\frac{f(U)}{K} < 0.$$

Hence Ω is positively invariant under (4.36). Now we verify the tangent vector to the unstable manifold of (4.36) at saddle point $(1, 0)$ pointing towards the interior of Ω. Let $\lambda_1 > 0$, $\lambda_2 < 0$ be the eigenvalues of the variational matrix of (4.36) at $(1, 0)$. Then the tangent vector $(x_1, x_2)^T$ to the unstable manifold at $(1, 0)$ satisfies

$$\begin{pmatrix} 0 & -1 \\ \frac{f'(1)}{K} & \frac{-c}{K} \end{pmatrix} \begin{pmatrix} x_1 \\ x_2 \end{pmatrix} = \lambda_1 \begin{pmatrix} x_1 \\ x_2 \end{pmatrix}.$$

Then $-x_2 = \lambda_1 x_1$. Thus the unstable manifold points to the interior of Ω. Since Ω is positively invariant and thus $\frac{dU}{d\xi} = -W < 0$ for all ξ and hence the unstable manifold will converge to $(0, 0)$ for all $c \geq c_0$. This establishes the existence of traveling wave solutions. We note that the wave is monotonic due to the fact $\frac{dU}{d\xi} = W > 0$ (see Fig. 4.13).

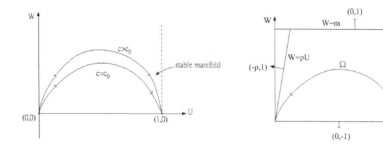

Fig. 4.13   Fig. 4.14

Bistable equations:

Consider the following scalar diffusion-reaction PDE

$$u_t = u_{xx} + f(u) \tag{4.38}$$

where $f(u)$ satisfies

$$
\begin{aligned}
&f(0) = f(1) = 0,\ f'(0) < 0,\ f'(1) < 0 \text{ and}\\
&\text{there exists } \alpha,\ 0 < \alpha < 1 \text{ such that}\\
&f(x) < 0 \text{ for } 0 < x < \alpha \text{ and } f(x) > 0, \alpha < x < 1.
\end{aligned}
\tag{4.39}
$$

Equation (4.38) with conditions (4.39) can be derived from nerve conduction (see [Kee1; AW; FMcL]). The typical examples of $f(u)$ are (see Fig. 4.15(a) and Fig. 4.15(b))

$$f(u) = au(u - 1)(\alpha - u), 0 < \alpha < 1,$$

and

$$f(u) = -u + H(u - \alpha), 0 < \alpha < 1,$$

where $H(v)$ is a Heaviside function,

$$H(v) = \begin{cases} 1, v \geq 0, \\ 0, v < 0. \end{cases}$$

We note that $u = 0$ and $u = 1$ are stable steady states of the ordinary differential equation

$$\frac{du}{dt} = f(u).$$

We look for the traveling wave solutions

$$u(x,t) = U(x + ct).$$

Let $\xi = x + ct$. Then from (4.38) we have

$$\frac{d^2U}{d\xi^2} - c\frac{dU}{d\xi} + f(U) = 0. \tag{4.40}$$

Set $W = \frac{dU}{d\xi}$, then (4.40) can be rewritten as

$$\frac{dU}{d\xi} = W, \tag{4.41}$$

$$\frac{dW}{d\xi} = cW - f(U).$$

We want to find traveling front solutions connecting $(0,0)$ to $(1,0)$, i.e.,

$$\lim_{\xi \to -\infty} (U(\xi), W(\xi)) = (0,0),$$

$$\tag{4.42}$$

$$\lim_{\xi \to +\infty} (U(\xi), W(\xi)) = (1,0).$$

From the variational matrix $M(U, W)$ of (4.41),

$$M(U, W) = \begin{pmatrix} 0 & 1 \\ -f'(U) & c \end{pmatrix}, \tag{4.43}$$

it is easy to verify that $(0,0)$ and $(1,0)$ are both saddle points of (4.41) due to the fact $f'(0) < 0, f'(1) < 0$. Multiplying W on both sides of (4.40) and integrating from $-\infty$ to ∞, we obtain from (4.42)

$$c\int_{-\infty}^{\infty} W^2(\xi)d\xi = \int_0^1 f(u)du. \tag{4.44}$$

From now on we assume

$$\int_0^1 f(u)du > 0 \tag{4.45}$$

then $c > 0$.

Consider the unstable manifold Γ_c of $(0,0)$. Obviously when $c = 0$, Γ_c cannot reach $U = 1$ otherwise from (4.44) we obtain

$$\int_0^1 f(u)du = 0$$

which is a contradiction to the assumption (4.45) (see Fig. 4.16(a)). Since the slope of the unstable manifold Γ_c at $(0,0)$ is the positive eigenvalue λ_+ of $M(0,0)$,

$$\lambda_+ = \frac{c + \sqrt{c^2 - 4f'(0)}}{2} > c.$$

Let $K = \sup_{0 < u < 1} \frac{f(u)}{u}$ and $\sigma > 0$ be a fixed number. Consider the line $W = \sigma U$ in the phase plane (see Fig. 4.16(a)). Claim $W(U) > \sigma U$ for c large. If not, there exists U_0 such that $W(U_0) = \sigma U_0$, $W'(U_0) < \sigma$, $W(U) > \sigma U > 0$ for $0 < U < U_0$, then for c large

$$\frac{dW}{dU}\bigg|_{U=U_0} = c - \frac{f(U_0)}{W(U_0)} = c - \frac{f(U_0)}{\sigma U_0} > c - \frac{K}{\sigma} > \sigma.$$

This is a contradiction.

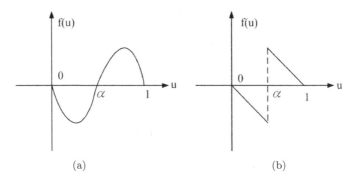

(a)   (b)

Fig. 4.15

Then for large c the unstable manifold Γ_c is above the line $W = \sigma U$. By continuous dependence on the parameter c, there exists c_0 such that the unstable manifold Γ_{c_0} of $(0, 0)$ reach the saddle point $(1, 0)$ (see Fig. 4.16(a)).

Next we show the uniqueness of traveling waves. Suppose we have two traveling waves connecting $(0, 0)$ and $(1, 0)$ with two different speeds c_1 and c_0. Assume $c_1 > c_0$. From (4.41) we have

$$\frac{dW}{dU} = c - \frac{f(U)}{W}. \tag{4.46}$$

It follows that $W(U, c_1) \geq W(U, c_0)$.

From (4.43), the slope of the stable manifold of $(1, 0)$ is

$$\lambda_- = \frac{c - \sqrt{c^2 - 4f'(1)}}{2} < 0.$$

Hence

$$|\lambda_-| = \frac{\sqrt{c^2 - 4f'(1)} - c}{2} \quad \text{and}$$

$$\frac{d|\lambda_-|}{dc} = \frac{1}{2}\left[\frac{c}{\sqrt{c^2 - 4f'(1)}} - 1\right] < 0. \tag{4.47}$$

However from Fig. 4.16(b), $|\lambda_(c_1)| > |\lambda_(c_0)|$, we obtain a contradiction.

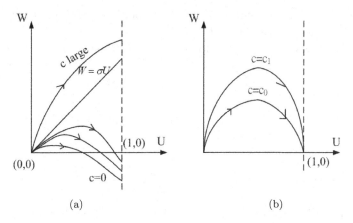

Fig. 4.16

4.6 Exercises

Exercise 4.1 For the equation $v''(t) + t\sin v(t) = 0$, show that $v(t) \equiv \pi$, $v'(t) \equiv 0$ is an unstable solution and $v(t) \equiv 0$, $v'(t) \equiv 0$ is asymptotically stable.

Exercise 4.2 Consider the equation $\dot{y} = (a\cos t + b)y - y^3$ in $\mathbb{R}$, where $a, b > 0$.

(1) Study the stability of the zero solution.
(2) Suppose it is known there is a unique 2π-periodic solution $y_p(t; \xi_0)$ satisfying $y_p(0; \xi_0) = \xi_0 \in [c, a+b+1]$, $c > 0$. Show that $y_p(t; \xi_0)$ is stable.

Exercise 4.3 Consider the spiral system

$$\dot{x}_1 = -x_2 + x_1(1 - r^2),$$
$$\dot{x}_2 = x_1 + x_2(1 - r^2)$$

where $r^2 = x_1^2 + x_2^2$. Let $\mathbf{x}(t, \tau, \xi)$ denote the solution which passes through ξ at time τ and note that $\mathbf{x}(t, \tau, (1, 0)) = (\cos(t - \tau), \sin(t - \tau))$.

(1) Explicitly calculate the linear system of differential equations for which the Jacobian matrix

$$\left(\frac{\partial \mathbf{x}}{\partial \xi}(t, 0, (1, 0)) \right)$$

is a fundamental matrix solution.

(2) Use the given information to exhibit a periodic solution of period 2π for the linear system in part (a).

(3) Find the characteristic multipliers of the linear system in part (a).

Exercise 4.4

(1) Check that $(x = 3\cos 3t, y = \sin 3t)$ is a periodic solution of the system

$$\dot{x} = -9y + x\left(1 - \frac{x^2}{9} - y^2\right),$$

$$\dot{y} = x + y\left(1 - \frac{x^2}{9} - y^2\right).$$

(2) Find the derivative (eigenvalue) of the Poincaré first return map to the real axis at $(3, 0)$.

Exercise 4.5 Prove the converse statement in Lemma 4.3.1.

Exercise 4.6

(1) Let $B \in \mathbb{R}^{2\times 2}$. Show that for the linear difference equation $x_{n+1} = Bx_n$, fixed point 0 is asymptotically stable if and only if $|\det B| < 1$ and $|\text{Trace}B| < 1 + \det B$.

(2) Do the stability analysis of the map

$$x_{n+1} = \frac{ay_n}{1 + x_n{}^2}, \quad a, b \in \mathbb{R}$$

$$y_{n+1} = \frac{bx_n}{1 + y_n{}^2}.$$

Exercise 4.7 Analyze the following Host-parasitoid system

$$\begin{cases} x_{n+1} = \sigma x_n \exp(-ay_n) \qquad a > 0, \sigma > 0 \\ y_{n+1} = \sigma x_n(1 - \exp(-ay_n)) \end{cases}$$

$x_0 > 0, \quad y_0 > 0.$

Exercise 4.8 Analyze the following Predator-Prey system with refugee.

$$\frac{dx}{dt} = ax - c_1y(x - k) \qquad a, c_1, c_2, e, k > 0$$

$$\frac{dy}{dt} = -ey + c_2y(x - k)$$

$$x(0) > 0, \quad y(0) > 0.$$

Exercise 4.9 Let $p(\lambda) = \lambda^3 + a_1\lambda^2 + a_2\lambda + a_3$ and $p(\lambda_i) = 0$, $i = 1, 2, 3$. Show that $|\lambda_i| < 1$, $i = 1, 2, 3$ if and only if

(1) $p(1) = 1 + a_1 + a_2 + a_3 > 0$
(2) $(-1)^3 p(-1) = 1 - a_1 + a_2 - a_3 > 0$
(3) $1 - a_3^2 > |a_2 - a_1 a_3|$.

Exercise 4.10 Do the stability analysis of the Brusselator equations

$$\frac{dx}{dt} = a - (b+1)x + x^2 y, \qquad a, b > 0$$

$$\frac{dy}{dt} = bx - x^2 y.$$

Exercise 4.11 Suppose a, b, c are nonnegative continuous functions on $[0, \infty)$, u is a nonnegative bounded continuous solution of the inequality

$$u(t) \le a(t) + \int_0^t b(t-s)u(s)ds + \int_0^\infty c(s)u(t+s)ds, \quad t \ge 0,$$

and $a(t) \to 0$, $b(t) \to 0$ as $t \to \infty$, $\int_0^\infty b(s)ds < \infty$, $\int_0^\infty c(s)ds < \infty$. Prove that $u(t) \to 0$ as $t \to \infty$ if

$$\int_0^\infty b(s)ds + \int_0^\infty c(s)ds < 1.$$

Exercise 4.12 Let $x' = f(x)$, $x \in \mathbb{R}^n$ with flow $\varphi_t(x)$ defined for all $t \in \mathbb{R}$, $x \in \mathbb{R}^n$. Show that if $\text{Trace} Df(x) = 0$ for all $x \in \mathbb{R}^n$, then the flow $\varphi_t(x)$ preserves volume, i.e., $vol(\Omega) = vol(\varphi_t(\Omega))$ for all $t > 0$.

Exercise 4.13 Consider

$$x_1' = -x_2 + x_1 \left(1 - x_1^2 - x_2^2\right),$$
$$x_2' = x_1 + x_2 \left(1 - x_1^2 - x_2^2\right).$$

Show that $(\cos t, \sin t)$ is a periodic solution. Compute the characteristic multipliers of the linearized system and discuss the orbital stability.

Exercise 4.14 Prove Lemma 4.4.1.

Exercise 4.15 Consider Fibonacci sequence $\{a_n\}$ satisfying $a_0 = 0, a_1 = 1, a_n = a_{n-1} + a_{n-2}, n \ge 2$. Find a formula for a_n. Show that a_n increases like a geometric progression and find $\lim\limits_{n \to \infty} \frac{\ln a_n}{n}$.

Exercise 4.16 Do the stability analysis for the following Henon map

$$x_{n+1} = A - By_n - x_n^2, \quad A, B \in \mathbb{R},$$
$$y_{n+1} = x_n.$$

Exercise 4.17 Consider the following Predator-Prey system

$$\begin{cases} \frac{dx}{dt} = x \left(\gamma \left(1 - \frac{x}{K} \right) - \alpha y \right), \\ \frac{dy}{dt} = (\beta x - d - \delta y) y, \qquad \gamma, K, \alpha, \beta, d, \delta > 0. \\ x(0) > 0, y(0) > 0. \end{cases}$$

For various possible cases, do the following:

(1) Find all equilibria with nonnegative components.
(2) Do stability analysis for each equilibrium.
(3) Find the stable manifold of each saddle point.
(4) Predict the global asymptotic behavior.

Exercise 4.18 Read pp. 273–275 of [Kee1] where the explicit form of $U(\xi)$ is shown as

$$U(\xi) = \frac{1}{2} \left[1 + \tan h \left(\frac{\xi}{2\sqrt{2}} \right) \right],$$
$$c = \frac{1}{\sqrt{2}} (1 - 2\alpha),$$

for (4.40).

Exercise 4.19 Let K be a closed and bounded subset of $\mathbb{R}^n$ and let $F : K \to K$ be a continuous map, not necessarily either one-to-one or onto. A point p in K is said to be a wandering point for F if there exists a neighborhood U of p for which $\{F^{-m}(U) : m \geq 0\}$ is a pairwise disjoint family of sets. Let p be a point of K.

(1) Given $M > 0$, prove that if $p \notin F^M(K)$, then

$$\{F^{-m}(\{p\}) : 0 \leq m < M\}$$

is a pairwise disjoint family of sets. (It is possible that some of the sets in this family are in fact empty.)

(2) Prove that if there exists $M > 0$ such that $p \notin F^M(K)$, then p is a wandering point for F.

Exercise 4.20 Consider a mathematical model describing the Pollen coupling of forest trees: Let $Y_i(t)$ be the non-dimensionalized energy

$$Y_i(t+1) = \begin{cases} Y_i(t) + 1 & \text{if } Y_i(t) \le 0 \\ -kP_i(y)Y_i(t) + 1 & \text{if } Y_i(t) > 0. \end{cases}$$

Where $P_i(t)$ is given by

$$P_i(t) = \left(\frac{1}{N-1} \sum_{j \ne i} [Y_j(t)]_+ \right)^{\beta} \quad i = 1, 2, \ldots, N$$

N is the number of trees, $\beta > 0$, $k > 0$, $[Y]_+ = Y$ if $Y > 0$, $[Y]_+ = 0$ if $Y \le 0$. Find the synchronized equilibrium $(Y^*, \ldots, Y^*) \in \Re^N$, $Y^* > 0$ and the condition that $(Y^*, \ldots, Y^*)$ is locally stable.

Exercise 4.21 Consider the system $x' = a + x^2 - xy, y' = y^2 - x^2 - 1$ where a is a parameter.

(1) Sketch the phase portrait for $a = 0$. Show that there is a trajectory connecting two saddle points.
(2) Sketch the phase portrait for $a < 0$ and $a > 0$ by computer simulations.

Exercise 4.22 The system $x' = xy - x^2y + y^3, y' = y^2 + x^3 - xy^2$ has a nasty higher-order fixed point at the origin. Using polar coordinates to sketch the phase portrait.

Exercise 4.23 Consider the equation $\dot{x} = Ax + h(t)$ in $\mathbb{R}^n$, where all eigenvalues of A have negative real parts and h is continuous and bounded on R.

(1) Prove that solutions are unique and defined for all $t \in (-\infty, \infty)$.
(2) Prove that if in addition $||h(t)|| \to 0$ as $t \to \infty$, then every solution tends to 0 as $t \to \infty$.

Exercise 4.24 (Kermack and McKendrick's Epidemics model)

Consider SIR model, $S \longrightarrow I \longrightarrow R$,

$$\frac{dS}{dt} = -rSI,$$

$$\frac{dI}{dt} = rSI - aI,$$

$$\frac{dR}{dt} = aI,$$

$$S(0) = S_0 > 0, \quad I(0) = I_0 > 0, \quad R(0) = 0$$

where $S(t), I(t), R(t)$ are populations of susceptible, infectives and removed respectively; $r > 0, a > 0$ are infective rate, removal rate respectively. Let $\rho = \frac{a}{r}$ be the relative removable rate, $N = S_0 + I_0$.
Show that

(1) $S(t) + I(t) + R(t) = N$ for all $t > 0$.
(2) If $S_0 < \rho$ then $I(t)$ goes monotonely to zero. No epidemics can occur.
(3) If $S_0 > \rho$, $I(t)$ increases as t increases then tends monotonely to zero. This is a threshold phenomenon — there is a critical value which the initial susceptible population must exceed for there to be an epidemic.
(4) $\lim_{t \to \infty} S(t) = S(\infty)$ exists and is the unique root of the transcendental equation

$$S_0 \, exp \left\{ -\frac{1}{\rho}(N - z) \right\} - z = 0.$$

Chapter 5

METHOD OF LYAPUNOV FUNCTIONS

5.1 An Introduction to Dynamical Systems

Example 5.1.1 Consider the autonomous system
$$x' = f(x), \quad f : D \subseteq \mathbb{R}^n \to \mathbb{R}^n. \tag{5.1}$$
Let $\varphi(t, x_0)$ be the solution of (5.1) with initial condition $x(0) = x_0$. Then it follows that

  (i) $\varphi(0, x_0) = x_0$
  (ii) $\varphi(t, x_0)$ is a continuous function of t and x_0
  (iii) $\varphi(t + s, x_0) = \varphi(s, \varphi(t, x_0))$.

Property (i) is obvious; property (ii) follows directly from the property of continuous dependence on initial conditions. Property (iii) which is called semi-group property, follows from uniqueness of ODE and the fact that (5.1) is an autonomous system. We call $\varphi : \mathbb{R} \times \mathbb{R}^n \to \mathbb{R}^n$ the flow induced by (5.1).

Remark 5.1.1 Property (ii) does not hold.

In general we shall consider flows defined on the metric space M. In Example 5.1.1, $M = \mathbb{R}^n$. In fact, in many applications of dynamical systems to partial differential equations, functional differential equations, the metric space M is a Banach space and the flow φ is a semiflow, i.e., $\varphi : \mathbb{R}^+ \times M \to M$ satisfying (i), (ii), (iii). Now let (M, ρ) be a matric space. In the rest of Section 5.1, we refer to the book in [NS].

Definition 5.1.1 *We call a map $\pi : \mathbb{R} \times M \to M$ a **continuous** dynamical system if*

(i) $\pi(0, x) = x$

(ii) $\pi(t, x)$ is continuous in x and t

(iii) $\pi(t + s, x) = \pi(t, \pi(s, x))$, $x \in M$, $t, s \in \mathbb{R}$.

We may interpret $\pi(t, x)$ as the position of a particle at time t when the initial (i.e. time $= 0$) position is x.

Remark 5.1.2 A discrete dynamical system is defined as a continuous map $\pi : \mathbb{Z} \times M \to M$ satisfying (i), (ii), (iii) where $\mathbb{Z} = \{n : n$ is an integer$\}$. The typical examples of discrete dynamical systems are $x_{n+1} = f(x_n)$, $x_n \in \mathbb{R}^d$ and the Poincaré map $x_0 \longmapsto \varphi(w, x_0)$ where $\varphi(t, x_0)$ is the solution of a periodic system $x' = f(t, x)$, $x(0) = x_0$ where $f(t, x) = f(t + w, x)$.

The next lemma says that property (ii) implies the property of continuous dependence on initial data.

Lemma 5.1.1 Given $T > 0$ and $p \in M$. For any $\varepsilon > 0$, there exists $\delta > 0$ such that

$$\rho(\pi(t, p), \pi(t, q)) < \varepsilon \quad for \quad 0 \le t \le T$$

whenever $\rho(p, q) < \delta$.

Proof. If not, then there exists $\{q_n\}$, $q_n \to p$ and $\{t_n\}$, $|t_n| \le T$ and $\alpha > 0$ with

$$\rho(\pi(t_n, p), \ \pi(t_n, q_n)) \ge \alpha.$$

Without loss of generality, we may assume $t_n \to t_0$. Then

$$\begin{aligned} 0 < \alpha &\le \rho\left(\pi(t_n, p), \ \pi(t_n, q_n)\right) \qquad\qquad (5.2) \\ &\le \rho\left(\pi(t_n, p), \ \pi(t_0, p)\right) + \rho\left(\pi(t_0, p), \pi(t_n, q_n)\right). \end{aligned}$$

From (ii), the right hand side of (5.2) approaches zero as $n \to \infty$. This is a contradiction. $\qquad\square$

Notations:

$\gamma^+(x) = \{\pi(t, x) : t \ge 0\}$ is the positive orbit through x.

$\gamma^-(x) = \{\pi(t, x) : t \le 0\}$ is the negative orbit through x.

$\gamma(x) = \{\pi(t, x), -\infty < t < \infty\}$ is the orbit through x.

Definition 5.1.2 *We say that a set $S \subseteq M$ is positively (negatively) invariant under the flow π if for any $x \in S$, $\pi(t, x) \in S$ for all $t \geq 0$ ($t \leq 0$), i.e., $\pi(t, S) \subseteq S$ for all $t \geq 0$ (i.e., $\pi(t, S) \subseteq S$ for all $t < 0$). S is invariant if S is both positively and negatively invariant, i.e., $\pi(t, S) = S$ for $-\infty < t < \infty$.*

Lemma 5.1.2 *The closure of an invariant set is invariant.*

Proof. Let S be an invariant set and $\bar{S}$ be its closure. If $p \in S$, then $\pi(t, p) \in S \subseteq \bar{S}$. If $p \in \bar{S} \setminus S$, then there exists $\{p_n\} \subset S$ such that $p_n \to p$. Then for each $t \in \mathbb{R}$, we have $\lim_{n \to \infty} \pi(t, p_n) = \pi(t, p)$. Since $\pi(t, p_n) \in S$, it follows that $\pi(t, p) \in \bar{S}$. Hence $\pi(t, \bar{S}) \subset \bar{S}$. □

Definition 5.1.3 *We say a point $p \in M$ is an equilibrium or a rest point of the flow π if $\pi(t, p) \equiv p$ for all t. If $\pi(T, p) = p$ for some $T > 0$ and $\pi(t, p) \neq p$ for all $0 < t < T$, then we say $\{\pi(t, p) : 0 \leq t \leq T\}$ is a periodic orbit.*

Example 5.1.2 Rest points and periodic orbits are (fully) invariant. We note that positively invariance does not necessarily imply negatively invariance. For instance, if an autonomous system $x' = f(x)$ satisfies $f(x) \cdot \vec{n}(x) < 0$ for all $x \in \partial\Omega$ where Ω is a bounded domain in $\mathbb{R}^n$ and $\vec{n}(x)$ is an outward normal vector at $x \in \partial\Omega$, then Ω is positively invariant, but not negatively invariant.

Lemma 5.1.3
  (i) *The set of rest points is a closed set.*
  (ii) *No trajectory enters a rest point in finite time.*

Proof. (i) Trivial.
  (ii) Suppose $\pi(T, p) = p^*$ where p^* is a rest point and $p \neq p^*$. Then $p = \pi(-T, p^*) = p^*$. This is a contradiction. □

Lemma 5.1.4
  (i) *If for any $\delta > 0$, there exists $p \in B(q, \delta)$ such that $\gamma^+(p) \subset B(q, \delta)$, then q is a rest point.*
  (ii) *If $\lim_{t \to \infty} \pi(t, p) = q$ then q is a rest point.*

Proof. We shall prove (i). (ii) follows directly from (i). Suppose q is not a rest point, then $\pi(t_0, q) \neq q$ for some $t_0 > 0$. Let $\rho(q, \pi(t_0, q)) = d > 0$. From continuity of the flow π, there exists $\delta > 0$ such that if $\rho(p, q) < \delta < d/2$ then $\rho(\pi(t, q), \pi(t, p)) < d/2$ for all $t, |t| \leq t_0$. By hypothesis of (i) there exists $p \in B(q, \delta)$ such that $\gamma^+(p) \subset B(q, \delta)$. Then

$$d = \rho(q, \pi(t_0, q)) \leq \rho(q, \pi(t_0, p)) + \rho(\pi(t_0, p), \pi(t_0, q))$$
$$< \delta + d/2 < d/2 + d/2 = d.$$

This is a contradiction. □

Next we introduce the notion of limit sets.

Definition 5.1.4 *We say that p is an omega limit point of x if there exists a sequence $\{t_n\}$, $t_n \to +\infty$ such that $\pi(t_n, x) \to p$. The set*

$$\omega(x) = \{p : p \text{ is an omega limit point of } x\}$$

is called the ω-limit set of x.

Similarly we say that p is an alpha limit point of x if there exists a sequence $\{t_n\}$, $t_n \to -\infty$ such that $\pi(t_n, x) \to p$. The set

$$\alpha(x) = \{p : p \text{ is an alpha limit point of } x\}$$

is called the α-limit set of x.

Remark 5.1.3 Note that $\omega(x)$ represents where the positive orbit $\gamma^+(x)$ ends up, while $\alpha(x)$ represents where the negative orbit $\gamma^-(x)$ started. We note that α, ω are the initial and final alphabets of Greek characters.

Remark 5.1.4

$$\omega(x) = \bigcap_{t \geq 0} cl \left(\bigcup_{s \geq t} \pi(s, x) \right)$$

$$\alpha(x) = \bigcap_{t \leq 0} cl \left(\bigcup_{s \leq t} \pi(s, x) \right).$$

Theorem 5.1.1 $\omega(x)$ *and* $\alpha(x)$ *are closed, invariant sets.*

Proof. We shall prove the case of $\omega(x)$ only. First we show that $\omega(x)$ is invariant. Let $q \in \omega(x)$ and $\tau \in \mathbb{R}$. We want to show $\pi(\tau, q) \in \omega(x)$. Let $\pi(t_n, x) \to q$ as $t_n \to +\infty$. By continuity of π, it follows that

$$\pi(t_n + \tau, x) = \pi(\tau, \pi(t_n, x)) \to \pi(\tau, q),$$

as $t_n \to +\infty$. Thus $\pi(\tau, q) \in \omega(x)$ and it follows that $\omega(x)$ is invariant.

Next we show $\omega(x)$ is a closed set. Let $q_n \in \omega(x)$ and $q_n \to q$ as $n \to \infty$. We want to show $q \in \omega(x)$. Since $q_n \in \omega(x)$, there exists $\tau_n \geq n$ such that $\rho(\pi(\tau_n, x), q_n) < 1/n$. From $q_n \to q$, it follows that for any $\varepsilon > 0$ there exists $N = N(\varepsilon)$, such that $\rho(q_n, q) < \varepsilon/2$ for $n \geq N_\varepsilon$. Therefore,

$$\rho(\pi(\tau_n, x), q) < \rho(\pi(\tau_n, x), q_n) + \rho(q_n, q)$$
$$< \varepsilon/2 + \varepsilon/2 = \varepsilon.$$

Thus we have $\lim_{n \to \infty} \pi(\tau_n, x) = q$ and $q \in \omega(x)$. $\qquad\square$

Theorem 5.1.2 *If the closure of $\gamma^+(p)$, $Cl(\gamma^+(p))$, is compact then $w(p)$ is nonempty, compact and connected. Furthermore $\lim_{t \to \infty} \rho(\pi(t, p), w(p)) = 0$.*

Proof. Let $p_k = \pi(k, p)$. Since $Cl(\gamma^+(p))$ is compact, then there exists a subsequence of $\{p_{k_j}\}$, $p_{k_j} \to q \in Cl(\gamma^+(p))$. Then $q \in w(p)$ and hence $w(p)$ is nonempty. The compactness of $w(p)$ follows directly from the facts that $Cl(\gamma^+(p)) = \gamma^+(p) \cup w(p)$, $w(p)$ is closed and $Cl(\gamma^+(p))$ is compact.

We shall prove that $w(p)$ is connected by contradiction. Suppose on the contrary, $w(p)$ is disconnected. Then $w(p) = A \cup B$, a disjoint union of two closed subsets A, B of $w(p)$. Since $w(p)$ is compact, then A, B are compact. Then $d = \rho(A, B) > 0$ where $\rho(A, B) = \inf_{x \in A, y \in B} \rho(x, y)$. Consider the neighborhoods $N(A, d/3)$ and $N(B, d/3)$ of A, B respectively. Then there exist $t_n \to +\infty$ and $\tau_n \to +\infty$, $\tau_n > t_n$ such that $\pi(t_n, p) \in N(A, d/3)$ and $\pi(\tau_n, p) \in N(B, d/3)$.

Since $\rho(\pi(p, t), A)$ is a continuous function of t, then from the inequalities

$$\rho(\pi(\tau_n, p), A) > \rho(A, B) - \rho(\pi(\tau_n, p), B) > d - d/3 = 2d/3,$$

and

$$\rho(\pi(t_n, p), A) < d/3$$

we have $\rho(\pi(t_n^*, p), A) = d/2$ for some $t_n^* \in (t_n, \tau_n)$. $\{\pi(t_n^*, p)\}$ contains a convergent subsequence, $\pi(t_{n_k}^*, p) \to q \in w(p)$. However $q \notin A$ and $q \notin B$ and we obtain a desired contradiction.

Next we shall prove that a trajectory is asymptotic to its omega limit set. If not, then there exists a sequence $\{t_n\}$, $t_n \to +\infty$ and $\alpha > 0$ such that $\rho(\pi(t_n, p), w(p)) \geq \alpha$. Then there exists a subsequence of $\{\pi(t_n, p)\}$, $\pi(t_{n_k}, p) \to q \in w(p)$ and we obtain the following contradiction:

$$0 = \rho(q, w(p)) = \lim_{n_k \to \infty} \rho(\pi(t_{n_k}, p), w(p)) \geq \alpha.$$

$\square$

Example 5.1.3 Consider $\begin{cases} x' = x + y - x(x^2 + y^2) \\ y' = -x + y - y(x^2 + y^2). \end{cases}$

Using the polar coordinate (r, θ), we have

$$r' = r(1 - r^2)$$
$$\theta' = -1.$$

If $r(0) = 0$, then $(0, 0) = w((0, 0))$.

If $0 < r(0) < 1$, then $w(x_0, y_0) = $ unit circle, $\alpha(x_0, y_0) = (0, 0)$.

If $r(0) = 1$, then $w(x_0, y_0) = \alpha(x_0, y_0) = $ unit circle.

If $r(0) > 1$, then $w(x_0, y_0) = $ unit circle, $\alpha(x_0, y_0) = \emptyset$.

Example 5.1.4 Consider

$$\begin{cases} x' = \frac{x-y}{1+(x^2+y^2)^{1/2}}, \\[2mm] y' = \frac{x+y}{1+(x^2+y^2)^{1/2}}. \end{cases} \tag{5.3}$$

Then, in terms of polar coordinate, we have

$$\begin{cases} \frac{dr}{dt} = \frac{r}{1+r}, \\[2mm] \frac{d\theta}{dt} = \frac{1}{1+r}. \end{cases}$$

From Fig. 5.1 it follows that "$w(x, y) = \emptyset$ for $(x, y) \neq (0, 0)$ and $\alpha(x, y) = \{(0, 0)\}$ for any (x, y)."

The following examples show that an unbounded trajectory may have disconnected or noncompact ω-limit set.

Example 5.1.5 Let $X = \frac{x}{1-x^2}$, $Y = y$, where $(X(t), Y(t))$ satisfies (5.3). Then

$$x' = \frac{x(1-x^2) - y(1-x^2)^2}{(1+x^2)\left(1 + \left(\left(\frac{x}{1-x^2}\right)^2 + y^2\right)^{1/2}\right)} = f(x, y)$$

$$y' = \frac{y}{1 + \left(\left(\frac{x}{1-x^2}\right)^2 + y^2\right)^{1/2}} + \frac{x}{1-x^2}\frac{1}{1 + \left(\left(\frac{x}{1-x^2}\right)^2 + y^2\right)^{1/2}} = g(x, y).$$

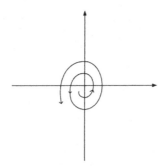

Fig. 5.1

Then $\lim_{x \to \pm 1} f(x, y) = 0$, $\omega((x_0, y_0)) = \{x = 1\} \bigcup \{x = -1\}$ is not connected (see Fig. 5.2).

Fig. 5.2

Example 5.1.6 Let $X = \log(1 + x)$, $Y = y$, $-1 < x < +\infty$, $y \in \mathbb{R}$. Where $(X(t), Y(t))$ satisfies (5.3). Then the equation

$$x' = f(x, y) = (1 + x)\frac{\log(1 + x) - y}{1 + \{(\log(1 + x))^2 + y^2\}^{1/2}}$$

$$y' = g(x, y) = \frac{y + \log(1 + x)}{1 + \{(\log(1 + x))^2 + y^2\}^{1/2}}$$

$f(x, y)$, $g(x, y)$ satisfies $\lim_{x \to -1} f(x, y) = 0$, $\lim_{x \to -1} g(x, y) = 1$. The ω-limit set $\omega(x) = \{(x, y) : x = -1\}$ is not compact (see Fig. 5.3).

5.2 Lyapunov Functions

Let $V : \Omega \subseteq \mathbb{R}^n \to \mathbb{R}$, $0 \in \Omega$, Ω is an open set in $\mathbb{R}^n$.

Fig. 5.3

Definition 5.2.1 $V : \Omega \subseteq \mathbb{R}^n \to \mathbb{R}$ *is said to be positive definite (negative definite) on* Ω *if* $V(x) > 0$ *for all* $x \neq 0$ $(V(x) < 0$ *for all* $x \neq 0)$ *and* $V(0) = 0$.

V is said to be semi-positive definite (semi-negative definite) on Ω *if* $V(x) \geq 0$ $(V(x) \leq 0)$ *for all* $x \in \Omega$.

Remark 5.2.1 In applications, we consider the autonomous system $x' = f(x)$ with x^* as an equilibrium. Let $y = x - x^*$, then $y' = g(y) = f(y + x^*)$ and 0 is an equilibrium of the system $y' = g(y)$. In general, $V(x)$ satisfies $V(x^*) = 0$, $V(x) > 0$ for all, $x \in \Omega$, $x \neq x^*$.

For the initial-value problem:

$$x' = f(x), \quad x(0) = x_0, \tag{5.4}$$

we assume that $x = 0$ is the unique equilibrium of (5.4) and the solution $\varphi(t, x_0)$ exists for all $t \geq 0$. We introduce "Lyapunov function" $V(x)$ to locate the ω-limit set $\omega(x_0)$. The function $V(x)$ satisfies the following: $V(0) = 0$, $V(x) > 0$ for all $x \in \Omega$ and $V(x) \to +\infty$ as $|x| \to \infty$. The level sets $\{x \in \Omega : V(x) = c\}$ is a closed surface. The surface $\{x \in \Omega : V(x) = c_1\}$ encloses the surface $\{x \in \Omega : V(x) = c_2\}$ if $c_1 > c_2$. We want to prove $\lim\limits_{t \to \infty} \varphi(t, x_0) = 0$, even though we don't have to know the exact location of the solution $\varphi(t, x_0)$. If we are able to construct a suitable Lyapunov function $V(x)$ such that $\frac{d}{dt}V(\varphi(t, x_0)) < 0$ as Fig. 5.4 shows, then $\lim_{t \to \infty} \varphi(t, x_0) = 0$.

Example 5.2.1 $m\ddot{x} + kx = 0$, $x(0) = x_0$, $x'(0) = x_0'$ describes the motion of a spring without friction according to Hooke's law.

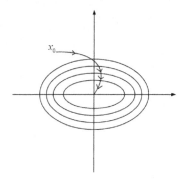

Fig. 5.4

Consider the total energy

$$V(t) = \text{kinetic energy} + \text{potential energy}$$

$$= \frac{1}{2}\, m(v(t))^2 + \int_0^{x(t)} ks\,ds$$

$$= \frac{1}{2}\, m(x'(t))^2 + \frac{k}{2}x^2(t).$$

Then, it follows that

$$\frac{d}{dt}V(t) = mx'(t)x''(t) + kx(t)x'(t) \equiv 0,$$

and hence the energy is conserved. In this case the Lyapunov function is $V(x, x') = \frac{m}{2}(x')^2 + \frac{k}{2}x^2$. It is easy to verify that $V(0,0) = 0$, $V(x, x') > 0$ for all $(x, x') \neq (0,0)$ and $V(x, x') \to +\infty$ as $|(x, x')| \to +\infty$.

Example 5.2.2 $mx'' + g(x) = 0$, $xg(x) > 0$, $x \neq 0$, $\int_0^x g(s)ds \to +\infty$ as $|x| \to \infty$. This describes the motion of a spring whose restoring force is a nonlinear function $g(x)$. The energy

$$V(t) = \frac{m}{2}\, (x'(t))^2 + \int_0^{x(t)} g(s)\,ds \qquad (5.5)$$

satisfies $\frac{dV}{dt} \equiv 0$, i.e., the energy is conserved. Hence the solution $(x(t), x'(t))$ is periodic. See Fig. 5.5.

Example 5.2.3 $mx'' + k(x)x' + g(x) = 0$, $k(x) \geq \delta > 0$ for all x. Use the energy function in (5.5), it follows that

$$\frac{dV}{dt} = x'(t)\left[-k(x(t))x'(t) - g(x(t))\right] + x'(t)g(x(t))$$

$$= -k(x(t))(x'(t))^2 < -\delta(x'(t))^2.$$

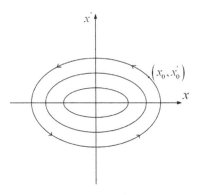

Fig. 5.5

Then, we expect $\lim_{t\to\infty} x(t) = 0$ and $\lim_{t\to\infty} x'(t) = 0$.

Let $x(t)$ be the solution of (5.4) and $V(x)$ be positive definite on Ω satisfying $V(x) \to \infty$ as $|x| \to \infty$. Compute the derivative of V along the trajectory of the solution $x(t)$,

$$\frac{d}{dt} V(x(t)) = grad\, V(x(t)) \cdot x'(t)$$

$$= \sum_{i=1}^{n} \frac{\partial V}{\partial x_i}(x(t)) f_i(x(t)) \overset{def}{=} \dot{V}(x(t)).$$

Definition 5.2.2 *A function* $V : \Omega \to \mathbb{R}$, $V \in C^1$ *is said to be a Lyapunov function for (5.4) if*

$$\dot{V}(x) = grad\, V(x) \cdot f(x) \leq 0 \text{ for all } x \in \Omega.$$

Remark 5.2.2 If V is merely continuous on Ω, we replace $\frac{d}{dt} V(x(t))$ by

$$\overline{\lim_{h\to 0}} \frac{1}{h} \left[V(x(t+h)) - V(x(t)) \right].$$

Let $\xi = x(t)$, then $x(t+h) = \varphi(h, \xi)$. Then we define

$$\dot{V}(\xi) = \overline{\lim_{h\to 0}} \frac{1}{h} \left[V(\varphi(h, \xi)) - V(\xi) \right].$$

Remark 5.2.3 In applications to many physical systems, $V(x)$ is the total energy. However for a mathematical problem, we may take V as a quadratic form, namely, $V(x) = x^T B x$, B is some suitable positive definite matrix.

Example 5.2.4 Consider the Lotka-Volterra predator-prey system,

$$\begin{aligned} \frac{dx}{dt} &= ax - bxy \\ \frac{dy}{dt} &= cxy - dy \end{aligned} \quad a, b, c, d > 0 \tag{5.6}$$

$$x(0) = x_0 > 0, \quad y(0) = y_0 > 0,$$

where $x = x(t)$, $y = y(t)$ are the densities of prey and predator species respectively. Let $x^* = \frac{d}{c}$, $y^* = \frac{a}{b}$. Then (5.6) can be rewritten as

$$\frac{dx}{dt} = -bx(y - y^*),$$

$$\frac{dy}{dt} = cy(x - x^*).$$

Consider the trajectory in phase plane, we have

$$\frac{dy}{dx} = \frac{cy(x - x^*)}{-bx(y - y^*)}. \tag{5.7}$$

Using separation of variables, from (5.7) it follows that

$$\frac{y - y^*}{y} dy + \frac{c}{b} \frac{x - x^*}{x} dx = 0$$

and

$$\int_{y(0)}^{y(t)} \frac{\eta - y^*}{\eta} d\eta + \frac{c}{b} \int_{x(0)}^{x(t)} \frac{\eta - x^*}{\eta} d\eta \equiv const.$$

Hence we define a Lyapunov function

$$V(x, y) = \int_{y^*}^{y} \frac{\eta - y^*}{\eta} d\eta + \frac{c}{b} \int_{x^*}^{x} \frac{\eta - x^*}{\eta} d\eta \tag{5.8}$$

$$= \left[y - y^* - y^* \ln\left(\frac{y}{y^*}\right) \right] + \frac{c}{b} \left[x - x^* - x^* \ln\left(\frac{x}{x^*}\right) \right].$$

Then it is straightforward to verify that $\dot{V}(x, y) \equiv 0$ and hence the system (5.6) is a conservative system. Each solution $(x(t, x_0, y_0), y(t, x_0, y_0))$ is a periodic solution. From the following Fig. 5.6, the system (5.6) has a family of "neutrally stable" periodic orbits.

Consider the system (5.4) and for simplicity we assume 0 is an equilibrium. To verify the stability property of the equilibrium 0, from previous chapter we can obtain the information regarding stability, from the eigenvalues of the variational matrix $D_x f(0)$. In the following, we present another method by using Lyapunov functions.

Theorem 5.2.1 (Stability, Asymptotic Stability [H1])
If there exists a positive definite function $V(x)$ on a neighborhood Ω of 0 such that $\dot{V} \leq 0$ on Ω, then the equilibrium 0 of (5.4) is stable. If, in addition, $\dot{V} < 0$ for all $x \in \Omega \setminus \{0\}$, then 0 is asymptotically stable.

Fig. 5.6

Proof. Let $r > 0$ such that $B(0, r) \subseteq \Omega$. Given ε, $0 < \varepsilon < r$ and let $k = \min\limits_{|x|=\varepsilon} V(x)$. Then $k > 0$. From continuity of V at 0, we choose $\delta > 0$, $0 < \delta \leq \varepsilon$ such that $V(x) < k$ when $|x| < \delta$. Then the solution $\varphi(t, x_0)$ satisfies $V(\varphi(t, x_0)) \leq V(x_0)$ because of $\dot{V} \leq 0$ on Ω. Hence for $x_0 \in B(0, \delta)$, $\varphi(t, x_0)$ stays in $B(0, \varepsilon)$. Therefore, 0 is stable.

Assume $\dot{V} < 0$ for all $x \in \Omega \backslash \{0\}$. To establish the asymptotic stability of the equilibrium 0, we need to show $\varphi(t, x_0) \to 0$ as $t \to \infty$ if x_0 is sufficiently close to 0. Let $r > 0$ such that $\overline{B}(0, r) \subset \Omega$. Let $H = \min\{V(x) : |x| = r\}$. Then $H > 0$. It suffices to show that $\lim\limits_{t \to \infty} V(\varphi(t, x_0)) = 0$ if $x_0 \in B(0, d)$ where $0 < d < r$ is a number such that $V(x) \leq \frac{H}{2}$ for every $x \in B(0, d)$. If not, since $\dot{V} < 0$, $V(\varphi(t, x_0))$ is decreasing in t to some $\eta > 0$. Then $V(\varphi(t, x_0)) \geq \eta$ for all $t \geq 0$. By continuity of V at 0, there is a $\delta > 0$ such that $0 \leq V(x) < \eta$ for every x with $|x| < \delta$. Hence $|\varphi(t, x_0)| \geq \eta$ for all $t \geq 0$. Set $S = \{x : \eta \leq |x| \leq r, V(x) \leq \frac{H}{2}\}$. Then $\varphi(t, x_0) \in S$ for all $t \geq 0$. Let $\gamma = -\min\{-\dot{V}(x) : x \in S\}$. By compactness of S and continuity of $\dot{V}$, $\gamma = \max\{\dot{V}(x) : x \in S\} > 0$ because $0 \notin S$, and it follows that $-\frac{d}{dt}V(\varphi(t, x_0)) \geq \gamma$. Integrating both sides from 0 to t yields $-[V(\varphi(t, x_0)) - V(x_0)] \geq \gamma t$ or $0 < V(\varphi(t, x_0)) \leq V(x_0) - \gamma t$.

Let $t \to +\infty$, we obtain a contradiction.   □

Theorem 5.2.2 *If there exists a neighborhood U of 0 and $V : \Omega \subseteq \mathbb{R}^n \to \mathbb{R}$, $0 \in \Omega$ such that V and $\dot{V}$ are positive definite on $U \cap \Omega \setminus \{0\}$, then the equilibrium 0 is completely unstable, i.e., 0 is a repeller. More specifically, if Ω_0 is any neighborhood of 0, $\overline{\Omega_0} \subseteq \Omega$ then any solution $\varphi(t, x_0)$ of (5.4) with $x_0 \in U \cap \Omega_0 \setminus \{0\}$ leaves Ω_0 in finite time.*

Proof. If not, there exists a neighborhood Ω_0 such that $\varphi(t, x_0)$ stays in $U \cap \Omega_0 \setminus \{0\}$ for some $x_0 \in U \cap \Omega_0 \setminus \{0\}$. Then $V(\varphi(t, x_0)) \geq V(x_0) > 0$ for all $t \geq 0$. Let $\alpha = \inf\{\dot{V}(x) : x \in U \cap \Omega_0, V(x) \geq V(x_0)\} > 0$. Then,

$$V(\varphi(t, x_0)) = V(x_0) + \int_0^t \dot{V}(\varphi(s, x_0))ds$$
$$\geq V(x_0) + \alpha t.$$

Since $\varphi(t, x_0)$ remains in $U \cap \Omega_0 \setminus \{0\}$ for all $t \geq 0$, $V(\varphi(t, x_0))$ is bounded for $t \geq 0$. Thus for t sufficiently large, we obtain a contradiction from the above inequality. $\qquad\square$

Example 5.2.5

$$\begin{cases} x' = -x^3 + 2y^3 \\ y' = -2xy^2 \end{cases} \tag{5.9}$$

$(0,0)$ is an equilibrium. The variational matrix of (5.9) at $(0,0)$ is $\begin{pmatrix} 0 & 0 \\ 0 & 0 \end{pmatrix}$ whose eigenvalues $\lambda_1 = \lambda_2 = 0$ are not hyperbolic. However $(0,0)$ is indeed asymptotically stable. Use the Lyapunov function $V(x, y) = \frac{1}{2}(x^2 + y^2)$, then we have

$$\dot{V} = x(-x^3 + 2y^3) + y(-2xy^2) = -x^4 \leq 0.$$

From Theorem 5.2.1, $(0,0)$ is stable. In fact from invariance principle (see Theorem 5.2.4), $(0,0)$ is globally asymptotically stable.

Linear Stability by Lyapunov Method

Consider linear system

$$x' = Ax, \quad A \in \mathbb{R}^{n \times n}. \tag{5.10}$$

From Chapter 3, the equilibrium $x = 0$ is asymptotically stable if $Re\lambda(A) < 0$. We recall that in order to verify $Re\lambda(A) < 0$ analytically, we first calculate the characteristic polynomial $f(z) = \det(\lambda I - A)$ and then apply the Routh-Hurwitz criteria. In the following we present another criterion for $Re\lambda(A) < 0$.

Theorem 5.2.3 *Let $A \in \mathbb{R}^{n \times n}$. The matrix equation*

$$A^T B + BA = -C, \tag{5.11}$$

has a positive definite solution B for every positive definite matrix C if and only if $Re\lambda(A) < 0$.

Proof. Consider the linear system (5.10) and let $V(x) = x^T Bx$, where B is a positive definite matrix to be determined. Then

$$\dot{V}(x) = \dot{x}^T Bx + x^T B\dot{x} = (Ax)^T Bx + x^T BAx$$
$$= x^T (A^T B + BA)x.$$

If for any positive definite matrix C there exists a positive definite matrix B satisfying (5.11), then $\dot{V}(x) = -x^T Cx < 0$ for $x \neq 0$. The asymptotic stability of $x = 0$, i.e., $Re\lambda(A) < 0$ follows directly from Theorem 5.2.1.

Assume $Re\lambda(A) < 0$. Let C be any positive definite matrix, we define

$$B = \int_0^\infty e^{A^T t} Ce^{At} dt.$$

First we claim that B is well-defined and positive definite. From $Re\lambda(A) < 0$, there exist $\alpha, K > 0$ such that $\| e^{At} \| \leq Ke^{-\alpha t}$ for $t \geq 0$. For any $s > 0$

$$\| \int_0^s e^{A^T t} Ce^{At} dt \| \leq \int_0^s \| e^{A^T t} \| \| C \| \| e^{At} \| dt$$
$$\leq \int_0^s K^2 e^{-2\alpha t} \| C \| dt < \infty \quad \text{for all} \quad s > 0.$$

Obviously B is positive definite for $x^T Bx > 0$ for $x \neq 0$. Compute

$$A^T B + BA = \int_0^\infty \left[A^T e^{A^T t} Ce^{At} + e^{A^T t} Ce^{At} A \right] dt$$
$$= \int_0^\infty \frac{d}{dt} \left(e^{A^T t} Ce^{At} \right) dt = 0 - ICI = -C.$$

Hence the proof is complete.   □

Linearization via Lyapunov Method

Consider the nonlinear system

$$x' = Ax + f(x), \ f(0) = 0, \ f(x) = o(|x|) \text{ as } x \to 0. \tag{5.12}$$

Now we shall apply Theorem 5.2.3 to obtain the stability property of the equilibrium $x = 0$ of (5.12).

Claim: If $Re\lambda(A) < 0$ then $x = 0$ is asymptotically stable.

From Theorem 5.2.3, there exists a positive definite matrix B satisfying $A^T B + BA = -I$. Let $V(x) = x^T Bx$. Then

$$\dot{V} = (Ax + f(x))^T Bx + x^T B(Ax + f(x))$$
$$= x^T(A^T B + BA)x + 2x^T Bf(x)$$
$$= -x^T x + 2x^T Bf(x).$$

Since $f(x) = o(|x|)$ as $x \to 0$, for any $\varepsilon > 0$ there exists $\delta > 0$ such that $|f(x)| < \varepsilon|x|$ for $|x| < \delta$. Then for $|x| < \delta$ it follows that

$$\dot{V} \leq -x^T x + 2\varepsilon x^T Bx \leq (-1 + 2\varepsilon \parallel B \parallel)x^T x < 0.$$

Hence $\dot{V}$ is negative definite for $|x| < \delta$ if ε is sufficiently small. Thus $x = 0$ is asymptotically stable.

If $Re\lambda(A) \neq 0$ and there exists an eigenvalue of A with positive real part, then $x = 0$ is an unstable equilibrium of (5.12). Without loss of generality we may assume that $A = diag(A_-, A_+)$ where $Re\lambda(A_-) < 0$, $Re\lambda(A_+) > 0$. Let B_1 be a positive definite matrix satisfying

$$A_-^T B_1 + B_1 A_- = -I$$

and B_2 be a positive definite matrix satisfying

$$\left(-A_+^T\right) B_2 + B_2 \left(-A_+\right) = -I.$$

Let $x = (u, v)$ where u, v has the same dimension as matrices B_1, B_2 respectively. Rewrite (5.12) as

$$u' = A_- u + f_1(u, v)$$
$$v' = A_+ v + f_2(u, v).$$

Introduce

$$V(x) = -u^T B_1 u + v^T B_2 v.$$

Then

$$\dot{V} = -u^T \left(A_-^T B_1 + B_1 A_-\right) u + v^T \left(A_+^T B_2 + B_2 A_+\right) v + O(|x|^2)$$
$$= x^T x + O(|x|^2) > c|x|^2 > 0 \text{ for some } c > 0 \text{ and for } |x| < \delta, \ \delta > 0.$$

On the region where V is positive definite, the conditions of Theorem 5.2.2 hold. Hence $x = 0$ is unstable.

Invariance Principle

The following invariance principle provides a method to locate ω-limit set $\omega(x_0)$ of the solution of I.V.P. of autonomous system $x' = f(x)$, $x(0) = x_0$. It is also a tool to estimate the domain of attraction of an asymptotically stable equilibrium x^*.

Definition 5.2.3 *We say that a scalar function V is a Lyapunov function on an open set $G \subseteq \mathbb{R}^n$ if V is continuous on $\overline{G}$ and $\dot{V}(x) = gradV(x) \cdot f(x) \leq 0$ for all $x \in G$.*

We note that V need not necessarily be positive definite on G. Let $S = \{x \in \overline{G} : \dot{V}(x) = 0\}$ and M be the largest invariant set in S with respect to the flow $\dot{x} = f(x)$.

Theorem 5.2.4 *(LaSalle's invariance principle)*
If V is a Lyapunov function on G and $\gamma^+(x_0)$ is a bounded orbit of $\dot{x} = f(x)$, $\gamma^+(x_0) \subseteq G$, then $\omega(x_0) \subseteq M$ and $\lim_{t \to \infty} dist(\varphi(t, x_0), M) = 0$.

Proof. Since $\gamma^+(x_0) \subseteq G$ is bounded and V is continuous on $\overline{G}$, $V(t) = V(\varphi(t, x_0))$ is bounded for all $t \geq 0$. From $\dot{V} \leq 0$ on G, it follows that $V(\varphi(t, x_0))$ is nonincreasing in t and hence $\lim_{t \to +\infty} V(\varphi(t, x_0)) = c$. Let $y \in \omega(x_0)$ then there exists a sequence $\{t_n\} \uparrow +\infty$ such that $\varphi(t_n, x_0) \to y$ as $n \to \infty$, then we have $V(y) = c$. From the invariance of $\omega(x_0)$, we have $V(\varphi(t, y)) = c$ for all $t \in \mathbb{R}^+$. Differentiating $V(\varphi(t, y)) = c$ with respect to t yields $\dot{V}(\varphi(t, y)) = 0$. Hence $\varphi(t, y) \subseteq S$ and $\omega(x_0) \subseteq S$. By the definition of M, it follows that $\omega(x_0) \subseteq M$. Thus the proof is complete. $\square$

Next Corollary provides us a way to estimate the domain of attraction of the asymptotic stable equilibrium O.

Corollary 5.2.1 *If $G = \{x \in \mathbb{R}^n : V(x) < \rho\}$, $\gamma^+(x_0) \subseteq G$ and G is bounded, then $\varphi(t, x_0) \to M$ as $t \to \infty$.*

Corollary 5.2.2 *If $V(x) \to \infty$ as $|x| \to \infty$ and $\dot{V} \leq 0$ on $\mathbb{R}^n$ then every solution of $x' = f(x)$ is bounded and approaches M. In particular, if $M = \{0\}$, then 0 is globally asymptotically stable.*

Example 5.2.6 Consider the nonlinear equation $x'' + f(x)x' + g(x) = 0$ where $xg(x) > 0$, $x \neq 0$, $f(x) > 0$, $x \neq 0$ and $G(x) = \int_0^x g(s)ds \to \infty$ as $|x| \to \infty$. The equation describes a spring motion with friction. It is equivalent to the system

$$\begin{cases} x' = y \\ y' = -g(x) - f(x)y. \end{cases} \tag{5.13}$$

Let $V(x, y)$ be the total energy of the system, that is, $V(x, y) = \frac{y^2}{2} + G(x)$. Then $\dot{V}(x, y) = -f(x)y^2 \leq 0$. For any $\rho > 0$, the function V is a Lyapunov function on the bounded set $\Omega = \{(x, y) : V(x, y) < \rho\}$. Also, the set S where $\dot{V} = 0$, belongs to the union of x-axis and y-axis. From (5.13), it is easy to check $M = \{(0, 0)\}$ and Corollary 5.2.2 implies that $(0, 0)$ is globally asymptotically stable.

Example 5.2.7 Consider the Lotka-Volterra Predator-Prey system

$$\frac{dx}{dt} = \gamma x \left(1 - \frac{x}{K}\right) - \alpha xy,$$

$$\frac{dy}{dt} = \beta xy - dy, \quad \gamma, \alpha, \beta, d, K > 0,$$

$$x(0) = x_0 > 0, \quad y(0) = y_0 > 0.$$

Fig. 5.7

Case 1: $x^* = d/\beta < K$
Let

$$V(x, y) = \int_{x^*}^x \frac{\eta - x^*}{\eta} d\eta + c \int_{y^*}^y \frac{\eta - y^*}{\eta} d\eta.$$

With the choice of $c = \frac{\alpha}{\beta}$, we have

$$\dot{V} = -\frac{\gamma}{K}(x - x^*)^2.$$

Then $S = \{(x, y) : x = x^*\}$ and it is easy to verify $M = \{(x^*, y^*)\}$ (see Fig. 5.7). Hence

$$\lim_{t \to \infty} x(t) = x^*, \quad \lim_{t \to \infty} y(t) = y^*.$$

Case 2: $x^* > K$

Then with Lyapunov function

$$V(x, y) = \int_K^x \frac{\eta - K}{\eta} d\eta + cy, \quad c = \frac{\alpha}{\beta},$$

it follows that

$$\dot{V} = -\frac{\gamma}{K}(x - K)^2 + c\beta y(K - x^*) \leq 0.$$

Then $S = \{(K, 0)\} = M$ and hence

$$\lim_{t \to \infty} x(t) = K, \quad \lim_{t \to \infty} y(t) = 0.$$

We note that in this example the Lyapunov function V does not satisfy the hypotheses of Theorem 5.2.4. We leave it as an exercise to show that the invariance principle holds.

Example 5.2.8 Consider Lotka-Volterra two species competition model (Example 4.2.4).

$$\frac{dx_1}{dt} = \left[r_1 \left(1 - \frac{x_1}{K_1} \right) - \alpha_1 x_2 \right] x_1$$

$$\frac{dx_2}{dt} = \left[r_2 \left(1 - \frac{x_2}{K_2} \right) - \alpha_2 x_1 \right] x_2.$$

If $\frac{r_1}{\alpha_1} > K_2$, $\frac{r_2}{\alpha_2} > K_1$, then the positive equilibrium (x_1^*, x_2^*) is globally asymptotically stable.

Proof. Let the Lyapunov function be

$$V(x_1, x_2) = \int_{x_1^*}^{x_1} \frac{\xi - x_1^*}{\xi} d\xi + c \int_{x_2^*}^{x_2} \frac{\xi - x_2^*}{\xi} d\xi,$$

for some $c > 0$ to be determined. Then

$$\dot{V}(x_1, x_2)$$

$$= (x_1 - x_1^*) \left[-\frac{r_1}{K_1}(x_1 - x_1^*) - \alpha_1(x_2 - x_2^*) \right]$$

$$+ c(x_2 - x_2^*) \left[-\frac{r_2}{K_2}(x_2 - x_2^*) - \alpha_2(x_1 - x_1^*) \right]$$

$$= \left(-\frac{r_1}{K_1} \right)(x_1 - x_1^*)^2$$

$$+ (-\alpha_1 - \alpha_2 c)(x_1 - x_1^*)(x_2 - x_2^*) + \left(-\frac{r_2}{K_2}c \right)(x_2 - x_2^*)^2.$$

The discriminant

$$P(c) = (\alpha_1 + \alpha_2 c)^2 - 4\frac{r_1}{K_1}\frac{r_2}{K_2}c$$

$$= \alpha_2 c^2 + 2c\left(\alpha_1\alpha_2 - 2\frac{r_1}{K_1}\frac{r_2}{K_2}\right) + \alpha_1^2.$$

Since $P(0) > 0$, the discriminant of quadratic polynomial $P(c)$ is

$$D = \left(\alpha_1\alpha_2 - 2\frac{r_1}{K_1}\frac{r_2}{K_2}\right)^2 - \alpha_1^2\alpha_2^2 = 4\frac{r_1}{K_1}\frac{r_2}{K_2}\left(\frac{r_1}{K_1}\frac{r_2}{K_2} - \alpha_1\alpha_2\right) > 0.$$

Hence there are two positive roots c_1^*, c_2^* of $P(c) = 0$ and there exists c, $c_1^* < c < c_2^*$ such that $P(c) < 0$. With this choice of c, we have $V(x_1, x_2) < 0$ for $(x_1, x_2) \neq (x_1^*, x_2^*)$. Thus we complete the proof that (x_1^*, x_2^*) is globally asymptotically stable. $\qquad\square$

Example 5.2.9 Consider the Van der Pol equation

$$x'' + \varepsilon(x^2 - 1)x' + x = 0$$

and its equivalent "Lienard form" (see Chapter 6)

$$\begin{cases} x' = y - \varepsilon\left(\frac{x^3}{3} - x\right), \\ y' = -x. \end{cases} \tag{5.14}$$

In the next chapter we shall show that $(0,0)$ is an unstable focus and the equation has a unique asymptotically stable limit cycle for every $\varepsilon > 0$. The exact location of this cycle in (x, y) plane is extremely difficult to obtain but the above theory allows one to determine a region near $(0,0)$ in which the limit cycle cannot lie. Such a region can be found by determining the domain of repulsion, i.e., the domain of attraction of $(0,0)$ with t replaced by $-t$. This has the same effect of $\varepsilon < 0$. Suppose $\varepsilon < 0$ and let $V(x, y) = \frac{x^2+y^2}{2}$. Then

$$\dot{V}(x, y) = -\varepsilon x^2\left(\frac{x^2}{3} - 1\right)$$

and $\dot{V}(x, y) \leq 0$ if $x^2 < 3$. Consider the region $G = \{(x, y) : V(x, y) < 3/2\}$. Then G is bounded and V is a Lyapunov function on G.

Furthermore $S = \{(x, y) : \dot{V} = 0\} = \{(x, 0) : x = \sqrt{3}, -\sqrt{3}, 0\}$. From (5.14), $M = \{(0, 0)\}$. Then every solution starting in the circle $x^2 + y^2 < 3$ approaches zero as $t \to \infty$. Finally, the limit cycle of (5.14) with $\varepsilon > 0$ must be outside this circle.

5.3   Simple Oscillatory Phenomena

Consider the second order scalar equation

$$u'' + g(u) = 0 \tag{5.15}$$

or the equivalent system

$$\begin{aligned} u' &= v \\ v' &= -g(u) \end{aligned} \tag{5.16}$$

where $g(u)$ is continuous on $\mathbb{R}$. System (5.16) is a Hamiltonian system with the Hamiltonian function or total energy given by $E(u,v) = \frac{v^2}{2} + G(u)$ where $G(u) = \int_0^u g(s)ds$. As we did for Example 5.2.2, it is easy to show that $\frac{dE}{dt} \equiv 0$. Hence the orbits of solutions of (5.16) in the (u,v)-plane must lie on the level curves of $E(u,v) = h$, h is a constant. Then we have $v = \pm\sqrt{2(h - G(u))}$. We note that $g(u)$ does not necessarily satisfy $ug(u) > 0$, $x \neq 0$, $G(u) \to +\infty$ as $|u| \to \infty$ as we did in Example 5.2.2.

Example 5.3.1 Simple Pendulum (for the derivation see Example 1.1.9)

$$\theta''(t) + \frac{g}{l}\sin\theta(t) = 0. \tag{5.17}$$

Let $k^2 = \frac{g}{l}$, $g(\theta) = k^2\sin\theta$, $G(\theta) = k^2(1 - \cos\theta)$. $G(u)$ has the graph shown in Fig. 5.8(a). The curves $E(\theta,\psi) = h$, $\psi = \theta'$, are shown in Fig. 5.8(b).

We note that (5.17) can be written as

$$\begin{aligned} \theta' &= \psi, \\ \psi' &= -k^2\sin\theta. \end{aligned} \tag{5.18}$$

The equilibria of (5.18) are $(\pm h\pi, 0)$, $h = 0, 1, 2, \ldots$. It is easy to verify from variational matrix

$$\begin{pmatrix} 0 & 1 \\ -k^2\cos\theta & 0 \end{pmatrix}_{\theta=\pm h\pi}$$

that the equilibria $(\pm 2h\pi, 0)$ are linear centers and $((2k+1)\pi, 0)$ are saddles. For $0 < h < 2k^2$, the level curves $E(\theta,\psi) = h$ are periodic orbits. For $h = 2k^2$, we obtain heteroclinic orbits connecting saddles from $((2k-1)\pi, 0)$ to $((2k+1)\pi, 0)$. If $h > 2k^2$ then we obtain level curves lying above and below the heteroclinic orbits. Physically it means "whirling", the pendulum will rotate around the axis $\theta = 0$ infinitely many times provided the initial velocity $\theta' = \psi$ is sufficiently large.

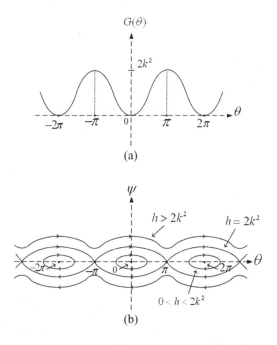

Fig. 5.8

Example 5.3.2 Suppose the potential $G(u)$ of (5.16) has the graph shown in Fig. 5.9(a) with A, B, C, D being extreme points of G. The orbits of the solution curves are sketched in Fig. 5.9(b), all curves of course being symmetric with respect to the u-axis. The equilibria corresponding to A, B, C, D are also labeled as A, B, C, D on the phase plane. The point A, C are centers, B is a saddle point and D is like the coalescence of a saddle and a center. The curves connecting B to B are homoclinic orbits.

Example 5.3.3 Simple Pendulum with Damping.
The equation of the motion is

$$\theta'' + \beta\theta' + k^2 \sin\theta = 0 , \qquad \beta > 0. \tag{5.19}$$

The equivalent system is

$$\begin{aligned} \theta' &= \psi \\ \psi' &= -k^2 \sin\theta - \beta\psi. \end{aligned} \tag{5.20}$$

If $E(\theta, \psi) = \frac{\psi^2}{2} + k^2 (1 - \cos\theta)$ is the energy of the system, then the derivative of E along the solution of (5.20) satisfies $\frac{dE}{dt} = -\beta\psi^2 \leq 0$.

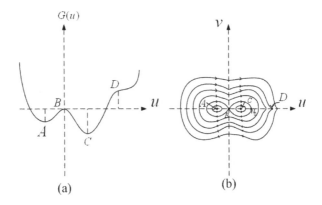

Fig. 5.9

Since $\frac{dE}{dt} \le 0$, $\psi(t)$ is bounded for all $t \ge 0$ and from second equation of (5.20) $\psi'(t)$ is also bounded $t \ge 0$. We claim $\lim_{t\to\infty} \psi(t) = 0$. If $\psi(t)$ does not approach zero as $t \to \infty$. Then there exist $\epsilon > 0$, $\delta > 0$ and a sequence $\{t_n\}$, $t_n \to \infty$ such that $\psi^2(t) > \epsilon$ on nonoverlapping intervals I_n, $I_n = [t_n - \delta, t_n + \delta]$. Let $t > 0$ sufficiently large and let $p(t) = \max\{n | t_n < t\}$. Then

$$E\left(\theta(t), \psi(t)\right) - E\left(\theta(0), \psi(0)\right) = \int_0^t \left(\frac{dE}{dt}\right) ds$$

$$\le -\int_0^t \beta\psi^2(s)ds \le -\sum_{t_n < t} \int_{t_n - \delta}^{t_n + \delta} \beta\psi^2(s)ds$$

$$\le -2\beta\epsilon\delta \sum_{n, t_n < t} 1 < -2\beta\epsilon\delta p(t).$$

Since $p(t) \to \infty$ as $t \to \infty$, from above we obtain a contradiction to the boundedness of $E\left(\theta(t), \psi(t)\right)$. Hence $\lim_{t\to\infty} \psi(t) = 0$.

The equilibria of (5.20) are $(\pm h\pi, 0)$, $h = 0, 1, 2, \ldots$. From the variational matrix

$$\begin{pmatrix} 0 & 1 \\ -k^2 \cos\theta & -\beta \end{pmatrix}_{\theta = \pm h\pi}.$$

It is easy to verify that $(\pm 2h\pi, 0)$ are stable spirals and $((2h - 1)\pi, 0)$ are saddle points. With the information $\lim_{t\to\infty} \psi(t) = 0$ together with the level curves of $E(\theta, \psi)$ allows one to sketch the orbits in the phase plane. These orbits are shown in Fig. 5.10.

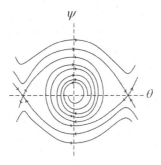

Fig. 5.10

5.4 Gradient Vector Fields

In physics especially in mechanics, gradient vector fields are the conservation systems:

Definition 5.4.1 *Let $F : \mathbb{R}^n \to \mathbb{R}$ be a C^2 function, the gradient vector field is*

$$-\nabla F(x) = -\left(\frac{\partial}{\partial x_1} F(x), \frac{\partial}{\partial x_2} F(x), \ldots, \frac{\partial}{\partial x_n} F(x) \right)$$

and the corresponding gradient system is

$$\frac{dx}{dt} = -\nabla F(x). \tag{5.21}$$

Definition 5.4.2 *A point $\bar{x}$ is said to be a critical point of F if $\nabla F(x) = 0$. A critical point $\bar{x}$ is called non-degenerate if the eigenvalues of the Hessian at $\bar{x}$, the symmetric matrix $\left(\frac{\partial^2 F}{\partial x_i \partial x_j} \right)(\bar{x})$ are nonzero.*

It is easy to prove the following lemma. (Exercises!)

Lemma 5.4.1 *An equilibrium point of a gradient system (5.21) is hyperbolic if and only if the corresponding critical point of F is non-degenerate. If $\bar{x}$ is a hyperbolic equilibrium of (5.21), then*

  (i) *$\bar{x}$ is an unstable node if and only if F has an isolated maximum at $\bar{x}$;*

  (ii) *$\bar{x}$ is asymptotically stable if and only if F has an isolated minimum at $\bar{x}$;*

  (iii) *$\bar{x}$ is a saddle point if and only if F has a saddle at $\bar{x}$.*

Theorem 5.4.1 [HK] *If $\gamma^+(x^0)$ is a bounded positive orbit of a gradient system (5.21), then the ω-limit set $\omega(x^0)$ belongs to the set of equilibria of equation (5.21). If the equilibrium points are isolated, then such an ω-limit set is a single equilibrium point. If $F(x) \to +\infty$ as $||x|| \to \infty$, then every positive orbit of equation (5.21) is bounded. Similarly, if $\gamma^-(x^0)$ is a bounded negative orbit of a gradient system (5.21), then the α-limit set $\alpha(x^0)$ belongs to the set of equilibria of equation (5.21). If the equilibrium points are isolated, then such an α-limit set is a single equilibrium point. If $F(x) \to -\infty$ as $||x|| \to +\infty$, then every negative orbit of equation (5.21) is bounded.*

Proof. Consider the positive orbit $\gamma^+(x^0)$. If $x(t)$ is a bounded solution of equation (5.21) for $t \geq 0$, then the derivative of F along such a solution satisfies

$$\frac{d}{dt}F(x(t)) = -|\nabla F(x(t))|^2 \leq 0.$$

Now, the Invariance Principle (Theorem 5.2.4) implies that $\omega(x(0))$ belongs to the set of equilibria. If the equilibria are isolated, then $\omega(x(0))$ is a single point because the limit set is connected. The statement on the boundedness of positive orbits is consequence of the fact that $F(x(t)) \leq F(x(0))$.   □

Example 5.4.1

$$\dot{x}_1 = 2(x_2 - x_1) + x_1(1 - x_1^2)$$
$$\dot{x}_2 = -2(x_2 - x_1) + x_2(1 - x_2^2).$$

This is a gradient system for the function

$$F(x_1, x_2) = \frac{1}{4}(x_1^4 + x_2^4 + 2x_2^2 + 2x_2^2 - 8x_1x_2).$$

The critical points of F are $(0,0)$, $(1,1)$, and $(-1,-1)$, all of which are non-degenerate. It is easy to verify that $(0,0)$ is a saddle point with stable manifold $x_1 = -x_2$, unstable manifold $x_1 = x_2$; $(1,1)$ and $(-1,-1)$ are asymptotically stable.

Example 5.4.2 Consider the system

$$\frac{dx_1}{dt} = -\lambda + x_1^2$$
$$\frac{dx_2}{dt} = -x_2.$$

It is a gradient system with the function
$$F(\lambda, x_1, x_2) = \lambda x_1 - \frac{1}{3}x_1^3 + \frac{1}{2}x_2^2.$$
The reader can sketch the orbits in the phase plane for the case $\lambda = 0$ and $\lambda < 0$.

The concept of gradient system can be extended to non-autonomous systems. When the vector field $F(t, x)$ is time-dependent, we still use the notation ∇ for the gradient with respect to the space variable x. The following theorem gives a criterion for a system to be locally a gradient system.

Theorem 5.4.2 Consider the system
$$\frac{dx}{dt} = f(t, x), \quad (t, x) \in I \times D,$$
where $I \subset \mathbb{R}$, $D \subset \mathbb{R}^n$ are open and f is of class C^1 with respect to x. If $Df(t, x) = D_x f(t, x)$ is symmetric for any (t, x), then the system is a locally gradient system in the sense that, for any $x_0 \in D$, there is a neighborhood U of x_0 and a function $F : I \times U \to \mathbb{R}$ such that $f(t, x) = \nabla F(t, x)$.

Proof. By translating the variable x, we may assume without loss of generality that $x_0 = 0$. Choose $\delta > 0$ such that $U = B_\delta(x_0) = B_\delta(0) \subset D$. For $x \in B_\delta(0)$, define
$$F(t, x) = \int_0^1 \langle f(t, sx), x \rangle \, ds.$$
It is a well-defined on $I \times U$ and, for $h \in \mathbb{R}^n$ close to 0,
$$F(t, x + h) - F(t, x)$$
$$= \int_0^1 \langle f(t, s(x + h)) - f(t, sx), x \rangle + \langle f(t, s(x + h)), h \rangle \, ds$$
$$= \int_0^1 \langle Df(t, sx)(sh), x \rangle + \langle f(t, sx), h \rangle \, ds + o(|h|)$$
$$= \int_0^1 s \langle Df(t, sx)x, h \rangle + \langle f(t, sx), k \rangle \, ds + o(|h|) \text{ as } h \to 0.$$
The last identity uses symmetry of $Df(t, x)$. The integrand in the last line is exactly
$$\left\langle \frac{d}{ds}(sf(t, sx)), h \right\rangle.$$
Therefore,
$$F(t, x + h) - F(t, x) = \langle f(t, x), h \rangle + o(|h|) \text{ as } h \to 0.$$
This implies $f(t, x) = \nabla F(t, x)$ for $(t, x) \in I \times U$. $\qquad\square$

5.5   Exercises

Exercise 5.1 Consider the second order system

$$\dot{x} = y - xf(x, y),$$
$$\dot{y} = -x - yf(x, y).$$

Discuss the stability properties of this system when f has a fixed sign.

Exercise 5.2 Consider the equation

$$\ddot{x} + a\dot{x} + 2bx + 3x^2 = 0, \quad a > 0, \ b > 0.$$

Determine the maximal region of asymptotic stability of the zero solution which can be obtained by using the total energy of the system as a Lyapunov function.

Exercise 5.3 Verify that the origin is asymptotically stable and its basin of attraction contains the unit disk for the system

$$\begin{cases} \dot{x} &= y, \\ \dot{y} &= -x - (1 - x^2)y. \end{cases}$$

(Basin of attraction for the origin := the set of all initial points that flow to the origin as $t \to \infty$.)

Exercise 5.4 Let $\mathbf{f} \in C^1(\mathbb{R}^n)$, $n \in \mathbb{N}$, and let $\mathbf{x}(t)$ be a solution of

$$\dot{\mathbf{x}} = \mathbf{f}(\mathbf{x}). \tag{5.22}$$

(1) Suppose $\lim_{t \to \infty} \mathbf{x}(t) = \bar{\mathbf{x}} \in \mathbb{R}^n$, show that $\bar{\mathbf{x}}$ must be an equilibrium of (1).
(2) Find an example of $\mathbf{f}$ so that there exists a solution $\mathbf{x}(t)$ satisfying $\lim_{t \to \infty} \dot{\mathbf{x}}(t) = 0$, yet $\lim_{t \to \infty} \mathbf{x}(t) \neq \bar{\mathbf{x}}$, for any equilibrium $\bar{\mathbf{x}}$ of (1). (Hint: Try $n = 1$.)
(3) Suppose every equilibrium of (1) is isolated, $\lim_{|\mathbf{x}| \to \infty} \mathbf{f}(\mathbf{x}) \neq 0$ and $\lim_{t \to \infty} \dot{\mathbf{x}}(t) = 0$, show that $\lim_{t \to \infty} \mathbf{x}(t) = \bar{\mathbf{x}}$, for some equilibrium $\bar{\mathbf{x}}$ of (1).

Exercise 5.5 Consider the system $\dot{x} = y$, $\dot{y} = z - ay$, $\dot{z} = -cx - F(y)$, $F(0) = 0$, $a > 0$, $c > 0$, $aF(y)/y > c$ for $y \neq 0$ and $\int_0^y [F(\xi) - c\xi/a]d\xi \to \infty$ as $|y| \to \infty$. If $F(y) = ky$ where $k > c/a$, verify that the characteristic roots of the linear system have negative real parts. Show that the origin is asymptotically stable even when F is nonlinear.

Hint: Choose V as a quadratic form plus the term $\int_0^y F(s)ds$.

Exercise 5.6 Suppose there is a positive definite matrix Q such that $J'(x)Q + QJ(x)$ is negative definite for all $x \neq 0$, where $J(x)$ is the Jacobian matrix of $f(x)$. Prove that the solution $x = 0$ of $\dot{x} = f(x)$, $f(0) = 0$, is globally asymptotically stable.

Hint: Prove and make use of the fact that $f(x) = \int_0^1 J(sx)x\,ds$.

Exercise 5.7 Suppose $h(x, y)$ is a positive definite function such that for any constant c, $0 < c < \infty$, the curve defined by $h(x, y) = c$ is a Jordan curve, $h(x, y) \to \infty$ as $x^2 + y^2 \to \infty$. Discuss the behavior in the phase plane of the solution of the equations

$$\dot{x} = \varepsilon x + y - xh(x, y),$$
$$\dot{y} = \varepsilon y - x - yh(x, y),$$

for all values of ε in $(-\infty, \infty)$.

Exercise 5.8 Consider the system

$$x' = x^3 + yx^2,$$
$$y' = -y + x^3.$$

Show that $(0, 0)$ is an unstable equilibrium by using Lyapunov function $V(x, y) = \frac{x^2}{2} - \frac{y^2}{2}$. We note that the variational matrix at $(0, 0)$ is

$$\begin{pmatrix} 0 & 0 \\ 0 & -1 \end{pmatrix}$$

and the eigenvalue $\lambda_1 = 0$ is not hyperbolic.

Exercise 5.9 Consider the n-dimensional system $\dot{x} = f(x) + g(t)$ where $x^T f(x) \leq -k|x|^2$, $k > 0$, for all x and $|g(t)| \leq M$ for all t. Find a sphere of sufficiently large radius so that all trajectories enter this sphere. Show this equation has a ω-periodic solution if g is ω-periodic. If, in addition, $(x - y)^T[f(x) - f(y)] < 0$ for all $x \neq y$ show there is a unique ω-periodic solution.

Hint: Use Brouwer's fixed point theorem (see p. 217).

Exercise 5.10 We say an autonomous system $x' = f(x)$, $f : D \overset{\text{open}}{\subseteq} \mathbb{R}^n \to \mathbb{R}^n$ is dissipative if there exists a bounded domain $\Omega \subseteq \mathbb{R}^n$ such that any trajectory of $x' = f(x)$, $\varphi(t, x_0)$, $x_0 \in \mathbb{R}^n$ will enter Ω and stay there, i.e. there exists $T = T(x_0)$ such that $\varphi(t, x_0) \in \Omega$ for all $t \geq T$.
Show that the Lorenz system

$$
\begin{aligned}
x' &= \sigma(y - x), \\
y' &= \rho x - y - xz, \qquad \sigma, \rho, \beta > 0 \\
z' &= xy - \beta z,
\end{aligned}
$$

is dissipative.
Hint: Consider the function

$$
V(x, y, z) = \frac{1}{2}(x^2 + \sigma y^2 + \sigma z^2) - \sigma \rho z.
$$

Prove that the ellipsoid $V(x, y, z) = c, c > 0$ sufficiently large, is the desired bounded domain Ω.

Exercise 5.11 Consider predator-prey system with Holling-Type II functional response

$$
\begin{aligned}
\frac{dx}{dt} &= rx\left(1 - \frac{x}{K}\right) - \frac{mx}{a + x}y, \\
\frac{dy}{dt} &= \left(\frac{mx}{a + x} - d\right)y, \\
x(0) &> 0, \ y(0) > 0.
\end{aligned}
$$

Prove that the equilibrium (x^*, y^*) is globally asymptotically stable if $x^* > \frac{K-a}{2}$ by using the following Lyapunov function

$$
V(x, y) = \int_{y^*}^{y} s^{(\theta-1)}(s - y^*)ds + cy^\theta \int_{x^*}^{x} \frac{\xi - x^*}{\xi}d\xi,
$$

for some $c > 0$, $\theta > 0$.

Exercise 5.12 Consider the predator-prey system of Gause type:

$$
\begin{aligned}
\frac{dx}{dt} &= xg(x) - yp(x), && (*) \\
\frac{dy}{dt} &= y[cp(x) - q(x)],
\end{aligned}
$$

where x and y represent the prey density and predator density respectively satisfying

(H1) $g(0) = 0$. There exists $K > 0$ such that $g(K) = 0$ and $(x - K)g(x) < 0$ for $x \neq K$.

(H2) $p(0) = 0$, $p'(x) > 0$.

(H3) $q(0) > 0$, $q'(x) \leq 0$ for $x \geq 0$, $\lim\limits_{x \to \infty} q(x) = q_\infty > 0$.

Let $x^* > 0$ satisfying $cp(x^*) = q(x^*)$ and $x^* < K$. Let $y^* = \frac{x^* g(x^*)}{p(x^*)}$ and $H(x^*) = x^* g(x^*) \frac{d}{dx} \ln \left[\frac{xg(x)}{p(x)} \right]_{x=x^*}$. Show that if $H(x^*) < 0$ then (x^*, y^*) is asymptotically stable; if $H(x^*) > 0$ then (x^*, y^*) is unstable focus. Hence the stability condition can be stated graphically as follows: if the prey isocline $y = \frac{xg(x)}{p(x)}$ is decreasing (increasing) at x^*, then (x^*, y^*) is asymptotically stable (an unstable focus).

Exercise 5.13 Show that for the predator-prey system $(*)$ if $\left(\frac{xg(x)}{p(x)} - y^* \right)(x - x^*) \leq 0$ for all $x \geq 0$ then (x^*, y^*) is globally stable in $\mathbb{R}_+^2$. (Hint: Use Lyapunov function $V(x, y) = \int_{x^*}^x \frac{cp(\xi) - q(\xi)}{p(\xi)} d\xi + \int_{y^*}^y \frac{\xi - y^*}{\xi} d\xi$).

Exercise 5.14 Let $A^i = (a_{jk}^i)$ be an $n \times n$ symmetric matrix for $i = 1, 2, \ldots, n$. Assume $a_{jk}^i = a_{ij}^k = a_{ik}^j$ for all $i, j, k = 1, 2, \ldots, n$. For the system $\frac{dx}{dt} = -\nabla V$ with $V(x) = \sum_{i=1}^n x_i \left(x^T A^i x \right)$. Prove that $\dot{V}(x) = 3 \sum_{i=1}^n \left(x^T A^{(i)} x \right)^2 \geq 0$.

Exercise 5.15 Prove Lemma 5.4.1.

Exercise 5.16 Sketch the orbits of the gradient system in Example 5.4.2 for $\lambda = 0$, $\lambda < 0$ and $\lambda > 0$.

Exercise 5.17 Consider the system of differential equations

$$(*) \qquad \frac{dx}{dt} = f(x), \quad f : \Omega \subseteq \mathbb{R}^n \to \mathbb{R}^n$$

is assumed to be continuous.

We call V a Lyapunov function on $G \subseteq \Omega$ for $(*)$ if

(1) V is continuous on G

(2) If V is not continuous at $\bar{x} \in cl(G) \setminus G$ then $\lim\limits_{x \in G, x \to \bar{x}} V(x) = +\infty$

(3) $\dot{V} = \text{grad} V \cdot f \leq 0$ on G.

Prove the following invariance principle:

Assume that V is a Lyapunov function of (1) on G. Define $S = \{x \in \bar{G} \cap \Omega : \dot{V}(x) = 0\}$. Let M be the largest invariant set in S. Then every bounded trajectory (for $t \geq 0$) of $(*)$ that remains in G for $t \geq 0$ approaches the set M as $t \to +\infty$.

Remark: In the example of Lotka-Volterra Predator-prey system, we did apply the above modified invariance principle.

Exercise 5.18 Consider the mathematical model of n microorganisms competing for a single-limited nutrient in the chemostat [SW; Hsu]

$$S'(t) = \left(S^{(0)} - S\right) D - \sum_{i=1}^{n} \frac{1}{y_i} \frac{m_i S}{a_i + S} x_i,$$

$$x_i' = \left(\frac{m_i S}{a_i + S} - d_i\right) x_i,$$

$$S(0) \geq 0, x_i(0) > 0, i = 1, 2, \ldots, n,$$

where $S(t)$ is the concentration of the single-limiting nutrient, $x_i(t)$ is the concentration of i-th microorganism, $S^{(0)}$ is the input concentration, D is the dilution rate, y_i is the yield constant of i-th microorganism, m_i, a_i, d_i are the maximal birth rate, Michaelis-Menten constant, death rate of i-th species respectively. Show that if

$$0 < \lambda_1 < \lambda_2 \leq \ldots \leq \lambda_n \leq S^{(0)}, \lambda_i = \frac{a_i}{\left(\frac{m_i}{d_i}\right) - 1} > 0, i = 1, \ldots, n$$

then

$$\lim_{t \to \infty} S(t) = \lambda_1, \quad \lim_{t \to \infty} x_1(t) = x_1^* > 0 \quad \text{and}$$

$$\lim_{t \to \infty} x_i(t) = 0, \quad i = 2, \ldots, n, \quad x_1^* = \frac{\left(S^{(0)} - \lambda_1\right) D y_1}{d_1}.$$

Hint: Construct Lyapunov function

$$V(S, x_1, \ldots, x_n) = \int_{\lambda_1}^{S} \frac{\xi - \lambda_1}{\xi} d\xi + c_1 \int_{x_1^*}^{x_1} \frac{\eta - x_1^*}{\eta} d\eta + \sum_{i=2}^{n} c_i x_i,$$

where $c_i > 0$ to be determined, $i = 1, 2, \ldots, n$ and apply Exercise 5.9.

Exercise 5.19 Consider $x'' + g(x) = 0$ or equivalent system

$$x' = y,$$

$$y' = -g(x).$$

Verify that this is a conservative system with the Hamiltonian function $H(x, y) = \frac{y^2}{2} + G(x)$, where $G(x) = \int_0^x g(s)ds$. Show that any periodic orbit of this system must intersect x-axis at two points $(a, 0)$, $(b, 0)$, $a < b$. Using symmetry of the periodic orbit with respect to x-axis, show that the minimal period T of a periodic orbit passing through two such points is given by

$$T = 2 \int_a^b \frac{du}{\sqrt{2[G(b) - G(u)]}}.$$

Exercise 5.20 Sketch the phase portrait for the unforced, undamped Duffing oscillator

$$x'' + x^3 - x = 0,$$

by considering the energy $E = \frac{y^2}{2} + V(x)$, where $V(x) = \frac{x^4}{4} - \frac{x^2}{2}$ is a double-well potential. Show that there is a homoclinic orbit. Also consider the same problem for unforced, damped oscillator $x'' + \delta x' + x^3 - x = 0$, $\delta > 0$.

Exercise 5.21 Sketch the phase portrait of

$$x'' - x + x^2 = 0,$$

and

$$x'' - (1 - x)e^{-x} = 0.$$

Exercise 5.22 (Transport Theorem)
Let ϕ_t denote the flow of the system $\dot{x} = f(x)$, $x \in \mathbb{R}^n$, and let Ω be a bounded region in $\mathbb{R}^n$. Define

$$V(t) = \int_{\phi_t(\Omega)} dx_1 dx_2 \cdots dx_n$$

and recall that the divergence of a vector field $f = (f_1, f_2, \ldots, f_n)$ on $\mathbb{R}^n$ with the usual Euclidean structure is

$$\operatorname{div} f = \sum_{i=1}^n \frac{\partial f_i}{\partial x_i}.$$

(1) Use Liouville's theorem and the change of variables formula for multiple integrals to prove that

$$\dot{V}(t) = \int_{\phi_t(\Omega)} \operatorname{div} f(x) dx_1 dx_2 \cdots dx_n.$$

(2) Prove: The flow of a vector field whose divergence is everywhere negative contracts volume.

(3) Suppose that $g : \mathbb{R}^n \times \mathbb{R} \to \mathbb{R}$ and, for notational convenience, let $dx = dx_1 dx_2 \cdots dx_n$. Prove the transport theorem:

$$\frac{d}{dt} \int_{\phi_t(\Omega)} g(x,t)dx = \int_{\phi_t(\Omega)} g_t(x,t) + \operatorname{div}(gf)(x,t)dx.$$

(4) Suppose that the mass in every open set remains unchanged as it is moved by the flow (that is, mass is conserved) and let ρ denote the corresponding mass-density. Prove that the density satisfies the equation of continuity

$$\frac{\partial \rho}{\partial t} + \operatorname{div}(\rho f) = 0.$$

(5) The flow of the system $\dot{x} = y$, $\dot{y} = x$ is area preserving. Show directly that the area of the unit disk is unchanged when it is moved forward two time units by the flow.

Exercise 5.23 Consider $\dot{x} = Ax + h(t,x)$, where A is a constant symmetric matrix with all of its eigenvalues positive, h is continuously differentiable and has period 1 in t, and $\|h(t,x)\| \leq \sqrt{\|x\|} + 2004$. Prove that there is a periodic solution.

Exercise 5.24 *Hamiltonian flows preserve area*: Given a C^1 function $H : \mathbb{R}^2 \to \mathbb{R}$, the planar system

$$\dot{q} = \frac{\partial H(q,p)}{\partial p},$$

$$\dot{p} = -\frac{\partial H(q,p)}{\partial q}$$

is called a *Hamiltonian system* with the Hamiltonian function H.

(1) Show that the total energy of a second-order conservative system is a Hamiltonian function.

(2) Let D_0 be a region, say, with a smooth boundary, in the plane and consider the image of D_0 under the flow of a planar differential equation. That is, consider the set $D(t) = \{\varphi(t, \mathbf{x}^0) : \mathbf{x}^0 \in D_0\}$, where $\varphi(t, \mathbf{x}^0)$ is the solution of the equation $\dot{\mathbf{x}} = \mathbf{f}(\mathbf{x})$ through $\mathbf{x}^0$. If $A(t)$ is the area of $D(t)$, prove that

$$\dot{A}(t) = \int_{D(t)} \operatorname{div} \mathbf{f}(\mathbf{x}) d\mathbf{x}.$$

Suggestion: Use the fact that

$$A(t) = \int_{D_0} \det \frac{\partial \varphi(t, \mathbf{x})}{\partial \mathbf{x}} d\mathbf{x}.$$

(3) Prove now that the flow of a Hamiltonian system preserves area.

Exercise 5.25 Find a Hamiltonian for the system

$$\dot{x}_1 = -\cos x_1 \sin x_2, \qquad \dot{x}_2 = \sin x_1 \cos x_2$$

and draw the flow.

Exercise 5.26 *Rotating pendulum*: Consider a pendulum of mass m and length l constrained to oscillate in a plane rotating with angular velocity w about a vertical line. If u denotes the angular deviation of the pendulum from the vertical and I is the moment of inertia, then

$$I\ddot{u} - mw^2l^2 \sin u \cos u + mgl \sin u = 0.$$

By changing the time scale, this is equivalent to

$$\ddot{u} - (\cos u - \lambda)\sin u = 0,$$

where $\lambda = g/(w^2 l)$. Discuss the flows for each $\lambda > 0$ paying particular attention to the bifurcations in the flow.

Exercise 5.27 Show that each of the systems below is a gradient system, determine the values of the parameters for which the vector field is generic, and discuss the bifurcations in the flows:

(1) $\dot{x}_1 = x_1 + \beta x_2$, $\quad \dot{x}_2 = x_2 + \beta x_1 \quad$ for $\beta \in \mathbb{R}$.
(2) $\dot{x}_1 = \mu(x_2 - x_1) + x_1(1 - x_1^2)$, $\quad \dot{x}_2 = -\mu(x_2 - x_1) + x_2(1 - x_2^2) \quad$ for $\mu > 0$.

Exercise 5.28 Analyze the flow of the gradient system

$$\dot{x}_1 = -x_1^3 - bx_1 x_2^2 + x_1, \qquad \dot{x}_2 = -x_2^3 - bx_1^2 x_2 + x_2$$

coming from vibrating membranes for the parameter values $b > 1$.

Exercise 5.29 Find the critical ($=$ equilibrium) points for the scalar equation

$$\ddot{x} + 2b\dot{x} + \sin(x^3) = 0$$

where b is a positive constant. The linearization of this equation about a stable critical point $(\xi, 0)$ in the $(x, \dot{x})$-plane with $\xi > 0$ has three qualitatively different kinds of phase portraits, one for small positive b, one for large positive b, and one for a single positive value of b, with these ranges depending on ξ. Give a careful sketch of such a phase portrait for a large value of b. Be sure to identify important directions with their slopes and include arrows.

Chapter 6

TWO-DIMENSIONAL SYSTEMS

6.1   Poincaré-Bendixson Theorem

Assume $f : \Omega \subseteq \mathbb{R}^2 \to \mathbb{R}^2$ and Ω is open in $\mathbb{R}^2$. Consider the two-dimensional autonomous system

$$x' = f(x). \tag{6.1}$$

Let $\varphi(t)$ be a solution of (6.1) for $t \geq 0$ and $\omega(\varphi)$ be the ω-limit set of $\varphi(t)$. We recall that if $\varphi(t)$ is bounded for $t \geq 0$, then $\omega(\varphi)$ is nonempty, compact, connected and invariant. The following Poincaré-Bendixson Theorem characterizes the ω-limit set $\omega(\varphi)$ of the solution $\varphi(t)$ of the two-dimensional autonomous system (6.1).

Theorem 6.1.1 (Poincaré-Bendixson Theorem)
If the solution $\varphi(t)$ of (6.1) is bounded for all $t \geq 0$, then either
(i) $\omega(\varphi)$ contains an equilibrium
or
(ii) (a) $\varphi(t)$ is periodic
or (b) $\omega(\varphi)$ is a periodic orbit.

Remark 6.1.1 In case (ii), we assume that $\omega(\varphi)$ contains no equilibria and in the case (b) we call $\omega(\varphi)$ a limit cycle. There is a difference between a periodic orbit and a limit cycle. A limit cycle must be a periodic orbit but not vice versa. In Example 5.2.4 the periodic orbits are not limit cycles.

Remark 6.1.2 The Poincaré-Bendixson Theorem is one of the most important results of nonlinear dynamical systems. It says that the dynamical possibilities in the phase plane are very limited: if a trajectory is confined to a closed, bounded region such that its ω-limit set contains no fixed points, then the trajectory must eventually approach a closed orbit. Nothing else is possible. This result depends crucially on the Jordan Curve Theorem in the plane. In higher dimensional system ($n \geq 3$, for example, Lorenz system), the Poincaré-Bendixson Theorem is no longer true and we may have "strange attractors". However, for three-dimensional competitive systems [Hir1; SW] and n-dimensional feedback systems [M-PS], the Poincaré-Bendixson Theorem holds. For Poincaré-Bendixson Theorem on two-dimensional manifold, the interested reader may find in the book [Ha].

Remark 6.1.3 To show the existence of a limit cycle by Poincaré-Bendixson Theorem, we usually apply the standard trick to construct a trapping region R which contains no equilibrium. Then there exists a closed orbit in R (see Fig. 6.1).

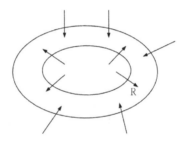

Fig. 6.1

Example 6.1.1 Consider the system

$$r' = r(1 - r^2) + \mu r \cos \theta$$
$$\theta' = 1.$$

Show that a closed orbit exists for $\mu \in (0, 1)$.

To construct an annulus $0 < r_{\min} \leq r \leq r_{\max}$ to be a desired trapping region. To find $r_{\min}$, we required $r' = r(1 - r^2) + \mu r \cos \theta > 0$ for all θ. Since $\cos \theta \geq -1$ then $r' \geq r(1 - r^2) - \mu r = r[(1 - r^2) - \mu]$. Hence any $r_{\min} < \sqrt{1 - \mu}$ will work as long as $\mu < 1$. By similar argument, the flow is inward on the outer circle if $r_{\max} > \sqrt{1 + \mu}$. Therefore, a closed orbit exists for all $\mu < 1$ and it lies somewhere in the annulus $0.999\sqrt{1 - \mu} < r < 1.001\sqrt{1 + \mu}$.

Example 6.1.2 Consider the following system of glycolytic oscillation arising from biochemistry.

$$x' = -x + ay + x^2y$$
$$y' = b - ay - x^2y.$$

From Fig. 6.2,

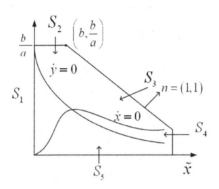

Fig. 6.2

(1) On S_1, $\vec{n} = (-1, 0)$,

$$\vec{n} \cdot \vec{f}(x) = -(-x + ay + x^2y)|_{x=0,\ 0 \le y \le b/a} = -ay \le 0.$$

(2) On S_2, $\vec{n} = (0, 1)$,

$$\vec{n} \cdot \vec{f}(x) = (b - ay - x^2y)|_{y=b/a, 0 \le x \le b} = -x^2 \cdot b/a \le 0.$$

(3) On S_3, $\vec{n} = (1, 1)$,

$$\vec{n} \cdot \vec{f}(x) = b - x|_{b \le x \le \tilde{x}} \le 0.$$

(4) On S_4, $\vec{n} = (1, 0)$,

$$\vec{n} \cdot \vec{f}(x) = -x + ay + x^2y|_{x=\tilde{x},\ 0 \le y \le \tilde{y}},$$

where

$$\tilde{x} + \tilde{y} = b + b/a, \quad \tilde{x} \text{ is close to } b + b/a, \quad \text{i.e. } \tilde{y} \text{ is close to } 0,$$

then

$$\vec{n} \cdot \vec{f}(x) \le \tilde{y}(\tilde{x}^2 + a) - \tilde{x} \le 0.$$

(5) On S_5, $\vec{n} = (0, -1)$,

$$\vec{n} \cdot \vec{f}(x) = -(b - ay - x^2 y)|_{0 \leq x \leq \tilde{x}} {}_{y=0} = -b < 0.$$

Thus the region is a trapping region. Next we verify that the equilibrium (x^*, y^*) where $x^* = b$, $y^* = \frac{b}{a+b^2}$ is an unstable focus if

$$\tau = -\frac{b^4 + (2a - 1)b^2 + (a + a^2)}{a + b^2} > 0.$$

Thus by Poincaré-Bendixson Theorem, there exists a limit cycle for $\tau > 0$.

Example 6.1.3 Consider the Lotka-Volterra Predator-Prey system

$$x' = ax - bxy,$$
$$y' = cxy - dy.$$

Each periodic orbit is neutrally stable. There is no limit cycle. We note that the system is a conservative system.

Example 6.1.4 Consider the predator-prey system with Holling type II functional response

$$\frac{dx}{dt} = \gamma x \left(1 - \frac{x}{K}\right) - \frac{mx}{a+x} y = f(x, y),$$
$$\frac{dy}{dt} = \left(\frac{mx}{a+x} - d\right) y = g(x, y),$$
$$x(0) > 0, \ y(0) > 0.$$

Assume $\frac{K-a}{2} > \lambda$ where $\lambda = x^*$, (x^*, y^*) is the equilibrium. We note that in Example 4.2.3, we proved that (x^*, y^*) is an unstable focus. First we show that the solution $(x(t), y(t))$ is positive and bounded. It is easy to verify that $x(t) > 0$, $y(t) > 0$ for all $t \geq 0$. Then from differential inequality

$$\frac{dx}{dt} \leq \gamma x \left(1 - \frac{x}{K}\right)$$

it follows that for small $\epsilon > 0$, $x(t) \leq K + \varepsilon$ for $t \geq T(\varepsilon)$. Consider

$$\frac{dx}{dt} + \frac{dy}{dt} \leq \gamma x - dy \leq (\gamma + d)K - d(x + y),$$

then $x(t) + y(t) \leq \frac{\gamma+d}{d} K$ for t large. Hence $x(t), y(t)$ are bounded.

From Poincaré-Bendixson Theorem, there exists a limit cycle in the first quadrant of $x - y$ plane (see Remark 6.1.4 and Fig. 6.3).

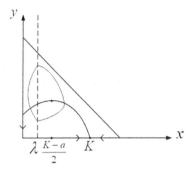

Fig. 6.3

To establish the global asymptotic stability of the equilibrium by using Poincaré-Bendixson's Theorem, it suffices to eliminate the possibility of the existence of periodic solutions.

Theorem 6.1.2 (Bendixson Negative Criterion)
Consider the system

$$\frac{dx}{dt} = f(x, y),$$
$$\frac{dy}{dt} = g(x, y). \tag{6.2}$$

If $\frac{\partial f}{\partial x} + \frac{\partial g}{\partial y}$ is of the same sign in a simple-connected domain D in $\mathbb{R}^2$, then there are no periodic orbits in D.

Proof. Suppose, on the contrary, there is a periodic orbit $C = \{(x(t), y(t))\}_{0 \le t \le T}$. Then, by Green's Theorem,

$$\int\int_D \left(\frac{\partial f}{\partial x} + \frac{\partial g}{\partial y}\right) dxdy = \oint_C (-g(x, y)) \, dx + f(x, y)dy$$

$$= \int_0^T \left[-g(x(t), y(t))x'(t) + f(x(t), y(t))y'(t)\right] dt$$

$$= 0.$$

But $\frac{\partial f}{\partial x} + \frac{\partial g}{\partial y}$ is of same sign, then $\int\int_D \left(\frac{\partial f}{\partial x} + \frac{\partial g}{\partial y}\right) dxdy$ is either positive or negative. This is a contradiction. □

Example 6.1.5 Show that $x'' + f(x)x' + g(x) = 0$ cannot have any periodic solution in a region where $f(x)$ is of the same sign (i.e. such region has only "positive damping" or "negative damping"). Write the equation as

$$\begin{pmatrix} x' \\ y' \end{pmatrix} = \begin{pmatrix} y \\ -f(x)y - g(x) \end{pmatrix} = \begin{pmatrix} F(x, y) \\ G(x, y) \end{pmatrix}.$$

Then

$$\frac{\partial F}{\partial x} + \frac{\partial G}{\partial y} = -f(x) \quad \text{is of same sign.}$$

We leave it as an exercise to generalize the Bendixson Negative criterion to the following Dulac's criterion which is much more powerful than Negative Bendixson Criterion.

Theorem 6.1.3 (Dulac's criterion)
Let $h(x,y)$ be continuously differentiable in a simply-connected region D. For the system (6.2), if $\frac{\partial(fh)}{\partial x} + \frac{\partial(gh)}{\partial y}$ is of the same sign in D, then (6.2) has no periodic orbit in D.

Example 6.1.6 Consider Lotka-Volterra two species competition model

$$x' = \gamma_1 x \left(1 - \frac{x}{K_1}\right) - \alpha xy = f(x,y),$$

$$y' = \gamma_2 y \left(1 - \frac{y}{K_2}\right) - \beta xy = g(x,y).$$

Show that there is no periodic orbit in the first quadrant.
If we try the Bendixson Negative criterion, then

$$\frac{\partial f}{\partial x} + \frac{\partial g}{\partial y} = \left(\gamma_1 - 2\frac{\gamma_1}{K_1}x - \alpha y\right) + \left(\gamma_2 - 2\frac{\gamma_2}{K_2}y - \beta x\right)$$

which is not of the same sign.

Choose $h(x,y) = x^\xi y^\eta$ where $\xi, \eta \in \mathbb{R}$ are to be determined. Then from routine computation,

$$\begin{aligned}
\Delta &= \frac{\partial(fh)}{\partial x} + \frac{\partial(gh)}{\partial y} \\
&= x^\xi y^\eta \left\{ (\gamma_1 + \gamma_2 + \xi\gamma_1 + \eta\gamma_2) + \left((-2-\xi)\frac{\gamma_1}{K_1} - \beta(1+\eta)\right)x \right. \\
&\quad \left. + \left[(-2-\eta)\frac{\gamma_2}{K_2} - \alpha(1+\xi)\right]y \right\}.
\end{aligned}$$

Choose $\xi = \eta = -1$, then $\Delta < 0$ for $x, y > 0$. Thus, from Dulac's criterion, we complete the proof.

Now we return to the proof of Poincaré-Bendixson Theorem [MM] for the system

$$x' = f(x), \quad f : \Omega \subseteq \mathbb{R}^2 \to \mathbb{R}^2.$$

The proof needs the following Jordan Curve Theorem.

Jordon Curve Theorem: A simple closed curve separates plane into two parts, Ω_i and Ω_e. The interior component Ω_i is bounded while the exterior component Ω_e is unbounded.

Given a closed line segment $L = \overrightarrow{\xi_1 \xi_2}$. Let the vector $\vec{b} = \xi_2 - \xi_1$ and $\vec{a}$ is the normal vector to L (see Fig. 6.4).

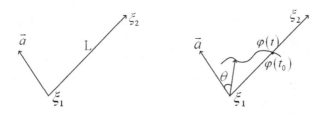

Fig. 6.4

We say that a continuous curve $\varphi : (\alpha, \beta) \to \mathbb{R}^2$ cross L at time t_0 if $\varphi(t_0) \in L$ and there exists $\delta > 0$ such that $(\varphi(t) - \xi_1, \vec{a}) > 0$ for $t_0 - \delta < t < t_0$ and (or for $t_0 < t < t_0 + \delta$) $(\varphi(t) - \xi_1, \vec{a}) < 0$ for $t_0 < t < t_0 + \delta$ (or for $t_0 - \delta < t < t_0$).

Definition 6.1.1 *We say that a closed segment L in $\mathbb{R}^2$ is a transversal with respect to continuous vector field $f : \mathbb{R}^2 \to \mathbb{R}^2$ if*

(i) *for each $\xi \in L$, ξ is a regular point of the vector field f, i.e. $f(\xi) \neq 0$.*

(ii) *The vector $f(\xi)$ is not parallel to the direction of $\vec{b}$.*

Assume that L is a transversal with respect to vector field $f(x)$.

Lemma 6.1.1 *Let ξ_0 be an interior point of a transversal L, then for any $\varepsilon > 0$ there exists $\delta > 0$ such that, for any $x_0 \in B(\xi_0, \delta)$, $\varphi(t, x_0)$ crosses L at some time $t \in (-\varepsilon, \varepsilon)$.*

Proof. Let the equation of L be

$$g(x) = a_1 x_1 + a_2 x_2 - c = 0.$$

Consider

$$G(t, \xi) = g(\varphi(t, \xi))$$

where $\varphi(t, \xi)$ is the solution of I.V.P. $x' = f(x)$, $x(0) = \xi$, then

$$G(0, \xi_0) = g(\xi_0) = 0,$$

$$\frac{\partial G}{\partial t}(t, \xi) = g'(\varphi(t, \xi))f(\varphi(t, \xi)) = \overrightarrow{a} \cdot f(\varphi(t, \xi)),$$

$$\frac{\partial G}{\partial t}(0, \xi_0) = \overrightarrow{a} \cdot f(\xi_0) \neq 0.$$

By Implicit Function Theorem, we may express t as a function of ξ, i.e., there exist $\delta > 0$ and a map $t : B(\xi_0, \delta) \to \mathbb{R}$ s.t. $G(t(\xi), \xi) = 0$, $t(\xi_0) = 0$. Hence $g(\varphi(t(\xi), \xi)) = 0$ and the proof of Lemma 6.1.1 is complete. $\square$

Lemma 6.1.2 *Let $\varphi(t)$ be an orbit of (6.1) and $S = \{\varphi(t) : \alpha \leq t \leq \beta\}$. If S intersects L, then $L \cap S$ consists of only finite points whose order is monotone with respect to t. If $\varphi(t)$ is periodic, then $L \cap S$ is a singleton set.*

Proof. First we show that $L \cap S$ is finite. If not, then there exists $\{t_m\} \subset [\alpha, \beta]$ such that $\varphi(t_m) \in L$. We may assume $t_m \to t_0 \in [\alpha, \beta]$. Then $\varphi(t_m) \to \varphi(t_0)$ and

$$\lim_{m \to \infty} \frac{\varphi(t_m) - \varphi(t_0)}{t_m - t_0} = \varphi'(t_0) = f(\varphi(t_0)).$$

Thus the vector $f(\varphi(t_0))$ is parallel to L and we obtain a contradiction to the transversality of L.

Next we prove the monotonicity. We may have two possibilities as Fig. 6.5 shows:

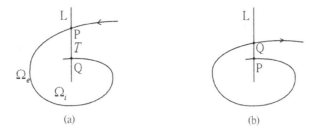

(a)   (b)

Fig. 6.5

We consider case (a) only. From the uniqueness of O.D.E., it is impossible for the trajectory to cross the arc $\widehat{PQ}$. Let $\sum$ be the simple closed curve made up of the curve $\widehat{PQ}$ and the segment T between P and Q. Let D be the closed region bounded by $\sum$. Since T is transverse to the flow, at each point of T, the trajectory either leaves or enters D. We assert that at any point of T the trajectory leaves D. Let T_- be the set of points whose trajectories leave D and T_+ be the set of points whose trajectories enter D. From the continuity of the flow, T_-, T_+ are open sets in T. Since T is a disjoint union of T_+ and T_- and T is connected, then we have T_+ is empty and $T = T_-$. Obviously the monotonicity of $L \cap S$ follows. It is obvious from the monotonicity that $L \cap S$ is singleton if $\varphi(t)$ is periodic. $\qquad\square$

Lemma 6.1.3 *A transversal L cannot intersect $\omega(\varphi)$, the ω-limit set of a bounded trajectory, in more than one point.*

Proof. Let $\omega(\varphi)$ interest L at ξ', then there exists $\{t'_m\} \uparrow +\infty$, $t'_{m+1} > t'_m + 2$, $\varphi(t'_m) \to \xi'$. By Lemma 6.1.1, there exists $M \geq 1$ sufficiently large, such that φ must cross L at some time t_m, $|t_m - t'_m| < 1$ for $m \geq M$. From Lemma 6.1.2, $\varphi(t_m) \downarrow \xi \in L \cap \omega(\varphi)$. From Lemma 6.1.1, we may assume $|t_m - t'_m| \to 0$ as $m \to \infty$. Then from Mean Value Theorem, we have
$$\varphi(t_m) = \varphi(t'_m) + (\varphi(t_m) - \varphi(t'_m))$$
$$= \varphi(t'_m) + (t_m - t'_m)f(\varphi(\eta_m)).$$
Since $\varphi(t)$ is bounded, $\lim_{m \to \infty} \varphi(t_m) = \xi'$ and $\xi = \xi'$. Hence there exists $\{t_m\} \uparrow \infty$ such that $\varphi(t_m) \downarrow \xi$.

If η is a second point in $L \cap \omega(\varphi)$, then there exists $\{s_m\} \uparrow \infty$ such that $\varphi(s_m) \in L$ and $\varphi(s_m) \downarrow \eta$ or $\varphi(s_m) \uparrow \eta$. Assume the sequences $\{s_m\}$, $\{t_m\}$ interlace i.e., $t_1 < s_1 < t_2 < s_2 < \cdots$, then $\{\varphi(t_1), \varphi(s_1), \varphi(t_2), \cdots\}$ is a monotone sequence. Thus $\xi = \eta$. $\qquad\square$

Lemma 6.1.4 *Let $\varphi(t)$ be the solution of (6.1) and $\omega(\varphi)$ be the ω-limit set of $\gamma^+(\varphi)$.*
(i) If $\omega(\varphi) \cap \gamma^+(\varphi) \neq \phi$, then $\varphi(t)$ is a periodic solution.
(ii) If $\omega(\varphi)$ contains a nonconstant periodic orbit Γ, then $\omega(\varphi) = \Gamma$.

Proof. (i) Let $\eta \in \omega(\varphi) \cap \gamma^+(\varphi)$, then η is a regular point and there exists a transversal L through η. Let $\eta = \varphi(\tau)$, then from the invariance of $\omega(\varphi), \gamma(\eta) \subset \omega(\varphi)$.

Fig. 6.6

Since $\eta \in w(\varphi)$, there exists $\{t'_m\} \uparrow +\infty$ s.t. $\varphi(t'_m) \to \eta$ (see Fig. 6.6).
By Lemma 6.1.1, there exists t_m near t'_m, $\varphi(t_m) \in L$, $|t_m - t'_m| \to 0$,
$\varphi(t_m) \to \eta$ as $m \to \infty$ as we did in the proof of Lemma 6.1.3. Since
$\eta = \varphi(\tau) \in w(\varphi)$ and $w(\varphi)$ is invariant, it follows that $\varphi(t_m) \in w(\varphi)$ for
m sufficiently large. From Lemma 6.1.3 and $\varphi(t_m) \in L \cap w(\varphi)$, we have
$\varphi(t_m) = \eta$ for m sufficiently large. This implies that $\varphi(t)$ is a periodic
solution.

(ii) Assume a periodic orbit Γ is contained in $w(\varphi)$. We claim that $\Gamma = w(\varphi)$.
If not, $\Gamma \neq w(\varphi)$. Since $w(\varphi)$ is _connected_, there exists $\{\xi_m\} \subseteq w(\varphi) \setminus \Gamma$ and
$\xi_0 \in \Gamma$ such that $\xi_m \to \xi_0$ (see Fig. 6.7).

Fig. 6.7

Let L be a transversal through ξ_0. By Lemma 6.1.1, for m sufficiently
large, the orbit through ξ_m must intersect L. Let $\varphi(\tau_m, \xi_m) = \xi'_m \in L \cap$
$w(\phi)$. From Lemma 6.1.3, $\{\xi'_m\}$ is a constant sequence and hence $\xi'_m = \xi_0$.
Then $\xi_m = \varphi(-\tau_m, \xi'_m) = \varphi(-\tau_m, \xi_0) \in \Gamma$. This is a contradiction. $\square$

Proof of Poincaré-Bendixson Theorem. Assume $w(\varphi)$ contains no equilib-
ria and φ is not a periodic solution. Let $y_0 \in w(\varphi)$ and $C^+ = \gamma^+(y_0)$, then
$C^+ \subseteq w(\varphi)$. Let $w(y_0)$ be the w-limit set of C^+. Obviously $w(y_0) \subseteq w(\varphi)$
and hence $w(y_0)$ contains no equilibria. Let $\eta \in w(y_0)$ and introduce a
transversal L through η.

Fig. 6.8

Since $\eta \in \omega(y_0)$, there exists a sequence $\{t_n\} \uparrow \infty$ such that $\varphi(t_n, y_0) \to \eta$ (see Fig. 6.8). By Lemma 6.1, there exists $\{t'_n\}$, $|t_n - t'_n| \to 0$, $\varphi(t'_n, y_0) \in L \cap \omega(y_0)$. But $C^+ \subseteq \omega(\varphi)$. C^+ meets L only once. Then $\eta \in C^+$. From Lemma 6.1.4(i), we have that C^+ is a periodic orbit. Since $C^+ \subseteq \omega(\varphi)$, from Lemma 6.1.4(ii), $\omega(\varphi) = C^+$. Thus we complete the proof. $\square$

In the following, we present a result related to the Poincaré-Bendixson's Theorem: A closed orbit in the plane must enclose an equilibrium. We shall prove it in Chapter 8 by using the index theory. Before that we present a result in $\mathbb{R}^n$ by using Brouwer's fixed point theorem.

Theorem 6.1.4 *If K is a positively invariant set of the system $x' = f(x), f : \Omega \subseteq \mathbb{R}^n \to \mathbb{R}^n$ and K is homeomorphic to the closed unit ball in $\mathbb{R}^n$, then there is at least one equilibrium in K.*

Proof. For any fixed $\tau_1 > 0$, consider the map $\varphi(\tau_1, \cdot) : K \to K$. From Brouwer's fixed point theorem, there is a $p_1 \in K$ such that $\varphi(\tau_1, p_1) = p_1$, thus, a periodic orbit of period τ_1. Choose $\tau_m \to 0$, $\tau_m > 0$ as $m \to \infty$ and corresponding points p_m such that $\varphi(\tau_m, p_m) = p_m$. Without loss of generality, we assume $p_m \to p^* \in K$. For any t and any integer m, there is an integer $k_m(t)$ such that $k_m(t)\tau_m \leq t < k_m(t)\tau_m + \tau_m$ and $\varphi(k_m(t)\tau_m, p_m) = p_m$ for all t. Furthermore,

$$|\varphi(t, p^*) - p^*| \leq |\varphi(t, p^*) - \varphi(t, p_m)| + |\varphi(t, p_m) - p_m| + |p_m - p^*|$$
$$= |\varphi(t, p^*) - \varphi(t, p_m)| + |\varphi(t - k_m(t)\tau_m, p_m) - p_m| + |p_m - p^*|$$
$$\to 0 \text{ as } m \to \infty.$$

Therefore, p^* is an equilibrium. $\square$

Remark 6.1.4 In the book of [H1] or [ASY] p. 341, we have further results on the Case (i) of Poincaré-Bendixson Theorem. If $\omega(\varphi)$ contains only finite number of equilibria of (6.1), then either
(a) $\omega(\varphi)$ is a single equilibrium of (6.1)

or

(b) $\omega(\varphi)$ consists of a finite number of equilibria $q_1 \cdots q_n$ together with a finite set of orbits $\gamma_1 \cdots \gamma_n$ such that the $\alpha(\gamma_j)$ and $\omega(\gamma_j)$ are equilibria, i.e. $\omega(\varphi) = \{q_1 \cdots q_n\} \cup \gamma_1 \cup \cdots \cup \gamma_n$, which is a closed contour.

6.2   Levinson-Smith Theorem

Consider the following Lienard equation

$$x'' + f(x)x' + g(x) = 0 \tag{6.3}$$

where $f(x)$, $g(x)$ satisfy

(H1) $f : \mathbb{R} \to \mathbb{R}$ is continuous and $f(x) = f(-x)$, i.e., f is even.
(H2) $g : \mathbb{R} \to \mathbb{R}$ is continuous, $xg(x) > 0$ for all $x \neq 0$, $g(x) = -g(-x)$, i.e., g is odd.
(H3) $F(x) = \int_0^x f(s)ds$ satisfies
$0 < x < a$, $F(x) < 0$
$x > a$, $F(x) > 0$, $f(x) > 0$.
(H4) $G(x) = \int_0^x g(s)ds \to \infty$ as $|x| \to \infty$;
$F(x) \to +\infty$ as $x \to +\infty$.

Example 6.2.1 One of the most important examples of Lienard equations is van der Pol equation,

$$x'' + \varepsilon(x^2 - 1)x' + x = 0, \quad \varepsilon > 0.$$

The derivation of the model can be consulted in [Keel] p. 481. In this example, $f(x) = \varepsilon(x^2 - 1)$, $F(x) = \varepsilon\left(\frac{x^3}{3} - x\right)$, $g(x) = x$, $G(x) = x^2/2$ with $a = \sqrt{3}$ in (H3).

Theorem 6.2.1 *(Levinson-Smith)*
There exists a unique of limit cycle of (6.3), which is globally orbital stable.

Proof. Rewrite (6.3) in the following Lienard form

$$\begin{cases} x' = y - F(x), \\ y' = -g(x). \end{cases} \tag{6.4}$$

Consider the energy function

$$v(x, y) = y^2/2 + G(x). \tag{6.5}$$

Then

$$\frac{dv}{dt} = y'y + g(x)x' = -g(x)F(x). \tag{6.6}$$

We note that F is odd, i.e., $F(-x) = -F(x)$, and the trajectory $(x(t), y(t))$ is symmetric with respect to origin because of the fact that $(x(t), y(t))$ is a solution of (6.3) if and only if $(-x(t), -y(t))$ is a solution of (6.3). Let $\alpha > 0$, $A = (0, \alpha)$ and $D = (0, y(t(\alpha)))$ be the points of the trajectory $(x(t), y(t))$ on the y-axis with $(x(0), y(0)) = A$ and $(x(t(\alpha)), y(t(\alpha))) = D$ (see Fig. 6.9). From (6.5) we have

$$\alpha^2 - y^2(t(\alpha)) = 2(v(A) - v(D)). \tag{6.7}$$

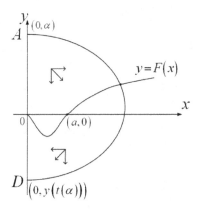

Fig. 6.9

Due to the symmetry of $(x(t), y(t))$ with respect to the origin, we prove the theorem in three steps.

Step 1: To show $v(A) - v(D)$ is a monotonically increasing function of α (see Fig. 6.12).

From (6.4), (6.6), we have

$$\frac{dv}{dx} = \frac{dv/dt}{dx/dt} = \frac{-g(x)F(x)}{y - F(x)}, \tag{6.8}$$

$$\frac{dv}{dy} = \frac{\frac{dv}{dt}}{\frac{dy}{dt}} = F(x). \tag{6.9}$$

From (6.8), (6.9), we have the following inequalities (see Fig. 6.10)

$$v(B) - v(A) = \int_0^a \left(\frac{dv}{dx}\right) dx = \int_0^a \frac{-g(x)F(x)}{y(x) - F(x)} dx$$

$$< \int_0^a \frac{-g(x)F(x)}{\tilde{y}(x) - F(x)} dx = v(B') - v(A'); \tag{6.10}$$

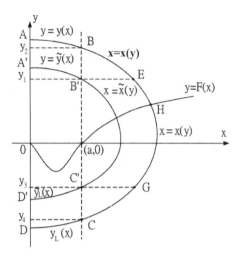

Fig. 6.10

$$v(E) - v(B) = -\int_{y_1}^{y_2} \left(\frac{dv}{dy}\right) dy = -\int_{y_1}^{y_2} F(x(y)) dy < 0; \qquad (6.11)$$

$$v(G) - v(E) = -\int_{y_3}^{y_1} \frac{dv}{dy} dy = -\int_{y_3}^{y_1} F(x(y)) dy \qquad (6.12)$$
$$< -\int_{y_3}^{y_1} F(\tilde{x}(y)) dy = v(C') - v(B')$$

(here we use the fact that $f(x) > 0$ for $x > a$ to deduce $F(x(y)) > F(\tilde{x}(y))$)

$$v(C) - v(G) = \int_{y_3}^{y_4} \frac{dv}{dy} dy = -\int_{y_4}^{y_3} \frac{dv}{dy} dy$$
$$= -\int_{y_4}^{y_3} F(x(y)) dy < 0 \qquad (6.13)$$

$$v(D) - v(C) = -\int_0^a \left(\frac{dv}{dx}\right) dx = -\int_0^a \frac{-g(x)F(x)}{y_L(x) - F(x)} dx \qquad (6.14)$$
$$= \int_0^a \frac{-g(x)F(x)}{F(x) - y_L(x)} dx < \int_0^a \frac{-g(x)F(x)}{F(x) - \tilde{y}_L(x)} dx = v(D') - v(C').$$

Summing the inequalities, (6.10)–(6.14), yields

$$v(D) - v(A) < v(D') - v(A') \text{ or } v(A') - v(D') < v(A) - v(D).$$

Thus we complete the proof of Step 1.

Step 2: For $\alpha > 0$ sufficiently small, $v(A) - v(D) < 0$. Since $\frac{dv}{dt} = -g(x)F(x) > 0$ for α small, Step 2 follows.

Step 3: For α sufficiently large, $v(A) - v(D) > 0$ (see Fig. 6.11)

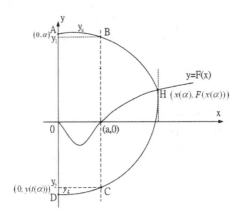

Fig. 6.11

$$v(A) - v(D) = (v(A) - v(B)) + (v(B) - v(C)) + (v(C) - v(D))$$
$$= \int_a^0 \frac{dv}{dx} dx + \int_{y_2}^{y_1} \frac{dv}{dy} dy + \int_0^a \frac{dv}{dx} dx \qquad (6.15)$$
$$= \int_0^a \frac{g(x)F(x)}{y_L(x) - F(x)} dx + \int_0^a \frac{-g(x)F(x)}{y_u(x) - F(x)} dx + \int_{y_2}^{y_1} F(x) dy.$$

Case 1: $x(\alpha)$ is bounded for all α.

Since the map $(x(\alpha), F(x(\alpha))) \to (0, y(t(\alpha)))$ is continuous in α and hence $y(t(\alpha))$ is bounded. Hence $v(A) - v(D) = \frac{1}{2}\left(\alpha^2 - y^2(t(\alpha))\right) > 0$ for α sufficiently large. (R is the region bounded by $\overline{BC}$ and $\overparen{BHC}$.)

Case 2: $x(\alpha) \to +\infty$ and $y(t(\alpha)) \to +\infty$ as $\alpha \to +\infty$.

From (6.4), we have

$$\frac{dy}{dx} = \frac{-g(x)}{y - F(x)}. \qquad (6.16)$$

For $0 < x < a$, consider I.V.P. (6.16) with different initial values $y(0) = \alpha$ and $y(0) = y(t(\alpha))$, it follows that for $0 \le x \le a$, $y_u(x) \to +\infty$ and $y_L(x) \to -\infty$ if $\alpha \to \infty$. Hence the first and second integrals in (6.15) go to zero as $\alpha \to \infty$. Consider the third integral in (6.15), from Green's theorem we have

$$\int_{y_2}^{y_1} F(x)dy = \int_{\widehat{CB}} F(x)dy + \int_{\overline{BC}} F(x)dy$$

$$= \oint F(x)dy = \int\int_R \frac{\partial}{\partial x}(F(x))dxdy$$

$$= \int\int_R f(x)dxdy \to +\infty \quad \text{as} \quad \alpha \to +\infty.$$

Hence $v(D) - v(A) > 0$ for α sufficiently large. (R is the region bounded by $\overline{BC}$ and $\widehat{BHC}$.)

Combining Step 1–Step 3 and the symmetry of the trajectory with respect to origin, there exists a unique solution α^* of $v(A) - v(D)$ (see Fig. 6.12).

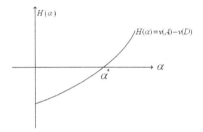

Fig. 6.12

And obviously the corresponding unique of limit cycle attracts every point $(x, y) \neq (0, 0)$. □

Remark 6.2.1 For a two-dimensional autonomous, dissipative system, we can prove the existence of limit cycle by Poincaré-Bendixson Theorem. To prove the uniqueness of limit cycles is a very difficult job. If we are able to show that every periodic solution is orbitally asymptotically stable, then uniqueness of limit cycles follows. Hence if we are able to prove that

$$\oint_\Gamma \left(\frac{\partial f}{\partial x} + \frac{\partial g}{\partial y}\right) dt < 0 \tag{6.17}$$

for system (6.2), then from Corollary 4.4.1, the periodic orbit Γ is orbitally asymptotically stable. Since the exact position of Γ is unknown, hence it is difficult to evaluate the integral (6.17). However, for some systems, like the predator-prey system with Holling type II functional response (Example 6.1.4), K. S. Cheng [Che] prove the uniqueness of limit cycle by the following method of reflection.

Consider the predator-prey system with Holling type II functional responses:

$$\begin{cases} S' = rS(1 - \frac{S}{K}) - (\frac{m}{y})\frac{Sx}{a+S} = \frac{m}{y}\frac{S}{a+S}(f(S) - x) = g(S,x), \\ x' = (\frac{mS}{a+S} - d)x = h(S,x) = (m-d)\frac{(S-\lambda)x}{a+S}, \end{cases} \quad (6.18)$$

where $x = f(S) = \frac{ry}{m}(a + S)(1 - \frac{S}{K})$ is the prey-isocline. Let $\Gamma = \{(S(t), x(t)), 0 \le t \le T\}$ be a periodic orbit of the predator-prey system (6.18).

To show the unique of limit cycle, it suffices to prove the Floque multiple ρ_2, satisfies $\rho_2 < 1$, or to show

$$\oint \frac{\partial g}{\partial S} + \frac{\partial h}{\partial x} = \oint \frac{\partial g}{\partial S} = \oint \frac{m}{y}\frac{Sf'(S)}{a+S} + \frac{m}{y}\frac{a(f(S) - x)}{(a+S)^2} < 0. \quad (6.19)$$

From symmetry of the graph of $x = f(S)$ with respect to the line $S = \frac{K-a}{2}$ (see Fig. 6.13)

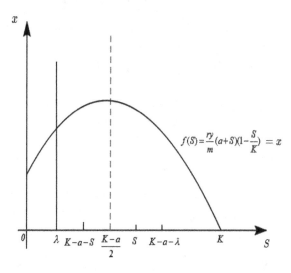

$$f(S) = \frac{ry}{m}(a+S)(1-\frac{S}{K}) = x$$

Fig. 6.13

we find that for $\frac{K-a}{2} < S < K - a - \lambda$,

$$f(S) = f(K - a - S) \tag{6.20}$$

$$f'(S) = -f'(K - a - S). \tag{6.21}$$

Since

$$0 = \int_0^T \frac{aS'}{a+S} = \int_0^T \frac{m}{y} \frac{a(f(S) - x)}{(a+S)^2}$$

from (6.19)

$$\oint_\Gamma \frac{\partial g}{\partial S} + \frac{\partial h}{\partial x} = \oint_\Gamma \frac{\partial g}{\partial S} = \oint_\Gamma \frac{m}{y} \frac{Sf'(S)}{a+S}.$$

Since $S'(t) = \frac{m}{y} \frac{S}{a+S}(f(S) - x)$, then $\frac{S'(t)}{f(S)-x} = \frac{m}{y} \frac{S}{a+S}$. To show

$$\oint_\Gamma \frac{\partial g}{\partial S} + \frac{\partial h}{\partial x} = \oint_\Gamma \frac{\partial g}{\partial S} = \oint_\Gamma \frac{m}{y} \frac{Sf'(S)}{a+S} < 0.$$

By Fig. 6.14, we reflect the arc $\overset{\frown}{BHQ'LP'JA}$ with respect to the line $S = \frac{K-a}{2}$, then

$$\oint_\Gamma \frac{m}{y} \frac{Sf'(S)}{a+S} = \int_{\overline{AP}} + \int_{\overset{\frown}{PRQ}} + \int_{\overline{QB}} + \int_{\overline{BQ'}} + \int_{\overset{\frown}{Q'LP'}} + \int_{\overline{P'A}}.$$

Since $\frac{S'(t)}{f(S)-x} = \frac{m}{y} \frac{S}{a+S}$, then from (6.20), (6.21)

$$\left(\int_{\overline{P'A}} + \int_{\overline{AP}} \right) \frac{m}{y} \frac{Sf'(S)}{a+S}$$

$$= \int_{S_P}^{S_A} \frac{f'(S)}{f(S) - x_1(S)} dS + \int_{S_A}^{S_P} \frac{f'(S)}{f(S) - x_2(S)} dS$$

$$= \int_{S_A}^{S_P} \frac{f'(K - a - S)}{f(K - a - S) - x_1(K - a - S)} dS + \int_{S_A}^{S_P} \frac{f'(S)}{f(S) - x_2(S)} dS$$

$$= \int_{S_A}^{S_P} f'(S) \left[\frac{x_2(S) - x_1(K - a - S)}{(f(S) - x_2(S))(f(S) - x_1(K - a - S))} \right] dS < 0$$

$$\tag{6.22}$$

where $x_2(S) > x_1(K - a - S)$ for $S_A < S < S_P$, and $f'(S) < 0$ for $S_A < S < S_P$.

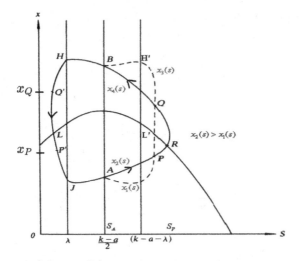

Fig. 6.14

Similarly

$$\left[\int_{\widehat{QB}} + \int_{\widehat{BQ'}}\right] \left(\frac{m}{y}\frac{Sf'(S)}{a+S}\right) dt$$

$$= \int_{S_B}^{S_Q} f'(S) \left[\frac{x_3(K-a-S) - x_4(S)}{(x_4(S) - f(S))(x_3(K-a-S) - f(S))}\right] dS < 0. \tag{6.23}$$

Since

$$x'(t) = x(m-d)\frac{S-\lambda}{a+S}$$

it follows that

$$\frac{1}{m-d}\frac{a+S}{(S-\lambda)x}dx = dt.$$

Let $\bar{\Omega}$ be the region bounded by arc $\widehat{Q'LP'}$ and line segment $\overline{P'Q'}$ and $\bar{\bar{\Omega}}$ be the region bounded by arc $\widehat{PRQ}$ and line segment $\overline{QP}$.

Then from Green's Theorem, we have

$$\int_{\widehat{Q'LP'}} \left(\frac{m}{y} \frac{Sf'(S)}{a+S} \right) dt = \frac{m}{y} \frac{1}{m-d} \int_{\widehat{Q'LP'}} \frac{Sf'(S)}{x(S-\lambda)} dx$$

$$= \frac{m}{y} \frac{1}{m-d} \left[\int_{\widehat{Q'LP'}} + \int_{\overline{P'Q'}} + \int_{\overline{Q'P'}} \frac{Sf'(S)}{x(S-\lambda)} dx \right]$$

$$= \frac{1}{m-d} \iint_{\Omega} -\frac{1}{x} \frac{\frac{\tau}{K}[2(S-\lambda)^2 + \lambda(K-a-2\lambda)]}{(S-\lambda)^2} dSdx$$

$$+ \frac{m}{y} \frac{1}{m-d} \int_{\overline{Q'P'}} \frac{Sf'(S)}{x(S-\lambda)} dx$$

$$< \frac{m}{y} \frac{1}{m-d} \int_{x_{p'}}^{x_{Q'}} \frac{S_1(x)f'(S_1(x))}{x(\lambda - S_1(x))} dx.$$

Similarly, we have

$$\int_{\widehat{PRQ}} \left(\frac{m}{y} \frac{Sf'(S)}{a+S} \right) dt = \frac{1}{m-d} \iint_{\bar{\Omega}} -\frac{1}{x} \frac{\frac{\tau}{K}[2(S-\lambda)^2 + \lambda(K-a-2\lambda)]}{(S-\lambda)^2} dSdx$$

$$+ \frac{m}{y} \frac{1}{m-d} \int_{\overline{PQ}} \frac{Sf'(S)}{x(S-\lambda)} dx$$

$$< \frac{m}{y} \frac{1}{m-d} \int_{x_p}^{x_Q} \frac{S_2(x)f'(S_2(x))}{x(S_2(x)) - \lambda} dx.$$

Then

$$\left(\int_{\widehat{Q'LP'}} + \int_{\widehat{PRQ}} \right) \frac{m}{y} \frac{Sf'(S)}{a+S} dt < \frac{m}{y} \frac{1}{m-d} \int_{x_{p'}}^{x_{Q'}} \frac{S_1(x)f'(S_1(x))}{x(\lambda - S_1(x))} dx$$

$$- \frac{m}{y} \frac{1}{m-d} \int_{x_{p'}}^{x_{Q'}} \frac{(K-a-S_1(x))f'(S_1(x))}{x(K-a-\lambda-S_1(x))} dx$$

$$= \frac{m}{y} \frac{1}{m-d} \int_{x_{p'}}^{x_{Q'}} \frac{f'(S_1(x))}{x} \frac{G(S_1(x))}{(\lambda - S_1(x))(K-a-\lambda-S_1(x))} dx$$

where

$$G(S') = (S-\lambda)(\lambda - S') \left(\frac{S'}{\lambda - S'} - \frac{S}{S-\lambda} \right). \tag{6.24}$$

Let $S = K - a - S'$, and $S' = S_1(x)$, from Lemma 6.2.1 and Lemma 6.2.2 (see below)

$$\begin{aligned} S_1(x) \le max\{S_{P'}, S_{Q'}\} \le S_-, && x_{P'} \le x \le x_{Q'} \\ G(S_1(x)) \le 0, && x_{P'} \le x \le x_{Q'}, \end{aligned} \tag{6.25}$$

then

$$\left(\int_{\widehat{Q'LP'}} + \int_{\widehat{PRQ}} \right) \left(\frac{m}{y} \frac{Sf'(S)}{a+S} \right) dt < 0. \tag{6.26}$$

From (6.22), (6.23), (6.24) it follow that $\oint_\Gamma \frac{m}{y} \frac{Sf'(S)}{a+S} < 0$ and (6.19) follows.

Lemma 6.2.1 From Fig. 6.14, $P'(S_{P'}, x_{P'})$, $Q'(S_{Q'}, x_{Q'})$ are respectively the mirror images of $P(S_P, x_P)$, $Q(S_Q, x_Q)$ with respect to the mirror $S = \frac{K-a}{2}$, then

$$0 < \frac{S_{Q'}}{\lambda - S_{Q'}} \leqslant \frac{S_Q}{S_Q - \lambda} \tag{6.27}$$

and

$$0 < \frac{S_{P'}}{\lambda - S_{P'}} \leqslant \frac{S_P}{S_P - \lambda}. \tag{6.28}$$

Proof. Let $(\frac{dx}{ds})'_Q$ and $(\frac{dx}{ds})_Q$ be the slopes of $\widehat{BH'QL'}$ and $\widehat{BRA}$ at point Q respectively. It is obvious from Fig. 6.14 that

$$0 > \left(\frac{dx}{ds}\right)_Q \geqslant \left(\frac{dx}{ds}\right)'_Q. \tag{6.29}$$

Since

$$\left(\frac{dx}{ds}\right)_Q = \frac{x_Q\left(\frac{mS_Q}{a+S_Q} - d\right)}{rS_Q\left(1 - \frac{S_Q}{K}\right) - \left(\frac{m}{y}\right)\left(\frac{x_Q S_Q}{a+S_Q}\right)}$$

$$= -(m-d)\left(\frac{y}{m}\right)\frac{x_Q(S_Q - \lambda)}{S_Q(x_Q - f(S_Q))}$$

and

$$\left(\frac{dx}{ds}\right)' = -(m-d)\left(\frac{y}{m}\right)\frac{x_{Q'}(\lambda - S_{Q'})}{S_{Q'}(x_{Q'} - f(S_{Q'}))}$$

$$= -(m-d)\left(\frac{y}{m}\right)\frac{x_Q(\lambda - S_{Q'})}{S_{Q'}(x_Q - f(S_Q))}.$$

(Recall that $x_Q = x_{Q'}$, $f(S_Q) = f(S_{Q'})$.) Thus from (6.27), (6.28), we have $0 < \frac{S_{Q'}}{\lambda - S_{Q'}} \leqslant \frac{S_Q}{S_Q - \lambda}$. Similarly we have $0 < \frac{S_{P'}}{\lambda - S_{P'}} \leqslant \frac{S_P}{S_P - \lambda}$. □

Lemma 6.2.2 $S_1(x) \leqslant max\{S_{P'}, S_{Q'}\} \leqslant S_-, x_{P'} \leqslant x \leqslant x_{Q'}$ and $G(S_1(x)) \leqslant 0$ for $x_{P'} \leqslant x \leqslant x_{Q'}$.

Proof. Consider the quadratic function $G(S')$ in (6.24). A straight calculation shows that

$$G(S') = -2(S')^2 + 2S'(K-a) - \lambda(K-a). \tag{6.30}$$

The two positive roots of $G(S') = 0$ are

$$S_\pm = \frac{K-a}{2} \pm \sqrt{\left(\frac{K-a}{2}\right)^2 - \lambda\left(\frac{K-a}{2}\right)}.$$

Hence $G(S') < 0$ if $S' < S_-$ or $S' > S_+$ and $G(S') > 0$ if $S_- < S' < S_+$. Since $S_{Q'} < \frac{K-a}{2} < S_+$, from (6.25), (6.27) we conclude that

$$S_{Q'} \leqslant S_- \quad or \quad S_Q \geqslant S_+. \tag{6.31}$$

Similarly

$$S_{P'} \leqslant S_- \quad or \quad S_P \geqslant S_+. \tag{6.32}$$

Then from (6.27), (6.28) we prove (6.26).   □

Remark 6.2.2 In the following we present another method to prove the uniqueness of limit cycle for van der Pol equation ([HK] p. 379)

$$x'' + \epsilon(x^2 - 1)x' + x = 0, \quad \epsilon > 0.$$

Let

$$x' = y = f(x, y)$$
$$y' = \epsilon(1 - x^2)y - x = g(x, y),$$

and $(x(t), y(t))$, $0 \leq t \leq T$, be a periodic solution. We want to show the periodic orbit $\Gamma = \{(x(t), y(t)) : 0 \leq t \leq T\}$ is orbitally asymptotically stable. It suffices to show

$$\int_0^T \frac{\partial f}{\partial x}(x(t), y(t)) + \frac{\partial g}{\partial y}(x(t), y(t))\, dt = \int_0^T \epsilon\left(1 - x^2(t)\right) dt < 0.$$

Consider

$$V(x, y) = \frac{1}{2}\left(x^2 + y^2\right).$$

Then

$$\dot{V}(x, y) = xx' + yy' = \epsilon\left(1 - x^2\right)y^2 = 2\epsilon\left(1 - x^2\right)\left[V(x, y) - \frac{x^2}{2}\right].$$

Assume $V(x(t), y(t))$ attains the minimum on the periodic orbit Γ at some point, say, $(x(\bar{t}), y(\bar{t}))$. Then $\dot{V}(x(\bar{t}), y(\bar{t})) = 0$ and it follows that either $y(\bar{t}) = 0$ or $x^2(\bar{t}) = 1$. First we prove that it is impossible that $y(\bar{t}) = 0$. Suppose $y(\bar{t}) = 0$ then from uniqueness of ODE, $y'(\bar{t}) \neq 0$, otherwise $y(\bar{t}) = 0$ and $y'(\bar{t}) = 0$ imply $(x(\bar{t}), y(\bar{t})) \equiv (0, 0)$. Hence $y'(\bar{t}) \neq 0$ i.e. $x''(\bar{t}) = y'(\bar{t}) \neq 0$. This says that $x(t)$ has either a maximum or a minimum at $t = \bar{t}$. Then $1 - (x(t))^2$ has a fixed sign for t near $\bar{t}$ and $\dot{V}(x(t), y(t))$ has a constant sign for t near $\bar{t}$. It follows that $V(x(t), y(t))$ is strictly monotone for t near $\bar{t}$. This leads to a contradiction that $V(x(t), y(t))$ attains a minimum at $\bar{t}$.

Hence we have $y(\bar{t}) = x'(\bar{t}) \neq 0$ and $x^2(\bar{t}) = 1$. Then $V(x(t), y(t)) \geq V(x(\bar{t}), y(\bar{t})) = \frac{1}{2}\left(1 + y^2(\bar{t})\right) > \frac{1}{2}$ for all t and

$$-\frac{\dot{V}\left(x(t), y(t)\right)}{V\left(x(t), y(t)\right) - \frac{1}{2}} + 2\epsilon\left(1 - x^2(t)\right) = 2\epsilon\left(1 - x^2(t)\right)\left[-\frac{V - \frac{x^2}{2}}{V - \frac{1}{2}} + 1\right]$$

$$= 2\epsilon\left(1 - x^2(t)\right)\frac{\left(\frac{x^2}{2} - \frac{1}{2}\right)}{V - \frac{1}{2}} = -\frac{\epsilon\left(1 - x^2(t)\right)^2}{V\left(x(t), y(t)\right) - \frac{1}{2}}.$$

Integrating both sides from 0 to T yields

$$0 + 2\epsilon\int_0^T \left(1 - x^2(t)\right) dt = -\epsilon\int_0^T \frac{\left(1 - x^2\right)^2}{V - \frac{1}{2}} dt < 0.$$

Hence $\int_0^T \frac{\partial f}{\partial x} + \frac{\partial g}{\partial y} dt = \epsilon\int_0^T \left(1 - x^2(t)\right) dt < 0$ and we complete the proof.

Relaxation of the van der Pol oscillator:

Consider van der Pol equation with $k \gg 1$

$$u'' - k(1 - u^2)u' + u = 0.$$

Let $\varepsilon = \frac{1}{k}$. Then $0 < \varepsilon \ll 1$ and

$$\varepsilon u' = v - F(u), \qquad (6.33)$$
$$v' = -\varepsilon u,$$

where $F(u) = -u + \frac{u^3}{3}$.
Consider the Jordan curve J as follows (see Fig. 6.15).

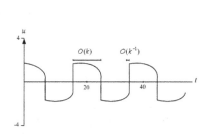

Fig. 6.15   Fig. 6.16

By Theorem 6.2.1, there is a unique limit cycle $\Gamma(\varepsilon)$ of (6.33). If the orbit is away from the isocline $v = F(u)$, then, from (6.24), $u' = \varepsilon^{-1}(v - F(u))$, $v' = -\varepsilon u$. It follows that $|u'| \gg 1$ and $|v'| \approx 0$, and the orbit has a tendency to jump except when it is closed to the curve $v = F(u)$.

Theorem 6.2.2 *As $\varepsilon \downarrow 0$, the limit cycle $\Gamma(\varepsilon)$ approaches the Jordan curve J* [H1].

Proof. We shall construct closed region U containing J such that the distance $\text{dist}(U, J)$ is any preassigned constant and for any $\varepsilon > 0$ sufficiently small, the vector field of (6.33) is directed inward U at the boundary ∂U, i.e. U is a trapped region. By Poincaré-Bendixson Theorem, U contains the limit cycle $\Gamma(\varepsilon)$.

Consider the following Fig. 6.17, due to the symmetry, we only need to check the flow on the following curves.

Fig. 6.17

(1) On the segment $\overline{12}$, the flow is inward since

$$\vec{n} \cdot (\dot{u}, \dot{v}) = (0, 1) \cdot (\dot{u}, \dot{v}) = -\varepsilon u < 0;$$

(2) On the segment $\overline{23}$

$$\vec{n} \cdot (\dot{u}, \dot{v}) = (1, 0) \cdot (\dot{u}, \dot{v}) = \frac{1}{\varepsilon}(v - F(u)) < 0;$$

(3) On the segment $\overline{13, 14}$

$$\vec{n} \cdot (\dot{u}, \dot{v}) = (0, 1) \cdot (\dot{u}, \dot{v}) = -\varepsilon u < 0;$$

(4) On the segment $\overline{11, 12}$

$$\vec{n} \cdot (\dot{u}, \dot{v}) = (-1, 0) \cdot (\dot{u}, \dot{v}) = -\frac{1}{\varepsilon}(v - F(u)) < 0;$$

(5) On the arc $\overparen{12, 13}$,

$$\vec{n} \cdot (\dot{u}, \dot{v}) = (-F'(u), 1) \cdot (0, \dot{v}) = -\varepsilon u < 0;$$

(6) On the arc $\overparen{3, 4}$,

$$\vec{n} \cdot (\dot{u}, \dot{v}) = (F'(u), -1) \cdot (\dot{u}, \dot{v})$$

$$= (u^2 - 1, -1) \cdot \left(\frac{-h}{\varepsilon}, -\varepsilon u\right)$$

$$= (u^2 - 1)\frac{-h}{\varepsilon} + \varepsilon u < 0, \quad \text{if } 0 < \varepsilon \ll 1;$$

(7) On the arc $\overparen{45}$, $n_1 > 0$, $n_2 < 0$

$$\vec{n} \cdot (\dot{u}, \dot{v}) = (n_1, n_2) \cdot \left(\frac{1}{\varepsilon}(v - F(u)), -\varepsilon u\right)$$

$$= n_1 \left(\frac{1}{\varepsilon}(v - F(u))\right) - \varepsilon u n_2 < 0$$

$$\text{if } 0 < \varepsilon \ll 1;$$

(8) On the arc $\overparen{10, 11}$, $n_1 < 0$, $n_2 < 0$,

$$\vec{n} \cdot (\dot{u}, \dot{v}) = n_1 \left(\frac{1}{\varepsilon}(v - F(u))\right) - \varepsilon u n_2 < 0, \quad \text{if } 0 < \varepsilon \ll 1.$$

$\square$

Remark 6.2.3 As $\varepsilon \ll 1$, we may evaluate the period of the periodic solution of the relaxation.

Since the periodic trajectory spends most of the time traveling on the isocline (see Fig. 6.16, Fig. 6.18), we need to consider the equation

$$\frac{dv}{dt} = -\varepsilon u \quad \text{or} \quad dt = -\frac{dv}{\varepsilon u}.$$

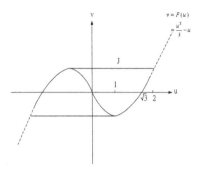

Fig. 6.18

Then

$$T(\varepsilon) \approx 2 \int dt = 2 \int_{u=2}^{u=1} \frac{dv}{-\varepsilon u}$$

$$= -2 \int_1^2 \frac{u^2 - 1}{-\varepsilon u} du = \frac{3 - 2\log 2}{\varepsilon}.$$

Generalized Lienard Equations

Consider generalized Lienard equation [Zh]:

$$\begin{cases} \dfrac{dx}{dt} = -\varphi(y) - F(x) \\ \dfrac{dy}{dt} = g(x). \end{cases} \tag{6.34}$$

Theorem 6.2.3 [Zh]
Assume

(i) $xg(x) > 0$, $x \neq 0$, $G(+\infty) = G(-\infty) = +\infty$. $G(x) = \int_0^x g(\xi)d\xi$.

(ii) $\frac{F'(x)}{g(x)}$ *is non-decreasing on* $(-\infty, 0), (0, +\infty)$, $\frac{F'(x)}{g(x)} \neq$ *constant in neighborhood of* $x = 0$.

(iii) $y\varphi(y) > 0$, $y \neq 0$, $\varphi(y)$ *is non-decreasing.* $\varphi(-\infty) = -\infty$, $\varphi(+\infty) = +\infty$. $\varphi(y)$ *has right and left derivative at* $y = 0$ *which are non-zero in case* $F'(0) = 0$.

Then the system (6.34) has at most one limit cycle.

We state the following Theorem 6.2.3 without proof.

In 1988, Kuang Yang and Freedman prove uniqueness of limit cycle for Gauss-type Predator-Prey System by reducing it to generalized Lienard equation [KF].

Consider Predator-Prey System:

$$\begin{cases} \dfrac{dx}{dt} = \phi(x)(F(x) - \pi(y)) \\ \dfrac{dy}{dt} = \rho(y)\psi(x). \end{cases} \tag{6.35}$$

For example, Gauss-type predator-prey system satisfies

$$\begin{cases} \pi(y) = y, \phi(x) = p(x). \\ F(x) = \dfrac{xg(x)}{p(x)} = prey\ isocline \\ \rho(y) = y, \psi(x) = p(x) - p(x^*). \end{cases}$$

Theorem 6.2.4 *Assume the system (6.35) satisfies*

(i) $\varphi(0) = \pi(0) = \rho(0) = 0.$ $\phi'(x) > 0,$ $\psi'(x) > 0$ *for* $x > 0,$ $\rho'(y) > 0,$ $\pi'(y) > 0$ *for* $y > 0,$ $\pi(+\infty) = +\infty.$

(ii) $\psi(x^*) = 0,$ $k > x^*,$ $F(K) = 0,$ $(x - K)F(x) < 0,$ $\forall x \neq K.$

(iii) $\dfrac{-F'(x)\phi(x)}{\psi(x)}$ *is non-decreasing for* $-\infty < x < x^*,$ $x^* < x < +\infty.$

Then there is at most one limit cycle for the system (6.35).

Proof. Let (x^*, y^*) be the equilibrium.
Consider change of variable $(x, y) \to (u, v).$

$$x = \xi(u) + x^*$$
$$y = \eta(v) + y^*$$

where $\xi(u), \eta(v)$ are to be determined.
Then

$$\begin{cases} \dfrac{du}{dt} = \dfrac{1}{\xi'(u)}\phi(\xi(u) + x^*)(F(\xi(u) + x^*) - \pi(\eta(v) + y^*)) \\ \dfrac{dv}{dt} = \dfrac{1}{\eta'(v)}\rho(\eta(v) + y^*)\psi(\xi(u) + x^*). \end{cases}$$

Let

$$\begin{cases} \xi'(u) = \phi(\xi(u) + x^*) \\ \xi(0) = 0 \end{cases}$$

and

$$\begin{cases} \eta'(v) = \rho(\eta(v) + y^*) \\ \eta(0) = 0. \end{cases}$$

Write

$$\frac{du}{dt} = -[\pi(\eta(v) + y^*) - y^*] - [-F(\xi(u) + x^*) + y^*]$$

$$= -\Phi(v) - A(u)$$

$$\frac{dv}{dt} = \psi(\xi(u) + x^*) = g(u).$$

Check: The conditions in Theorem 6.2.3.

Since $\xi(0) = 0$, $\xi'(u) = \phi(\xi(u) + x^*) > 0$, we have $\xi(u) > 0$ for $u > 0$. Thus u and $\xi(u)$ have the same sign. $ug(u) > 0$ if and only if $\xi(u)g(u) > 0$. Since $(x - x^*)\psi(x) > 0$ for all x,

$$\xi(u)g(u) = \xi(u)\psi(\xi(u) + x^*) = ((\xi(u) + x^*) - x^*)\psi(\xi(u) + x^*) > 0$$

and consequently $ug(u) > 0$.

$$G(u) = \int_0^u g(s)ds \to +\infty \quad as \ |u| \to +\infty.$$

$$v\Phi(v) = v(II(\eta(v) + y^*) - y^*) > 0, v \neq 0.$$

Check: $\frac{F'(u)}{g(u)}$ is non-decreasing on $(-\infty, 0), (0, +\infty)$.

$$A'(u) = -F'(\xi(u) + x^*)\xi'(u) = -F'(\xi(u) + x^*)\phi(\xi(u) + x^*).$$

$$\frac{A'(u)}{g(u)} = \frac{-F'(\xi(u) + x^*)\phi(\xi(u) + x^*)}{\psi(\xi(u) + x^*)}$$

is non-decreasing for $-\infty < u < 0, 0 < u < +\infty$, since $\frac{-F'(x)\phi(x)}{\psi(x)}$ in nondecreasing in $x, -\infty < x < x^*, x^* < x < +\infty$. $\therefore$ (iii) holds.   □

Example 6.2.2 Prove uniqueness of limit cycle for Rosenzweig-McArthur model (Example 6.1.4).

Here

$$F(x) = \frac{r}{m}(a + x)\left(1 - \frac{x}{K}\right) \text{ be the prey isocline.}$$

$$\phi(x) = \frac{mx}{a + x} = p(x)$$

$$\psi(x) = \frac{mx}{a + x} - d = (m - d)\frac{x - x^*}{a + x}.$$

Show that $\frac{-F'(x)\phi(x)}{\psi(x)}$ is nondecreasing on $(-\infty, x^*), (x^*, +\infty)$.
Compute the following:

Proof. $F'(x) = -\frac{r}{m}\frac{2}{K}\left(\frac{K-a}{2} - x\right)$, $\frac{\phi(x)}{\psi(x)} = \frac{m}{m-d}\frac{x}{x-x^*}$,

$$\left(-F'(x)\frac{\phi(x)}{\psi(x)}\right)' = -\frac{r}{m}\frac{2}{K}\frac{m}{m-d}\left(\left(\frac{K-a}{2} - x\right)\left(\frac{x}{x-x^*}\right)\right)' > 0$$

$$\Leftrightarrow \left(\left(\frac{K-a}{2} - x\right)\left(\frac{x}{x-x^*}\right)\right)' < 0 \text{ for } x \neq x^*$$

$$\Leftrightarrow \left(\frac{K-a}{2} - x\right)\frac{(-x^*)}{(x-x^*)^2} + (-1)\frac{x}{x-x^*} < 0 \text{ for } x \neq x^*$$

$$\Leftrightarrow \left(\frac{K-a}{2} - x\right)(-x^*) - (x - x^*)x < 0 \text{ for } x \neq x^*$$

$$\Leftrightarrow \frac{K-a}{2}(-x^*) - x^2 + 2xx^* < 0 \text{ for } x \neq x^*$$

$$\Leftrightarrow \frac{K-a}{2}(-x^*) + (x^*)^2 < (x - x^*)^2 \text{ for } x \neq x^*$$

$$\Leftrightarrow \left(-\frac{K-a}{2} + x^*\right)x^* < (x - x^*)^2 \text{ for } x \neq x^*$$

since $0 < x^* < \frac{K-a}{2}$, the above holds.

Hence we complete the proof of uniqueness of limit cycle by Theorem 6.2.4.

□

6.3 Hopf Bifurcation

Consider one-parameter family of differential equations

$$x' = f(x, \mu) = f_\mu(x) \tag{H_μ}$$

with $x \in \mathbb{R}^2$, $\mu \in \mathbb{R}$.

Let $f(x(\mu), \mu) = 0$, i.e., $x(\mu)$ is an equilibrium. As μ varies, we assume the stability property of the equilibrium $x(\mu)$ changes from asymptotic stability to complete instability at $\mu = \mu_0$, i.e., the variational matrix $D_x f(x(\mu), \mu)$ with eigenvalues $\alpha(\mu) \pm i\beta(\mu)$ satisfies

$$\alpha(\mu) < 0 \ (\alpha(\mu) > 0) \text{ for } \mu < \mu_0, \ \alpha(\mu_0) = 0, \ \beta(\mu_0) \neq 0$$
$$\text{and } \alpha(\mu) > 0 \ (\alpha(\mu) < 0) \text{ for } \mu > \mu_0.$$

The Hopf Bifurcation Theorem says that the equation (H_μ) has a one-parameter family of nontrivial periodic solution $x(t, \epsilon)$ with period $T(\epsilon) = \frac{2\pi}{\beta(\mu_0)} + O(\epsilon)$, $\mu = \mu(\epsilon)$ and $\mu_0 = \mu(0)$.

Let $\tilde{x} = x - x(\mu)$, $\tilde{\mu} = \mu - \mu_0$,

$$\tilde{x}' = f(\tilde{x} + x(\tilde{\mu} + \mu_0), \tilde{\mu} + \mu_0) = F(\tilde{x}, \tilde{\mu}).$$

Hence we assume (H_μ) satisfies

(H1) $f(0, \mu) = 0$, for all μ;

(H2) The eigenvalues of $Df_\mu(0)$ are $\alpha(\mu) \pm i\beta(\mu)$ with $\alpha(0) = 0$, $\beta(0) = \beta_0 \neq 0$ and $\alpha'(0) \neq 0$, so the eigenvalues are crossing the imaginary axis.

Example 6.3.1 Consider the predator-prey system

$$\begin{cases} x' = \gamma x \left(1 - \frac{x}{K}\right) - \frac{mx}{a+x} y, \\ y' = \left(\frac{mx}{a+x} - d\right) y. \end{cases}$$

Hopf Bifurcation occurs at $\lambda_0 = \frac{K-a}{2}$, where $\lambda = \frac{a}{(m/d)-1}$ is the parameter. In order to prove Hopf-Bifurcation Theorem, we need to introduce the concept of "normal form".

Normal form for Hopf Bifurction

Consider (H_μ) under the assumption (H1), (H2), with the form

$$\begin{pmatrix} x' \\ y' \end{pmatrix} = \begin{pmatrix} \alpha(\mu) & -\beta(\mu) \\ \beta(\mu) & \alpha(\mu) \end{pmatrix} \begin{pmatrix} x \\ y \end{pmatrix} + \begin{pmatrix} f^1(x, y, \mu) \\ f^2(x, y, \mu) \end{pmatrix}.$$

With the scaling $\tau = \frac{t}{\beta(\mu)}$ and let $\tilde{\mu} = \frac{\alpha(\mu)}{\beta(\mu)}$, we may assume $\alpha(\mu) = \mu$, $\beta(\mu) = 1$. Then we have

$$\begin{aligned} x' &= \mu x - y + \sum a_{mn} x^m y^n, \\ y' &= x + \mu y + \sum b_{mn} x^m y^n, \end{aligned} \qquad m + n \geq 2. \tag{6.36}$$

Since $\mu \pm i$ are eigenvalues of $\begin{pmatrix} \mu & -1 \\ 1 & \mu \end{pmatrix}$, we set

$$u = x + iy, \ v = x - iy.$$

Then we have

$$
\begin{aligned}
u' &= (\mu + i)u + \sum A_{mn}u^m v^n, \\
v' &= (\mu - i)v + \sum B_{mn}u^m v^n.
\end{aligned}
\tag{6.37}
$$

Claim $B_{mn} = \overline{A_{nm}}$.

Since $\bar{u} = v$, then

$$
\begin{aligned}
v' = \bar{u}' &= (\mu - i)v + \sum \overline{A_{mn}}\,\bar{u}^m \bar{v}^n \\
&= (\mu - i)v + \sum \overline{A_{mn}}\,v^m u^n.
\end{aligned}
$$

Hence $B_{mn} = \overline{A_{nm}}$.

Now we look for the following change of variables

$$
\begin{aligned}
u &= \xi + \sum \alpha_{mn}\xi^m \eta^n, \\
v &= \eta + \sum \beta_{mn}\xi^m \eta^n.
\end{aligned}
\tag{6.38}
$$

Write

$$
\begin{aligned}
\xi' &= (\mu + i)\xi + \sum A'_{mn}\xi^m \eta^n, \\
\eta' &= (\mu - i)\eta + \sum B'_{mn}\xi^m \eta^n.
\end{aligned}
\tag{6.39}
$$

Now we substitute (6.38) into (6.37) and use (6.39),

$$
\begin{aligned}
u' &= \xi' + \sum \alpha_{mn}\left(m\xi'\xi^{m-1}\eta^n + n\xi^m \eta^{n-1}\eta'\right) \\
&= (\mu + i)\left(\xi + \sum \alpha_{mn}\xi^m \eta^n\right) \\
&\quad + \sum A_{\ell k}\left(\xi + \sum \alpha_{mn}\xi^m \eta^n\right)^\ell \left(\eta + \sum \beta_{mn}\xi^m \eta^n\right)^k
\end{aligned}
\tag{6.40}
$$

$$
\begin{aligned}
&= \left(1 + \sum \alpha_{mn}m\xi^{m-1}\eta^n\right)\xi' + \left(\sum \alpha_{mn}n\xi^m \eta^{n-1}\right)\eta' \\
&= \left(1 + \sum \alpha_{mn}m\xi^{m-1}\eta^n\right)\left((\mu + i)\xi + \sum A'_{mn}\xi^m \eta^n\right) \\
&\quad + \left(\sum \alpha_{mn}n\xi^m \eta^{n-1}\right)\left((\mu - i)\eta + \sum B'_{mn}\xi^m \eta^n\right).
\end{aligned}
\tag{6.41}
$$

Compare the coefficients $\xi^m \eta^n$, $m + n = 1, 2$ in (6.40) and (6.41). For $m + n = 1$, we have

$$(\mu + i)\xi = (\mu + i)\xi.$$

For $m + n = 2, 3$

$$(\mu + i)\alpha_{mn}\xi^m\eta^n + A_{mn}\xi^m\eta^n \tag{6.42}$$
$$= \alpha_{mn}m(\mu + i)\xi^m\eta^n + A'_{mn}\xi^m\eta^n + \alpha_{mn}n(\mu - i)\xi^m\eta^n.$$

We want $A'_{mn} = 0$ as many terms as possible. From (6.42),

$$A'_{mn} = A_{mn} - \alpha_{mn}\left((\mu + i)(m - 1) + (\mu - i)n\right). \tag{6.43}$$

If $A'_{mn} = 0$, then

$$\alpha_{mn} = \frac{A_{mn}}{(\mu + i)(m - 1) + (\mu - i)n}. \tag{6.44}$$

For $m + n = 2$, we obviously have $m - n - 1 \neq 0$. For $\mu = 0$ or μ near 0 we choose α_{mn} as in (6.44) and hence we are able to kill all quadratic terms in (6.39). Now for $m + n = 3$, $\mu = 0$ or μ near 0, we want to check $m - n - 1 \neq 0$ or not. Then we find that except $m = 2$, $n = 1$, we are able to kill all other cubic terms. For $m = 2$, $n = 1$

$$\alpha_{21} = \frac{A_{21}}{3\mu}.$$

Then (6.39) becomes

$$\xi' = (\mu + i)\xi + A'_{21}\xi^2\eta + \text{4th-order terms},$$
$$\eta' = (\mu - i)\eta + \overline{A'_{21}}\xi\eta^2 + \text{4th-order terms}.$$

The last change of variables is

$$\xi = \rho e^{i\varphi}, \quad \eta = \rho e^{-i\varphi}.$$

Then

$$\begin{pmatrix} \xi' \\ \eta' \end{pmatrix} = \begin{pmatrix} e^{i\varphi} & i\rho e^{i\varphi} \\ e^{-i\varphi} & -i\rho e^{-i\varphi} \end{pmatrix} \begin{pmatrix} \rho' \\ \varphi' \end{pmatrix}$$

or

$$\begin{pmatrix} \rho' \\ \varphi' \end{pmatrix} = \begin{pmatrix} e^{i\varphi} & i\rho e^{i\varphi} \\ e^{-i\varphi} & -i\rho e^{-i\varphi} \end{pmatrix}^{-1} \begin{pmatrix} \xi' \\ \eta' \end{pmatrix}$$
$$= \frac{1}{(-2i\rho)} \begin{bmatrix} -i\rho e^{-i\varphi}, & -i\rho e^{i\varphi} \\ -e^{-i\varphi}, & e^{i\varphi} \end{bmatrix} \begin{bmatrix} (\mu + i)\rho e^{i\varphi} + A'_{21}\rho^3 e^{i\varphi} + \cdots \\ (\mu - i)\rho e^{-i\varphi} + \overline{A'_{21}}\rho^3 e^{-i\varphi} + \cdots \end{bmatrix}$$
$$= \begin{bmatrix} \mu\rho + G\rho^3 + \cdots \\ 1 + (Im A'_{21})\rho^2 + \cdots \end{bmatrix}.$$

Hence we have

$$\dot{\rho} = \mu\rho + G\rho^3 + O(\rho^4), \qquad (6.45)$$
$$\dot{\varphi} = 1 + K\rho^2 + O(\rho^3).$$

Neglecting the higher order terms in (6.45) we have

$$\dot{\rho} = \mu\rho + G\rho^3, \qquad (6.46)$$
$$\dot{\varphi} = 1 + O(\rho^2).$$

Let $G \neq 0$. If $\mu/G < 0$, then $\dot{\rho} = \rho G(\rho + \sqrt{-\mu/G})(\rho - \sqrt{-\mu/G})$. If $G < 0$, $\mu > 0$, then $(0,0)$ is an unstable spiral and $\rho = \sqrt{-\mu/G}$ is asymptotically stable (see Fig. 6.19).

Fig. 6.19

This is the case of supercritical Hopf Bifurcation. If $G > 0$, $\mu < 0$ then $(0,0)$ is asymptotically stable and the limit cycle $\rho = \sqrt{-\mu/G}$ is unstable (see Fig. 6.20).

Fig. 6.20

This is the case of subcritical Hopf Bifurcation. When $G = 0$, we call the Hopf Bifurcation degenerated.

Now we return to (6.46). Consider the annulus $A = \{(\rho, \varphi) : \rho_1 \le \rho \le \rho_2\}$ where $\rho_1 < \sqrt{-\mu/G} < \rho_2$, ρ_1, ρ_2 are sufficiently close to $\sqrt{-\mu/G}$. Then, for $G < 0$, the annulus is a trapped region for μ sufficiently small. For $G > 0$, the annulus region A is a trapped region in reverse time. Apply Poincaré-Bendixson Theorem, we obtain a closed orbit.

Remark 6.3.1 Return to original parameter,

$$\dot{\rho} = \alpha(\mu)\rho + G\rho^3 + O(\rho^4),$$
$$\dot{\theta} = \beta(\mu) + O(\rho^2).$$

Hence, for μ sufficiently small, from the scaling $\tau = \frac{t}{\beta(\mu)}$, the period of the closed orbit is $2\pi/\beta_0 + O(\mu)$.

Remark 6.3.2 To check the supercritical or subcritical Hopf Bifurcation, we need to evaluate G. It is a complicated matter. According to [WHK], if we have

$$\begin{pmatrix} \dot{x} \\ \dot{y} \end{pmatrix} = \begin{pmatrix} 0 & -\beta_0 \\ \beta_0 & 0 \end{pmatrix} \begin{pmatrix} x \\ y \end{pmatrix} + \begin{pmatrix} f(x,y,0) \\ g(x,y,0) \end{pmatrix}$$

at $\mu = 0$, then the formula is

$$G = \frac{1}{16} \left(f_{xxx} + f_{xyy} + g_{xxy} + g_{yyy} \right)$$
$$+ \frac{1}{16\beta_0} \left(f_{xy} \left(f_{xx} + f_{yy} \right) - g_{xy} \left(g_{xx} + g_{yy} \right) - f_{xx}g_{xx} + f_{yy} \cdot g_{yy} \right)$$

where all partial derivatives are evaluated at the bifurcation point $(0, 0, 0)$.

Example 6.3.2 $x'' + \mu x' + \sin x = 0$ has degenerated Hopf bifurcation at $\mu = 0$.

Example 6.3.3 Consider the following system

$$x_1' = x_2 + F(\lambda, r^2)x_1,$$
$$x_2 = -x_1 + F(\lambda, r^2)x_2,$$

where $r^2 = x_1^2 + x_2^2$. Then the polar equations are

$$\frac{dr}{dt} = F(\lambda, r^2)r,$$
$$\frac{d\theta}{dt} = 1.$$

Case 1: If $F(\lambda, r^2) \equiv \lambda$, then $r(t) = ae^{\lambda t}$. There is no nontrivial periodic orbit except $\lambda = 0$.

If $\lambda < 0$, then $(0,0)$ is stable equilibrium.

If $\lambda > 0$ then $(0,0)$ is a repeller.

If $\lambda = 0$ then $r(t) \equiv a > 0$ is a periodic orbit for any $a > 0$.

This is degenerated Hopf bifurcation.

Case 2: If $F(\lambda, r^2) = \lambda - r^2$ then

$$\frac{dr}{dt} = r(\lambda - r^2).$$

If $\lambda < 0$ then $(0,0)$ is stable equilibrium.

If $\lambda > 0$ then $r(t) \equiv \sqrt{\lambda}$ is a stable limit cycle.

This is supercritical Hopf bifurcation. The Hopf bifurcation occurs at $\lambda = 0$.

Case 3: If $F(\lambda, r^2) = -(r^2 - c)^2 + c^2 + \lambda$ where $c > 0$, then

(i) $\lambda < -c^2$

Then $F(\lambda, r^2) < 0$ and it follows that $(0,0)$ is a stable equilibrium and $r(t) \to 0$ as $t \to \infty$.

(ii) $-c^2 < \lambda < 0$

Then $\lambda + c^2 > 0$, $\lambda < 0$.

$$\frac{dr}{dt} = \left[\sqrt{\sqrt{c^2 + \lambda} + c} + r\right]\left[\sqrt{\sqrt{c^2 + \lambda} + c} - r\right]$$
$$\cdot \left[r - \sqrt{c - \sqrt{c^2 + \lambda}}\right]\left[r + \sqrt{c - \sqrt{c^2 + \lambda}}\right] r.$$

It follows that $(0,0)$ is stable; $r = \sqrt{c - \sqrt{c^2 + \lambda}}$ is a unstable limit cycle; $r = \sqrt{\sqrt{c^2 + \lambda} + c}$ is a stable limit cycle. This is a subcritical Hopf bifurcation. The bifurcation occurs at $\lambda = -c^2$.

(iii) $\lambda > 0$

$$\frac{dr}{dt} = \left[\sqrt{\sqrt{c^2 + \lambda} + c} - r\right]\left[\sqrt{\sqrt{c^2 + \lambda} + c} + r\right]\left[r^2 + (\sqrt{c^2 + \lambda} - c)\right] r$$

$r = \sqrt{\sqrt{c^2 + \lambda} + c}$ is a stable limit cycle.

Hopf Bifurcation in $\mathbb{R}^n$.

Consider the autonomous system

$$x' = f(x, \mu), \quad x \in \mathbb{R}^n \tag{6.47}$$

under the assumption that $f \in C^2(\mathbb{R}^n \times \mathbb{R}, \mathbb{R}^n)$ and

$$f(0, \mu) \equiv 0. \tag{6.48}$$

From (6.48), the system (6.47) possesses for any $\mu \in \mathbb{R}$ the trivial solution $x(t) \equiv 0$ and we want to find the possible values of μ where there is a branching off nonconstant periodic solution of (6.47). Let $A_\mu := Df_x(0, \mu)$. We shall assume there exists μ_0 such that, A_{μ_0}, satisfies

(H1) A_{μ_0} is non-singular and has a pair of simple purely imaginary eigenvalues $\pm i\omega_0$, $\omega_0 > 0$.
(H2) A_{μ_0} has no other eigenvalues of the form $\pm ik\omega_0$, $k \in \mathbb{N}$, $k \neq 1$.
(H3) $\alpha'(\mu_0) \neq 0$ where the eigenvalues of A_μ for μ near μ_0 are $\alpha(\mu) \pm i\beta(\mu)$, $\beta(\mu_0) = \omega_0$.

In the following, we state without proof the Hopf Bifurcation Theorem in $\mathbb{R}^n$ ([AP], p. 144).

Theorem 6.3.1 Let $f \in C^2(\mathbb{R}^n \times \mathbb{R}, \mathbb{R}^n)$ be such that $f(0, \mu) = 0$ for all $\mu \in \mathbb{R}$ and suppose that for $\mu = \mu_0$ (H1)–(H3) hold. Then from μ_0 there bifurcates a branch of periodic solutions of (6.47) with period close to $\frac{2\pi}{\omega_0}$.

More precisely, there exist a neighborhood I of $s = 0$, function $\omega(s)$, $\mu(s) \in C'(I)$ and a family x_s of non-constant, periodic solutions of (6.47) such that

(i) $\omega(s) \to \omega_0$, $\mu(s) \to \mu_0$ as $s \to s_0$,

(ii) x_s has period $T_s = \frac{2\pi}{w(s)}$,

(iii) the amplitude of the orbit x_s tends to 0 as $s \to 0$.

Example 6.3.4 Lorenz equation ([Str] p. 136)

$$\begin{aligned}
x' &= \sigma(y - x), \\
y' &= rx - y - xz, \quad \sigma, \ r, \ b > 0 \\
z' &= xy - bz
\end{aligned} \tag{6.49}$$

or the system (6.49), the equilibrium $E_0 = (0, 0, 0)$ always exists. If $r > 1$ then we have two equilibria,

$$\begin{aligned}
E_+ &= (\sqrt{b(r-1)}, \sqrt{b(r-1)}, r-1) = (x^*, y^*, z^*), \\
E_- &= (-\sqrt{b(r-1)}, -\sqrt{b(r-1)}, r-1) = (-x^*, -y^*, z^*).
\end{aligned}$$

It is easy to show E_0 is a stable node if $0 < r < 1$. When $r > 1$, E_0 is a saddle point and E_+, E_- are stable if $1 < r < r_H = \sigma\left(\frac{\sigma+b+3}{\sigma-b-1}\right)$ provided $\sigma > b + 1$. Let r be the bifurcation parameter and B^+, B^- be the bifurcation branches of E^+, E^- respectively.

We note that it can be shown that Hopf bifurcation occurs at $r = r_H$ and the bifurcation is subcritical, i.e., for r slightly less than r_H, the limit cycles are unstable and exist only for $r < r_H$. For $r < r_H$. the phase portrait near B^+ is shown schematically in Fig. 6.21.

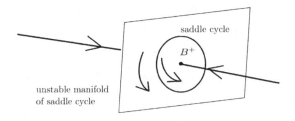

Fig. 6.21

The fixed point is stable. It is encircled by a saddle cycle, a new type of unstable limit cycle that is possible only in the phase space of three or more dimensions. The cycle has a two-dimensional unstable manifold (the sheet in Fig. 6.21), and a two-dimensional stable manifold. As $r \to r_H$ from below, the cycle shrinks down around the fixed point. At the Hopf bifurcation, the fixed point absorbs the saddle cycle and changes into a saddle point. For $r > r_H$ there are no attractors in the neighborhood (see Fig. 6.22).

Fig. 6.22

For the details related to chaotic behavior of the solutions of Lorenz equations (6.49), the reader may consult Section 9.3 in [Str].

6.4   Exercises

Exercise 6.1 Write the van der Pol equation $x'' + \alpha(x^2 - 1)x' + x = 0$ in the form

$$\begin{cases} x' = y, \\ y' = -x - \alpha(x^2 - 1)y. \end{cases} \tag{1}$$

Compare (1) with auxiliary system

$$\begin{cases} x' = y \\ y' = -x - \alpha \, sgn(|x| - \sqrt{2})y. \end{cases} \tag{2}$$

Show that system (2) has a unique limit cycle C and the vector field of (1) is directed inside C.

Exercise 6.2 Is it possible to have a two-dimensional system such that each orbit in an annulus is a periodic orbit and yet the boundaries of the annulus are limit cycles?

Exercise 6.3 Show that the system of equations

$$x' = x - xy^2 + y^3,$$
$$y' = 3y - yx^2 + x^3,$$

has no nontrivial periodic orbit in the region $x^2 + y^2 \leq 4$.

Exercise 6.4 Show that there exist $\mu \neq 0$ for which the system

$$\dot{x} = \mu x + y + xy - xy^2,$$
$$\dot{y} = -x + \mu y - x^2 - y^3$$

has periodic solutions.

Exercise 6.5 The second-order equation $y'' + (y')^3 - 2\lambda y' + y = 0$ where λ is a small scalar parameter, arises in the theory of sound and is known as Rayleigh's equation. Convert this into a first-order system and investigate the Hopf bifurcation.

Exercise 6.6 Investigate the Hopf bifurcation for the following predator-prey system

$$x' = rx\left(1 - \frac{x}{K}\right) - \frac{mx}{a+x}y,$$
$$y' = sy\left(1 - \frac{y}{\nu x}\right), \quad r, K, m, a, s, \nu > 0.$$

Exercise 6.7 Consider the two-dimensional system $x' = Ax - r^2 x$ where A is a 2×2 constant real matrix with complex eigenvalues $\alpha \pm iw$ and $x \in \mathbb{R}^2$, $r = \| x \|$. Prove that there exists at least one limit cycle for $\alpha > 0$ and that none for $\alpha < 0$.

Exercise 6.8 Show that the planar system

$$x' = x - y - x^3,$$
$$y' = x + y - y^3,$$

has a periodic orbit in some annulus region centered at the origin.

Exercise 6.9 Is it true that the solution $(x(t), y(t))$ of

$$x' = -x + ay + x^2 y,$$
$$y' = b - ay - x^2 y,$$

is bounded for any initial condition $x(0) > 0$, $y(0) > 0$?

Exercise 6.10 Consider Example 6.1.4. Show that, if $\frac{K-a}{2} < \lambda$, then (x^*, y^*) is globally asymptotically stable in $Int\left(\mathbb{R}_+^2\right)$. Hint: Apply Dulac's criterion with $h(S, x) = (a + S)^\alpha x^\beta$ for $\alpha, \beta \in \mathbb{R}$ to be determined.

Exercise 6.11 Consider the differential equations

$$\frac{dx}{dt} = a - x - \frac{4xy}{1 + x^2},$$
$$\frac{dy}{dt} = bx\left(1 - \frac{y}{1 + x^2}\right),$$

for $a, b > 0$.

(1) Show that $x^* = \frac{a}{5}$, $y^* = 1 + (x^*)^2$ is the only equilibrium point.
(2) Show that the equilibrium is repelling for $b < \frac{3a}{5} - \frac{25}{a}$, $a > 0$.
(3) Let x_1 be the value of x where the isocline $\{\dot{x} = 0\}$ crosses the x-axis. Let $y_1 = 1 + x_1^2$. Prove that the rectangle $\{(x, y) : 0 \le x \le x_1, 0 \le y \le y_1\}$ is positively invariant.

(4) Prove that there is a periodic orbit in the first quadrant for $a > 0$, $0 < b < \frac{3a}{5} - \frac{25}{a}$.

Exercise 6.12 Show that subcritical Hopf bifurcation occurs at $\mu = 0$ for the equations

$$\dot{r} = \mu r + r^3 - r^5,$$
$$\dot{\theta} = w + br^2.$$

Exercise 6.13 Consider a model of the cell division cycle

$$\frac{du}{dt} = b(v - u)(\alpha + u^2) - u,$$
$$\frac{dv}{dt} = c - u,$$

where $b \gg 1$ and $\alpha \ll 1$ are fixed and satisfy $8\alpha b < 1$, $c > 0$. Show that the system exhibits relaxations for $c_1 < c < c_2$, where c_1 and c_2 are to be determined approximately.

Exercise 6.14 Consider the equation $x'' + \mu(|x| - 1)x' + x = 0$. Find the approximate period of the limit cycle for $\mu \gg 1$.

Exercise 6.15 Let Ω be an annulus in $\mathbb{R}^2$. Assume the hypothesis of Dulac's criterion. Using Green's theorem, show that there exists at most one closed orbit in Ω.

Exercise 6.16 (Predator-prey system with Beddington-DeAngles type [Hw1; Hw2])
Let $x(t)$, $y(t)$ be population of prey and predator

$$\frac{dx}{dt} = rx\left(1 - \frac{x}{K}\right) - \frac{mxy}{a + x + by},$$
$$\frac{dy}{dt} = \left(\frac{emx}{a + x + by} - d\right)y,$$
$$x(0) > 0, \ y(0) > 0,$$

where the parameters r, K, a, b, m, e, $d > 0$

(1) Find the conditions for the existence of limit cycle.
(2) Find the conditions for the global stability of interior equilibrium (x^*, y^*).

Exercise 6.17 Show that the system

$$x' = y,$$
$$y' = -x + y(1 - x^2 - 2y^2),$$

is positively invariant in the annulus $\frac{1}{2} < x^2 + y^2 < 1$ and there exists at least one periodic solution in the annulus.

Exercise 6.18 Show that there is no periodic orbit for the quadratic system

$$x' = -y + x^2 - xy,$$
$$y' = x + xy,$$

by Dulac criterion with function $h(x, y) = (1 + y)^{-3}(-1 - x)^{-1}$.

Exercise 6.19 Verify the stability properties of E_0, E_+, E_- in Example 6.3.4.

Exercise 6.20 Verify that there exists a periodic orbit for the system

$$\begin{cases} \dot{x} &= -z + (1 - x^2 - z^2)x, \\ \dot{y} &= x - z + y, \\ \dot{z} &= x. \end{cases}$$

Exercise 6.21 Given a smooth autonomous ordinary differential equation on the plane, prove that for a non-periodic point when the intersection of its alpha-limit set and omega-limit set is not empty, it contains only fixed points.

Exercise 6.22 Consider the two-dimensional system:

$$\dot{x} = px + y - 2x + x^2 + y^2,$$
$$\dot{y} = py + x - 2y + 2xy.$$

(1) Find those real values of p such that the origin is a hyperbolic fixed point. For each such p, determine the type and stability of the origin.
(2) For those values of p such that the origin is *not* a hyperbolic fixed point, sketch the global phase portrait of the system.

Exercise 6.23 Consider the system:

$$\dot{x} = x - x\sqrt{x^2 + y^2} - y\sqrt{x^2 + y^2} + xy,$$
$$\dot{y} = y - y\sqrt{x^2 + y^2} + x\sqrt{x^2 + y^2} - x^2.$$

Observe that the positive x-axis is invariant. Find the w-limit set $w(p)$ for every point p in the plane.

Exercise 6.24 Prove the uniqueness of limit cycle for the predator-prey system [Hxc]:

If $0 < \lambda < \left(\frac{K-a}{x}\right)$, where $(x^*)^n = \lambda = \frac{a}{(\frac{m}{d})-1}$

$$x' = rx\left(1 - \frac{x}{K}\right) - \frac{mx^n}{a + x^n}y$$
$$y' = \left(\frac{mx^n}{a + x^n} - d\right)y$$

for $n = 2$.

Exercise 6.25

(1) For the Lorenz equations, show that the characteristic equation for the eigenvalues of the Jacobian matrix at C^+, C^- is

$$\lambda^3 + (\sigma + b + 1)\lambda^2 + (r + \sigma)b\lambda + 2b\sigma(r - 1) = 0.$$

(2) By seeking the solution of the form $\lambda = i\omega$ where ω is real, show that there is a pair of pure imaginary eigenvalues when $r = r_H = \sigma(\frac{\sigma+b+3}{\sigma-b-1})$. Explain why we need to assume $\sigma > b + 1$.

(3) Find the third eigenvalue.

Exercise 6.26 Consider a C^1 system $\dot{x} = f(x)$ in $\mathbb{R}^2$. Suppose Γ is a periodic orbit. Let us denote $w(x)$ the $w-$limit set of x. Prove that the set of points x which do not belong to Γ and satisfy $w(x) = \Gamma$ is open.

Chapter 7

SECOND ORDER LINEAR EQUATIONS

7.1 Sturm's Comparison Theorem and Sturm-Liouville Boundary Value Problem

Consider the general second order linear equation

$$u'' + g(t)u' + f(t)u = h(t), \ a \le t \le b. \tag{7.1}$$

Multiplying $\exp(\int_0^t g(s)ds) \equiv p(t)$ on both sides of (7.1) yields

$$(p(t)u')' + q(t)u = H(t), \tag{7.2}$$

with $p(t) > 0$, $p(t)$ and $q(t)$ are continuous on $[a, b]$. (7.3)

The advantage of the form (7.2) over (7.1) is that the linear operator $Lu = (p(t)u')' + q(t)u$ is self-adjoint (see Lemma 7.1.2).

Prüfer Transformation: Let $u(t) \not\equiv 0$ be a real-valued solution of

$$(p(t)u')' + q(t)u = 0. \tag{7.4}$$

Since $u(t)$ and $u'(t)$ cannot vanish simultaneously at any point t_0 in $J = [a, b]$, we introduce the following "polar" coordinate (see Fig. 7.1)

$$\rho = \left[u^2 + (pu')^2 \right]^{1/2}, \tag{7.5}$$

$$\varphi = \tan^{-1} \tfrac{u}{pu'}.$$

Then we have

$$pu' = \rho \cos \varphi, \tag{7.6}$$

$$u = \rho \sin \varphi.$$

From (7.4), (7.5), it follows that

$$\varphi' = \frac{\frac{pu' \cdot u' - u(pu')'}{(pu')^2}}{1 + \left(\frac{u}{pu'} \right)^2} = \frac{(pu')u' + q(t)u^2}{\rho^2}$$

$$= \frac{1}{p(t)} \left(\frac{(pu')^2}{\rho^2} \right) + q(t) \left(\frac{u}{\rho} \right)^2,$$

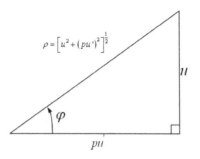

Fig. 7.1

i.e.,

$$\varphi' = \frac{1}{p(t)} \cos^2 \varphi + q(t) \sin^2 \varphi. \qquad (7.7)$$

We note that, in (7.6), the R.H.S. is independent of ρ.

A straightforward computation shows

$$\rho' = -\left[q(t) - \frac{1}{p(t)} \right] \rho \sin \varphi \cos \varphi. \qquad (7.8)$$

From (7.5), we have

$$\varphi(t_0) = 0 \ (\text{mod } \pi) \text{ if and only if } u(t_0) = 0.$$

Lemma 7.1.1 *Let $u(t) \not\equiv 0$ be a real-valued solution of (7.3) with $p(t) > 0$ and $q(t)$ continuous on $J = [a, b]$. Assume that $u(t)$ has exactly n zeros, $n \geq 1$, $t_1 < t_2 < \cdots < t_n$ in $[a, b]$. Let $\varphi(t)$ be a continuous solution of (7.6), (7.7) with $0 \leq \varphi(a) < \pi$. Then $\varphi(t_k) = k\pi$ and $\varphi(t) > k\pi$ for $t_k < t \leq b$.*

Proof. Note that $u(t_0) = 0$ iff $\varphi(t_0) = 0 \ (\text{mod } \pi)$. From (7.6) it follows that $\varphi'(t_k) = \frac{1}{p(t_k)} > 0$, if $\varphi(t_k) = 0$. Hence Lemma 7.1.1 follows. □

Definition 7.1.1 *Consider*

$$(p_1(t)u')' + q_1(t)u = 0, \quad t \in J = [a, b]; \qquad (I)$$

$$(p_2(t)u')' + q_2(t)u = 0, \quad t \in J = [a, b], \qquad (II)$$

we say that equation (II) is a Sturm majorant of (I) on J if $p_1(t) \geq p_2(t)$, $q_1(t) \leq q_2(t)$, $t \in J$. In addition, if $q_1(t) < q_2(t)$ or $p_1(t) > p_2(t) > 0$ at some point $t \in J$, we say that (II) is a strict Sturm majorant of (I) on J.

Theorem 7.1.1 (Sturm's 1st Comparison Theorem)
Let $p_i(t), q_i(t)$ be continuous on $J = [a, b]$, $i = 1, 2$, and (II) be a Strum majorant of (I). Assume that $u_1(t) \not\equiv 0$ is a solution of (I) and $u_1(t)$ has exactly n zeros $t_1 < t_2 < \cdots < t_n$ on J and $u_2(t) \not\equiv 0$ is a solution of (II) satisfying

$$\frac{p_1(t)u_1'(t)}{u_1(t)} \geq \frac{p_2(t)u_2'(t)}{u_2(t)}, \quad at \ t = a \qquad (7.9)$$

$$\left(if \ u_i(a) = 0, \quad set \ \frac{p_i(a)u_i'(a)}{u_i(a)} = +\infty \right).$$

Then $u_2(t)$ has at least n zeros on $(a, t_n]$. Furthermore, if either the inequality holds in (7.8) or (II) is a strict Sturm majorant of (I), then $u_2(t)$ has at least n zeros in (a, t_n).

Proof. Let $\varphi_i(t) = \tan^{-1} \frac{u_i(t)}{p_i(t)u_i'(t)}$, $i = 1, 2$. Then $\varphi_i(t)$ satisfies (7.6), i.e. $\varphi_i'(t) = \frac{1}{p_i(t)} \cos^2 \varphi_i(t) + q_i(t) \sin^2 \varphi_i(t) = f_i(t, \varphi_i)$. From (7.8), it follows that $0 \leq \varphi_1(a) \leq \varphi_2(a) < \pi$. Since (II) is a Sturm majorant of (I), it follows that $f_1(t, \varphi) \leq f_2(t, \varphi)$ on J. From differential inequality, we have $\varphi_1(t) \leq \varphi_2(t)$ on J. Since $\varphi_1(t_n) = n\pi \leq \varphi_2(t_n)$, $u_2(t)$ has at least n zeros on $(a, t_n]$. If $\varphi_1(a) < \varphi_2(a)$ or $f_1(t, \varphi) < f_2(t, \varphi)$ on J, then we have $\varphi_1(t) < \varphi_2(t)$ on J. Hence $u_2(t)$ has at least n zeros on (a, t_n). □

Corollary 7.1.1 (Sturm's Separation Theorem)
Let (II) be a Sturm majorant of (I) on J and $u_1(t)$, $u_2(t)$ be nonzero solutions of (I) and (II) respectively. Let $u_1(t)$ vanish at t_1, $t_2 \in J$, $t_1 < t_2$. Then $u_2(t)$ has at least one zero in $[t_1, t_2]$. In particular if $p_1(t) \equiv p_2(t) \equiv p(t)$, $q_1(t) \equiv q_2(t) \equiv q(t)$ and $u_1(t)$, $u_2(t)$ are two linearly independent solutions of (7.3), then the zeros of u_1 separate and are separated by those zeros of u_2.

Proof. We may assume $\varphi_1(t_1) = 0$, $\varphi_1(t_2) = \pi$ and $0 < \varphi_2(t_1) < \pi$. From the above proof it follows that $\varphi_1(t) < \varphi_2(t)$ for $t_1 \leq t \leq t_2$. Then there exists t_1^*, $t_1 < t_1^* < t_2$, such that $\varphi_2(t_1^*) = \pi$, i.e., there exists at least one zero of $u_2(t)$ in $[t_1, t_2]$. Hence we finish the proof of first part of the corollary. If $p_1(t) \equiv p_2(t) \equiv p(t)$, $q_1(t) \equiv q_2(t) \equiv q(t)$, then, from the first part of the corollary, the zeros of two linearly independent solutions $u_1(t)$ and $u_2(t)$ are separated by each other. □

Example 7.1.1 $\sin t$ and $\cos t$ are two linearly independent solutions of $u'' + u = 0$. Their zeros are separated by each other.

Example 7.1.2 Consider Airy's equation $u''(t) + tu(t) = 0$. Comparing with $u'' + u = 0$, it is easy to show that Airy's equation is oscillatory.

Remark 7.1.1 Sturm's Comparison Theorem is used to show that a second order linear equation (sometimes nonlinear) is oscillatory. For example in [HH] we consider the equation arising from the deformation of heavy cantilever

$$v''(x) + x \sin v = 0,$$
$$v'(0) = 0, \quad v(K) = 0, \tag{7.10}$$

where K is a parameter.

Let $v(x, a)$ be the solution of

$$v''(x) + x \sin v = 0,$$
$$v'(0) = 0, \quad v(0) = a. \tag{7.11}$$

From the uniqueness of solution of ordinary differential equations, we have

$$v(x, 2\pi + a) = 2\pi + v(x, a),$$
$$v(x, a) = -v(x, -a),$$
$$v(x, 0) \equiv 0, \quad v(x, \pi) \equiv \pi.$$

Hence we only need to consider the solution $v(x, a)$ with $0 < a < \pi$. We next show that $v(x, a)$ is oscillatory over $[0, \infty)$.

Let

$$V(x) = (1 - \cos v(x)) + \frac{1}{2} \frac{(v'(x))^2}{x}.$$

Then

$$V'(x) = -\frac{1}{2} \left(\frac{v'(x)}{x} \right)^2 \le 0$$

and

$$1 - \cos v(x) \le V(x) \le V(0) = 1 - \cos a.$$

Since $|v(x)| \leq \pi$, then $|v(x)| \leq a$ for all $x \geq 0$. Rewrite (7.9) as

$$v''(x) + x \left(\frac{\sin v(x)}{v(x)} \right) v(x) = 0.$$

Let $0 < \delta < \min\limits_{0 \leq v \leq a} \left(\frac{\sin v}{v} \right)$. Apply Sturm's Comparison Theorem to compare the above differential equation with

$$v'' + \delta v = 0$$

which is oscillatory over $[0, \infty)$, the solution $v(x, a)$ of (7.9) is oscillatory over $[0, \infty)$ with zeros $y_1(a) < y_2(a) < \dots$ (see Fig. 7.2).

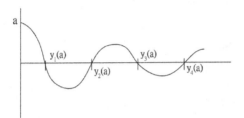

Fig. 7.2

To see how the number of solutions of (7.9) varies with the parameter K, we plot the curves $z = y_i(a)$, $i = 1, 2, \dots, n$ and find the numbers of intersections of these curves with $z = K$ (see Fig. 7.3).

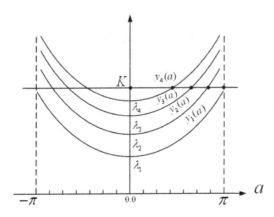

Fig. 7.3

Sturm-Liouville boundary value problem
Let $Lu = (p(t)u')' + q(t)u$, $p(t) > 0$, $q(t)$ continuous on $[a, b]$. Consider

the following eigenvalue problem:

$$Lu + \lambda u = 0,$$
$$u(a)\cos\alpha - p(a)u'(a)\sin\alpha = 0 \qquad (P_\lambda)$$
$$u(b)\cos\beta - p(b)u'(b)\sin\beta = 0.$$

Obviously the boundary conditions in (P_λ) are $\varphi(a) = \alpha$, $\varphi(b) = \beta$, where φ is the angle in the "polar" coordinate (7.4) or (7.5).

Let $y, z : [a, b] \to \mathbb{C}$ and define the inner product

$$\langle y, z \rangle = \int_a^b y(t)\overline{z(t)}dt.$$

Then we have the following

(i) $\langle y + z, w \rangle = \langle y, w \rangle + \langle z, w \rangle$,
(ii) $\langle \alpha y, z \rangle = \alpha \langle y, z \rangle, \alpha \in \mathbb{C}$,
(iii) $\langle z, y \rangle = \overline{\langle y, z \rangle}$,
(iv) $\langle y, y \rangle > 0$ when $y \neq 0$.

Lemma 7.1.2 *The linear operator L is self-adjoint, i.e., $\langle Ly, z \rangle = \langle y, Lz \rangle$ for all y, z satisfy the boundary condition in (P_λ).*

Proof.

$$\langle Ly, z \rangle - \langle y, Lz \rangle$$
$$= \int_a^b \left[(p(t)y')' + q(t)y \right] \overline{z}(t)dt - \int_a^b y(t) \left[(p(t)\overline{z}'(t))' + q(t)\overline{z}(t) \right] dt$$
$$= p(t)y'(t)\overline{z}(t) \Big|_a^b - \int_a^b \overline{z}'(t)p(t)y'(t)dt - p(t)\overline{z}'(t)y(t) \Big|_a^b + \int_a^b y'(t)p(t)\overline{z}'(t)dt$$
$$= p(b)y'(b)\overline{z}(b) - p(a)y'(a)\overline{z}(a) - p(b)\overline{z}'(b)y(b) + p(a)\overline{z}'(a)y(a)$$
$$= y(b)\cot\beta\overline{z}(b) - y(a)\cot\alpha\overline{z}(a) - \overline{z}(b)\cot\beta y(b) + \overline{z}(a)\cot\alpha y(a)$$
$$= 0.$$

$\square$

Lemma 7.1.3

(i) *All eigenvalues of (P_λ) are real.*
(ii) *The eigenfunctions ϕ_m and ϕ_n corresponding to distinct eigenvalues λ_m, λ_n are orthogonal, i.e. $\langle \phi_n, \phi_m \rangle = 0$.*
(iii) *All eigenvalues are simple.*

Proof. Let

$$L\varphi_m = -\lambda_m \varphi_m, \quad L\varphi_n = -\lambda_n \varphi_n,$$

then we have

$$\lambda_m \langle \varphi_m, \varphi_n \rangle = \langle \lambda_m \varphi_m, \varphi_n \rangle = -\langle L\varphi_m, \varphi_n \rangle$$
$$= -\langle \varphi_m, L\varphi_n \rangle = \langle \varphi_m, \lambda_n \varphi_n \rangle = \overline{\lambda_n} \langle \varphi_m, \varphi_n \rangle.$$

If $m = n$ then $\lambda_m = \overline{\lambda_m}$. Hence (i) follows. If $m \neq n$, then $(\lambda_m - \lambda_n)\langle \varphi_m, \varphi_n \rangle = 0$ and it follows that $\langle \varphi_m, \varphi_n \rangle = 0$ and we complete the proof of (ii). To show that each eigenvalue is simple, we note that the eigenfunction $\phi(t)$ satisfies the boundary condition $\phi(a) \cos \alpha - p(a)\phi'(a) \sin \alpha = 0$. If $\cos \alpha = 0$ then $\phi'(a) = 0$. If $\cos \alpha \neq 0$, then $\frac{\phi(a)}{p(a)\phi'(a)} = \tan \alpha$ or $\phi'(a) = \frac{\phi(a)}{p(a)\tan \alpha}$. In either case $\phi'(a)$ is determined uniquely by $\phi(a)$. Hence the eigenvalues are simple by the uniqueness theorem of ODE. $\square$

Now we return to Sturm-Liouville boundary value problem.

Example 7.1.3 $\begin{cases} u'' + \lambda u = 0 \\ u(0) = 0, \quad u(\pi) = 0 \end{cases}$ has eigenvalues $\lambda_n = (n+1)^2$ with eigenfunctions

$$u_n(x) = \sin(n+1)x, \quad n = 0, 1, 2, \ldots.$$

Theorem 7.1.2 [Ha] *There is a sequence of eigenvalues* $\lambda_0, \lambda_1, \cdots$, *forming a monotone increasing sequence with* $\lambda_n \to \infty$ *as* $n \to \infty$. *Moreover, the eigenfunction corresponding to* λ_n *has exactly n zeros on* (a, b).

Proof. Let $u(t, \lambda)$ be the solution of initial value problem

$$(p(t)u')' + q(t)u + \lambda u = 0, \tag{7.12}$$
$$u(a) = \sin \alpha,$$
$$u'(a) = \frac{\cos \alpha}{p(a)}.$$

We note that (P_λ) has a nontrivial solution if and only if $u(t, \lambda)$ satisfies the second boundary condition. For fixed λ, we define

$$\varphi(t, \lambda) = \tan^{-1} \frac{u(t, \lambda)}{p(t)u'(t, \lambda)}.$$

Then $\varphi(a, \lambda) = \alpha$ and $\varphi(t, \lambda)$ satisfies

$$\varphi'(t, \lambda) = \frac{1}{p(t)} \cos^2 \varphi(t, \lambda) + (q(t) + \lambda) \sin^2 \varphi(t, \lambda),$$

$$\varphi(a, \lambda) = \alpha.$$

The Sturm comparison theorem or the differential inequality shows that $\varphi(b, \lambda)$ is strictly increasing in λ. Without loss of generality, we may assume that $0 \le \alpha < \pi$. We shall show that

$$\lim_{\lambda \to \infty} \varphi(b, \lambda) = +\infty, \tag{7.13}$$

and

$$\lim_{\lambda \to -\infty} \varphi(b, \lambda) = 0. \tag{7.14}$$

We note that the second boundary condition in (P_λ) is equivalent to $\varphi(b, \lambda) = \beta + n\pi$ for some $n \ge 0$. Then if (7.12), (7.13), hold then there exist eigenvalues $\lambda_0, \lambda_1, \cdots, \lambda_n, \lambda_n \to +\infty$ such at $\varphi(b, \lambda_n) = \beta + n\pi$, $n \ge 0$. From Lemma 7.1.1, the corresponding eigenfunction $u_n(x)$ has exactly n zeros. Now we prove (7.12). Introduce scaling, $s = \int_a^t \frac{1}{p(\tau)} d\tau$ and set $U(s) = u(t)$. Then

$$\dot{U}(s) = \frac{du}{dt} \frac{dt}{ds} = p(t) \frac{du}{dt}$$

$$\ddot{U}(s) = p(t) \left(p(t) \frac{du}{dt} \right)'.$$

And (7.12) becomes

$$\ddot{U}(s) + p(t)(q(t) + \lambda)U = 0. \tag{7.15}$$

From arbitrary fixed $n > 0$ and for any fixed $M > 0$, choose λ sufficiently large such that

$$p(t)(q(t) + \lambda) \ge M^2, \quad a \le t \le b.$$

Compare (7.15) with

$$\ddot{u} + M^2 u = 0. \tag{7.16}$$

Then the solution $u(s)$ of (7.16) has at least n zeros provided M is sufficiently large. By Sturm Comparison Theorem, $U(s)$ has at least n zeros on $0 \le s \le \int_a^b \frac{dt}{p(t)}$, or equivalently, $u(t)$ has at least n zeros on $[a, b]$. Then $\varphi(b, \lambda) \ge n\pi$ if λ is sufficiently large. Hence we complete the proof of (7.13). Next we show that (7.14) holds. Obviously $\varphi(b, \lambda) \ge 0$ for all $\lambda \in \mathbb{R}$.

Choose $\lambda < 0$ with $|\lambda|$ sufficiently large and $p(t) \cdot (q(t) + \lambda) \le -M^2 < 0$. Compare (7.15) with

$$\ddot{u} - M^2 u = 0,$$
$$u(0) = \sin \alpha,$$
$$u'(0) = \cos \alpha.$$

Then

$$u(s) = \sin \alpha \cos h(Ms) + \frac{1}{M} \cos \alpha \sin h(Ms).$$

Let $\psi(s, M) = \tan^{-1} \frac{u(s)}{\dot{u}(s)}$, $\psi(0, M) = \alpha$. Since it can be verified that

$$\frac{u(s)}{\dot{u}(s)} \to 0 \quad \text{as} \quad M \to \infty,$$

it follows that $\psi(b_0, M) \to 0$ as $M \to \infty$ where $b_0 = \int_a^b \frac{dt}{p(t)}$. By Sturm Comparison Theorem, we have $0 \le \varphi(b, \lambda) \le \psi(b_0, M) \to 0$. Thus (7.14) holds. $\qquad \square$

7.2 Distributions

In 1930 famous physicist Paul Dirac introduced the concept of δ-function, $\delta_\xi(x)$ which satisfies

$$\delta_\xi(x) = 0, \text{ for } x \ne \xi,$$
$$\int_{-\infty}^{\infty} \delta_\xi(x) dx = 1,$$
$$\int_{-\infty}^{\infty} \delta_\xi(x) \varphi(x) dx = \varphi(\xi), \qquad \varphi \in C^\infty.$$

Physically $\delta_\xi(x)$ represents that a unit force is applied at the position $x = \xi$. In 1950 L. Schwartz introduced the definition of distribution to interpret the mathematical meaning of δ-function.

Definition 7.2.1 *A test function is a function $\varphi \in C^\infty(\mathbb{R})$ with compact support, i.e. there exists a finite interval $[a, b]$ such that $\varphi(x) = 0$ outside $[a, b]$. We denote $\varphi \in C_0^\infty(\mathbb{R})$.*

Example 7.2.1 Let $\varphi(x) = \begin{cases} \exp\left(\frac{1}{x^2-1}\right), |x| < 1; \\ 0 \qquad\qquad , |x| \ge 1. \end{cases}$ Then it is easy to verify that $\varphi \in C_0^\infty(\mathbb{R})$.

Definition 7.2.2 *We say that $t : C_0^\infty(\mathbb{R}) \to \mathbb{R}$ is a linear functional if $t(\alpha \varphi_1 + \beta \varphi_2) = \alpha t(\varphi_1) + \beta t(\varphi_2), \alpha, \beta \in \mathbb{R}, \varphi_1, \varphi_2 \in C_0^\infty(\mathbb{R})$. We denote $t(\varphi) = \langle t, \varphi \rangle$.*

Definition 7.2.3 *Let $\{\varphi_n\} \subseteq C_0^\infty(\mathbb{R})$. We say that $\{\varphi_n\}$ is a zero sequence if*

(i) $\cup_n \operatorname{supp} \varphi_n$ *is bounded,*

(ii) $\lim_{n \to \infty} \left(\max_x \left| \frac{d^k \varphi_n}{dx^k}(x) \right| \right) = 0$, $k = 0, 1, 2, \cdots$.

Definition 7.2.4 *A linear functional $t : C_0^\infty(\mathbb{R}) \to \mathbb{R}$ is continuous if $\langle t, \varphi_n \rangle \to 0$ as $n \to \infty$ for any zero sequence $\{\varphi_n\}$. A distribution is a continuous linear functional.*

Example 7.2.2 Let $f \in C_0^\infty(\mathbb{R})$ be locally integrable, i.e., $\int_I |f(x)|\, dx$ exists and bounded for any finite interval I. Then we define a distribution t_f associated with f by $\langle t_f, \varphi \rangle = \int_{-\infty}^{\infty} f(x)\varphi(x)dx$.

It is an exercise to verify that t_f is a distribution.

Example 7.2.3

(i) Heaviside distribution t_H. Let $H(x)$ be the Heaviside function:
$$H(x) = \begin{cases} 1, & x \geq 0; \\ 0, & x < 0. \end{cases}$$

Then
$$\langle t_H, \varphi \rangle = \int_{-\infty}^{\infty} H(x)\varphi(x)dx = \int_0^{\infty} \varphi(x)dx.$$

(ii) Delta distribution δ_ξ
$$\langle \delta_\xi, \varphi \rangle \overset{def}{=} \varphi(\xi)$$
Symbolically we write $\int \delta_\xi(x)\varphi(x)dx = \varphi(\xi)$.

(iii) Dipole distribution Δ
$$\langle \Delta, \varphi \rangle \overset{def}{=} \varphi'(0).$$

Definition 7.2.5 *We say that a distribution t is regular if $t = t_f$ for some locally integrable function f. Otherwise t is singular.*

Remark 7.2.1 δ_ξ is a singular distribution.

Proof. If not, $\delta_\xi = t_f$ for some locally integrable $f(x)$.

Consider $\varphi_a(x) = \begin{cases} \exp\left(\frac{a^2}{x^2-a^2}\right) & ,|x| < a, \\ 0 & ,|x| \geq a. \end{cases}$

$\varphi_a(0) = \frac{1}{e} = \max\limits_x |\varphi_a(x)|$.

Then $\langle t_f, \varphi_a \rangle \to 0$ as $a \to 0$ for

$|\langle t_f, \varphi_a \rangle| = \left| \int_{-\infty}^{\infty} f(x)\varphi_a(x)dx \right| \leq \frac{1}{e} \int_{-a}^{a} |f(x)|\, dx \to 0$.

However, if $\delta_\xi = t_f$, then $\langle t_f, \varphi_a \rangle = \varphi_a(0) = \frac{1}{e}$.

This is a contradiction. $\qquad\qquad\qquad\qquad\qquad\qquad\qquad\qquad\square$

Next we introduce some algebraic operations of distributions. Let $f \in C^\infty(\mathbb{R})$ and t be a distribution. Then we define ft to be a distribution defined as

$$\langle ft, \varphi \rangle = \langle t, f\varphi \rangle, \varphi \in C_0^\infty(\mathbb{R}).$$

We note that if t is a regular distribution, then

$$\langle ft, \varphi \rangle = \int_{-\infty}^{\infty} f(x)t(x)\varphi(x)dx = \int_{-\infty}^{\infty} t(x)f(x)\varphi(x)dx.$$

To define the derivative of a distribution t, we let t be a regular distribution. Then $\langle t', \varphi \rangle = \int_{-\infty}^{\infty} t'(x)\varphi(x)dx = \varphi(x)t(x) \mid_{-\infty}^{\infty} - \int \varphi'(x)t(x)dx = -\langle t, \varphi' \rangle$. Hence we have the following definition:

Definition 7.2.6 $\langle t', \varphi \rangle = -\langle t, \varphi' \rangle$.

It is an easy exercise to verify that t' is a distribution.

Example 7.2.4 $H' = \delta_0 = \delta$ in distribution sense.

$\langle H', \varphi \rangle = -\langle H, \varphi' \rangle = -\int_{-\infty}^{\infty} H(x)\varphi'(x)dx = -\int_0^{\infty} \varphi'(x)dx$
$= \varphi(0) = \langle \delta, \varphi \rangle, \varphi \in C_0^\infty(\mathbb{R})$.

Example 7.2.5 $\delta' = -\Delta$ in distribution sense.

$\langle \delta', \varphi \rangle = -\langle \delta, \varphi' \rangle = -\varphi'(0) = \langle -\Delta, \varphi \rangle$.

7.3 Green's Function

Let $Lu = a_n(x)\frac{d^n u}{dx^n} + a_{n-1}(x)\frac{d^{n-1}u}{dx^{n-1}} + \cdots + a_1(x)\frac{du}{dx} + a_0(x)u$ be a linear differential operator.

Consider the boundary value problem:

$(Lu)(x) = f(x)$, $0 \leq x \leq 1$ with boundary conditions at $x = 0$ and $x = 1$.

We say that u is a distribution solution of $Lu = f$ if $\langle Lu, \varphi \rangle = \langle f, \varphi \rangle$ for all test functions φ. To motivate the concept of Green's function, we consider the following example.

Example 7.3.1 Let $Lu = u''$. Consider two-point boundary value problem $Lu = f$,

$$u''(x) = f(x),$$
$$u(0) = 0, \quad u(1) = 0.$$

From direct integrations and the boundary conditions, we obtain

$$u(x) = -\int_0^1 x(1-\xi)f(\xi)d\xi + \int_0^x (x-\xi)f(\xi)d\xi$$

$$= \int_0^1 g(x,\xi)f(\xi)d\xi, \tag{7.17}$$

where

$$g(x,\xi) = \begin{cases} x(\xi-1) \, , 0 \leq x \leq \xi \leq 1; \\ \xi(x-1) \, , 0 \leq \xi \leq x \leq 1. \end{cases}$$

Write $u(x) = (L^{-1}f)(x) = \int_0^1 g(x,\xi)f(\xi)d\xi$.

Applying L on both sides of the above identity, we have

$$f(x) = L\left(\int_0^1 g(x,\xi)f(\xi)d\xi\right) = \int_0^1 Lg(x,\xi)f(\xi)d\xi.$$

Then $Lg(x,\xi) = \delta_x(\xi) = \delta(x-\xi)$ and $g(x,\xi)$ is called the Green's function of differential operator L subject to homogeneous boundary condition $u(0) = 0, u(1) = 0$.

Physical Interpretation of Green's Function

Suppose a string is stretched between $x = 0$ and $x = 1$ in equilibrium state. Let $u(x)$ be the vertical displacement from zero position and $u(0) = u(1) = 0$, i.e., two ends of the string are fixed. Since the string is in equilibrium state, the horizontal and vertical forces must be in balance. Then (see Fig. 7.4)

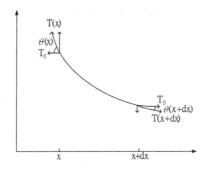

Fig. 7.4

$$T_0 = T(x)\cos\theta(x) = T(x+dx)\cos\theta(x+dx)$$

$$T(x+dx)\sin\theta(x+dx) - T(x)\sin\theta(x) = \rho(x)gds$$

where g is the constant of gravity, $\rho(x)$ is the density per unit length of the string at x and ds is the arclength between x and $x+dx$. Then

$$T_0\frac{\tan\theta(x+dx) - \tan\theta(x)}{dx} = \rho(x)g\frac{ds}{dx}.$$

Since $\frac{du}{dx} = \tan\theta(x)$, let $dx \to 0$ and it follows that $T_0\frac{d^2u}{dx^2} = \rho(x)g(1 + (\frac{du}{dx})^2)^{1/2}$.

Assume that $\left|\frac{du}{dx}\right|$ is sufficiently small, then it follows that

$$\frac{d^2u}{dx^2} = f(x), \quad f(x) = \frac{\rho(x)g}{T_0},$$
$$u(0) = u(1) = 0.$$

From Example 7.3.1, $u(x) = \int_0^1 g(x,\xi)f(\xi)d\xi$, where $g(x,\xi)$ satisfies

$$\frac{d^2}{dx^2}g(x,\xi) = \delta_\xi(x) = \delta(x-\xi),$$
$$g(0,\xi) = 0, \quad g(1,\xi) = 0.$$

Hence, $g(x,\xi)$ is the vertical displacement of a string fixed at $x = 0$ and $x = 1$ subject to the unit force at ξ. If we partition $[0,1]$ at points $\{\xi_k\}_{k=0}^n$ and apply force with magnitude $f(\xi_k)$ at point ξ_k, $k = 1, \cdots, n$. Then, by superposition principle , we have $u(x) = \sum_{k=1}^n f(\xi_k)g(x,\xi_k)\triangle\xi_k$.

Let $n \to \infty$, then we obtain (7.16).

Now we consider the following two-point boundary value problem

$$Lu = f \quad \text{on} \quad a < x < b, \quad \text{where}$$
$$Lu = a_n(x)\frac{d^n u}{dx^n} + a_{n-1}(x)\frac{d^{n-1} u}{dx^{n-1}} + \cdots + a_1(x)\frac{du}{dx} + a_0(x)u \qquad (7.18)$$

with homogeneous boundary conditions $B_1 u(a) = 0$, $B_2 u(b) = 0$.

To solve (7.17), we introduce Green's function $g(x, \xi)$ which satisfies

$$Lg(x, \xi) = \delta_\xi(x), \qquad \text{for all } x \in (a, b),$$
$$B_1 g(a, \xi) = 0, \quad B_2 g(b, \xi) = 0. \qquad (7.19)$$

Then we expect the solution $u(x)$ of (7.16) to be written as $u(x) = \int_a^b g(x, \xi) f(\xi) d\xi$.

We shall find Green's function $g(x, \xi)$ by the following:

(i)

$$Lg(x, \xi) = 0 \quad \text{for} \quad x \neq \xi$$
$$B_1 g(a, \xi) = 0, \quad B_2 g(b, \xi) = 0;$$

(ii) $\frac{d^k g}{dx^k}(x, \xi)$ is continuous at $x = \xi$ for $k = 0, 1, \cdots, n - 2$;

(iii) Jump condition: $\frac{d^{n-1} g}{dx^{n-1}}(\xi^+, \xi) - \frac{d^{n-1} g}{dx^{n-1}}(\xi^-, \xi) = \frac{1}{a_n(\xi)}$.

To see why the jump condition (iii) holds, we integrate (7.18) from ξ^- to ξ^+. Then $\int_{\xi^-}^{\xi^+} [a_n(x)\frac{d^n g}{dx^n}(x, \xi) + a_{n-1}(x)\frac{d^{n-1} g}{dx^{n-1}}(x, \xi) + \cdots + a_1(x)\frac{dg}{dx}(x, \xi) + a_0(x)g(x, \xi)]dx = 1$.

From condition (ii), the above identity becomes

$$a_n(\xi)\frac{d^{n-1} g}{dx^{n-1}}(x, \xi) \mid_{x=\xi^-}^{x=\xi^+} = 1.$$

Hence (iii) holds.

Example 7.3.2 Solve $\begin{cases} u''(x) = f(x), & 0 < x < 1, \\ u'(0) = a, & u(1) = b. \end{cases}$

Let $u_1(x)$, $u_2(x)$ be the solution of

$$u''(x) = f(x), \ 0 < x < 1,$$
$$u'(0) = 0, \ u(1) = 0, \qquad (7.20)$$

and

$$u''(x) = 0, \ 0 < x < 1,$$
$$u'(0) = a, \ u(1) = b, \qquad (7.21)$$

respectively.

Then $u(x) = u_1(x) + u_2(x)$, $u_1(x) = \int_0^1 g(x, \xi) f(\xi) d\xi$, $u_2(x) = ax + (b - a)$.

Find Green's function $g(x, \xi)$ of (7.19) which satisfies

(i)

$$\frac{d^2}{dx^2} g(x, \xi) = 0 \quad \text{for} \quad x \neq \xi,$$
$$\frac{dg}{dx}(0, \xi) = 0, \quad g(1, \xi) = 0;$$

(ii) $g(x, \xi)$ is continuous at $x = \xi$;

(iii) $\frac{dg}{dx}(x, \xi) \mid_{x=\xi^-}^{x=\xi^+} = 1.$

From (i),

$$g(x, \xi) = \begin{cases} Ax + B, & 0 \leq x < \xi \leq 1; \\ Cx + D, & 0 < \xi < x \leq 1. \end{cases}$$

$\frac{dg}{dx}(0, \xi) = 0$ implies $A = 0$.

$g(1, \xi) = 0$ implies $D = -C$.

From (ii), we have $A\xi + B = C\xi + D$. Hence,

$$g(x, \xi) = \begin{cases} C(\xi - 1), & 0 \leq x \leq \xi \leq 1; \\ C(x - 1), & 0 < \xi \leq x \leq 1. \end{cases}$$

From the jump condition (iii) $\frac{dg}{dx}(\xi^+, \xi) - \frac{dg}{dx}(\xi^-, \xi) = 1$, we have $C = 1$ and $g(x, \xi) = \begin{cases} \xi - 1, & 0 \leq x < \xi \leq 1 \\ x - 1, & 0 \leq \xi < x \leq 1. \end{cases}$

Next we want to show that $u_1(x)$ satisfies (7.19).

$$u_1'(x) = \frac{d}{dx}\left[\int_0^{x^-} g(x,\xi)f(\xi)d\xi + \int_{x^+}^1 g(x,\xi)f(\xi)d\xi\right]$$

$$= g(x,x^-)f(x^-) + \int_0^{x^-} \frac{dg}{dx}(x,\xi)f(\xi)d\xi$$

$$-g(x,x^+)f(x^+) + \int_{x^+}^1 \frac{dg}{dx}(x,\xi)f(\xi)d\xi$$

$$= \int_0^1 \frac{dg}{dx}(x,\xi)f(\xi)d\xi.$$

$$u_1'(0) = \int_0^1 \frac{dg}{dx}(0,\xi)f(\xi)d\xi = 0.$$

$$u_1(1) = \int_0^1 g(1,\xi)f(\xi)d\xi = 0.$$

$$u_1''(x) = \frac{dg}{dx}(x,x^-)f(x^-) + \int_0^{x^-} \frac{d^2g}{dx^2}(x,\xi)f(\xi)d\xi$$

$$-\frac{dg}{dx}(x,x^+)f(x^+) + \int_{x^+}^1 \frac{d^2g}{dx^2}(x,\xi)f(\xi)d\xi$$

$$= f(x)\left[\frac{dg}{dx}(x,x^-) - \frac{dg}{dx}(x,x^+)\right]$$

$$= f(x)\left[\frac{dg}{dx}(\xi^+,\xi) - \frac{dg}{dx}(\xi^-,\xi)\right] = f(x).$$

Example 7.3.3

$$\begin{cases} Lu = a_2(x)u''(x) + a_1(x)u'(x) + a_0(x)u(x) = f(x), \ a < x < b, \\ \alpha_1 u(a) + \beta_1 u'(a) = 0, \\ \alpha_2 u(b) + \beta_2 u'(b) = 0. \end{cases}$$

Let $u_1(x)$ be a solution of $Lu = 0$, $a < x < b$, $\alpha_1 u(a) + \beta_1 u'(a) = 0$, and $u_2(x)$ be a solution of $Lu = 0$, $a < x < b$, $\alpha_2 u(b) + \beta_2 u'(b) = 0$.

Let $g(x,\xi) = \begin{cases} Au_1(x)u_2(\xi), & a \leq x < \xi \leq b; \\ Au_2(x)u_1(\xi), & a \leq \xi < x \leq b. \end{cases}$

Obviously, $g(x,\xi)$ is continuous at $x = \xi$.

From jump condition $\frac{dg}{dx}\big|_{x=\xi^-}^{x=\xi^+} = \frac{1}{a_2(\xi)}$, it follows that

$$Au_2'(\xi)u_1(\xi) - Au_1'(\xi)u_2(\xi) = \frac{1}{a_2(\xi)},$$

i.e., $A = \frac{1}{a_2(\xi)W(u_1,u_2)(\xi)}$, where $W(u_1,u_2)(\xi)$ is the Wronskian of u_1, u_2.

We leave it as an exercise to verify that $u(x) = \int_a^b g(x, \xi) f(\xi) d\xi$ satisfies $Lu = f$.

In the following, we use Green's function to convert a nonlinear two-point boundary value problem

$$\begin{cases} y'' = f(x, y), & a < x < b, \\ y(a) = 0, & y(b) = 0, \end{cases} \tag{7.22}$$

into the following integral equation

$$y(x) = \int_a^b g(x, \xi) f(\xi, y(\xi)) d\xi \tag{7.23}$$

where

$$g(x, \xi) = \begin{cases} \frac{(x-a)(\xi-b)}{b-a}, & a \leq x \leq \xi \leq b \\ \frac{(x-b)(\xi-a)}{b-a}, & a \leq \xi \leq x \leq b. \end{cases} \tag{7.24}$$

To show the existence and uniqueness of the solution of (7.20), we shall apply the contraction mapping principle. Let $C[a, b] = \{\varphi : [a, b] \to \mathbb{R}.\}$ is continuous with supreme norm $\|\varphi\|_\infty = \sup\limits_{a \leq x \leq b} |\varphi(x)|$.

Define $T : C[a, b] \to C[a, b]$ by

$$(T\varphi)(x) = \int_a^b g(x, \xi) f(\xi, \varphi(\xi)) d\xi.$$

Theorem 7.3.1 *Assume $f(x, y)$ satisfies Lipschitz condition in y with Lipschitz constant K, i.e., $|f(x, y_1) - f(x, y_2)| \leq K |y_1 - y_2|$ for each x. If $K \frac{(b-a)^2}{4} < 1$ then there exists a unique solution of (7.22).*

Proof.

$$|T\varphi_1(x) - T\varphi_2(x)|$$

$$= \left| \int_a^b g(x, \xi)(f(\xi, \varphi_1(\xi)) - f(\xi, \varphi_2(\xi))) d\xi \right|$$

$$\leq K \left(\int_a^b |g(x, \xi)| \, d\xi \right) \|\varphi_1 - \varphi_2\|_\infty$$

$$\leq K \left[\max_{a \leq x \leq b} \left(\int_a^b |g(x, \xi)| \, d\xi \right) \right] \|\varphi_1 - \varphi_2\|_\infty.$$

It is easy to verify directly from (7.24) that $\max_{a \leq x \leq b} \left(\int_a^b |g(x,\xi)|\, d\xi \right) = \frac{(b-a)^2}{4}$.

Then

$$\|T\varphi_1 - T\varphi_2\|_\infty \leq K \frac{(b-a)^2}{4} \|\varphi_1 - \varphi_2\|_\infty.$$

Since $K \frac{(b-a)^2}{4} < 1$, from contraction mapping principle, T has a unique fixed point φ. Thus there is a unique solution of (7.22).   □

7.4   Fredholm Alternative

Let's recall the Fredholm Alternative in linear algebra. Let A be an $m \times n$ real matrix and $b \in \mathbb{R}^m$. For the solutions of the linear system $Ax = b$. We have two alternatives:

(i) If $Ax = 0$ has only trivial solution, i.e., the null space $N(A) = \{0\}$, then the solution of $Ax = b$ is unique;

(ii) If $Ax = 0$ has nontrivial solutions, then $A^T v = 0$ has nontrivial solutions and $Ax = b$ is solvable if and only if $\langle b, v \rangle = 0$ for all $v \in N(A^T)$.

Let $Lu = a_2(x)u'' + a_1(x)u' + a_0(x)u$.

Consider

$$\begin{cases} Lu = f(x) , \ a < x < b, \\ B_1(u) = 0 , \ B_2(u) = 0. \end{cases} \tag{7.25}$$

Let D be Domain $L = \{u \in C^2 : B_1(u) = 0, B_2(u) = 0\}$.

We shall find $L^* v$, the adjoint system of L, and D^*, the domain of L^*, and the corresponding boundary conditions $B_1^*(v) = 0$, $B_2^*(v) = 0$.

Compute

$$\langle Lu, v \rangle = \int_a^b (Lu)(x)v(x)dx$$

$$= \int_a^b [a_2(x)u''v + a_1(x)u'v + a_0(x)uv]dx$$

$$= (a_2v)u' \mid_a^b - u(a_2v)' \mid_a^b + u(a_1v) \mid_a^b$$

$$+ \int_a^b [(a_2v)'' - (a_1v)' + (a_0v)]udx$$

$$= J(u, v) \mid_a^b + \langle u, L^*v \rangle.$$

Definition 7.4.1 *We say that L is self-adjoint if $L = L^*$ and $D = D^*$, where*

$$L^*v = (a_2v)'' - (a_1v)' + (a_0v),$$
$$J(u, v) \mid_a^b = (a_2v)u' \mid_a^b - u(a_2v)' \mid_a^b + u(a_1v) \mid_a^b.$$

We define

$$D^* = domain \ of \ L^*$$
$$= \{v \in C^2 : J(u, v) \mid_a^b = 0 \ for \ all \ u \in D\}$$
$$= \{v \in C^2 : B_1^*(v) = 0 \ , \ B_2^*(v) = 0\}.$$

Then for $u \in D$, $v \in D^*$ we have

$$\langle Lu, v \rangle = \langle u, L^*v \rangle. \tag{7.26}$$

Example 7.4.1 If $Lu = a_2(x)u'' + a_1(x)u' + a_0(x)u$ and $B_1(u) = u(a) = 0$, $B_2(u) = u'(b) = 0$, then

$$J(u, v) \mid_a^b = a_2(vu' - uv') + (a_1 - a_2')uv \mid_a^b$$
$$= a_2(b)(-u(b)v'(b)) + (a_1(b) - a_2'(b))u(b)v(b) - a_2(a)v(a)u'(a) = 0,$$

for all $u \in$ Domain L.

Then $B_1^*(v) = v(a) = 0$, $B_2^*(v) = 0$ if and only if $v(b)(a_1(b) - a_2'(b)) = a_2(b)v'(b)$.

Remark 7.4.1 If $Lu = a_n(x)\frac{d^n u}{dx^n} + \cdots + a_0(x)u$, then $L^*u = \sum_{k=0}^n (-1)^k \frac{d^k}{dx^k}(a_k(x)v)$.

Now we are in a position to state Fredholm Alternatives for second order linear equations.

Theorem 7.4.1 (Fredholm Alternatives)
Let (1) $Lu = f(x)$, $B_1(u) = 0$, $B_2(u) = 0$, $a < x < b$;
  (2) $Lu = 0$, $B_1(u) = 0$, $B_2(u) = 0$;
  (3) $L^*v = 0$, $B_1^*(v) = 0$, $B_2^*(v) = 0$. *Then there are two alternatives*

(i) *If* (2) *has only trivial solution, then* (1) *has a unique solution;*
(ii) *If* (2) *has nontrivial solutions, then* (3) *has nontrivial solutions. Furthermore* (1) *is solvable if and only if* $\langle f, v \rangle = 0$ *for all solution v of* (3).

Proof. (i) is trivial. For part (ii) we only show the necessary condition.
  Let $Lu = f$ for some u. From (7.25), we have $\langle Lu, v \rangle = \langle f, v \rangle = \langle u, L^*v \rangle = 0$ for all v satisfying $L^*v = 0$. For the sufficient condition, reader may consult [Stak] Chapter 3, p. 254.   □

Remark 7.4.2 For second order linear operator L, we may assume L is self-adjoint, i.e., $Lu = (p(x)u')' + q(x)u$. From Theorem 7.1.2, there is a sequence of eigenvalues $\{\lambda_n\}$ such that $Lu_n = \lambda_n u_n$. In the following we express the solution u of $Lu = f$ in terms of the eigenvalues and eigenfunctions.

Let $\lambda_0 = 0$, then $Lu_0 = 0$ and $\langle f, u_0 \rangle = 0$. For $n \neq 0$, if $Lu = f$ then

$$\langle f, u_n \rangle = \langle Lu, u_n \rangle = \langle u, Lu_n \rangle = \lambda_n \langle u, u_n \rangle .$$

From the eigen-expansion, $u = u_0 + \sum \langle u, u_n \rangle u_n$ we find the solution u has the form

$$u = u_0 + \sum \frac{1}{\lambda_n} \langle f, u_n \rangle u_n. \tag{7.27}$$

Then

$$Lu = Lu_0 + \sum \frac{1}{\lambda_n} \langle f, u_n \rangle Lu_n = 0 + \sum \langle f, u_n \rangle u_n = f. \tag{7.28}$$

Hence, from (7.26), (7.27), u is the solution of $Lu = f$.

Remark 7.4.3 In Chapter 9, we shall apply Fredholm Alternatives in the regular perturbation method.

7.5 Exercises

Exercise 7.1

(1) Consider the differential equation

$$(*) \qquad \ddot{u} + q(t)u = 0.$$

Let $q(t)$ be real-valued, continuous, and satisfy $0 < m \leq q(t) \leq M$. If $u = u(t) \neq 0$ is a solution with a pair of zeros $t = t_1, t_2(> t_1)$, then $\pi/m^{1/2} \geq t_2 - t_1 \geq \pi/M^{1/2}$.

(2) Let $q(t)$ be continuous for $t \geq 0$ and $q(t) \to 1$ as $t \to \infty$. Show that if $u = u(t) \neq 0$ is a real-valued solution of $(*)$, then the zeros of $u(t)$ form a sequence $0 \leq t_1 < t_2 < \cdots$ such that $t_n - t_{n-1} \to \pi$ as $n \to \infty$.

(3) Consider the Bessel equation

$$(**) \qquad v'' + \frac{v'}{t} + \left(1 - \frac{\mu^2}{t^2}\right) v = 0$$

where μ is a real parameter. The change of variable $u = t^{1/2}v$ transforms $(**)$ into

$$u'' + \left(1 - \frac{\alpha}{t^2}\right) u = 0 \qquad \text{where} \quad \alpha = \mu^2 - \frac{1}{4}.$$

Show that the zeros of a real-valued solution $v(t)$ of $(**)$ on $t > 0$ form a sequence $t_1 < t_2 < \cdots$ such that $t_n - t_{n-1} \to \pi$ as $n \to \infty$.

Exercise 7.2 Show that all solutions of $y'' + (|t + \frac{1}{t}|)y + y^3 = 0$ are oscillatory on $[1, \infty)$. (I.e. they have infinitely many zeros in $[1, \infty)$.)

Exercise 7.3 Suppose $a(t) > 0$ increases to ∞ as $t \to \infty$, show that all solutions of $u'' + a(t)u = 0$ are bounded as $t \to \infty$. (Hint: Multiply the equation by u' and integrate, then try to apply some inequality.)

Exercise 7.4 Consider the second order equation

$$\ddot{y} + p(t)\dot{y} + q(t)y = 0$$

where $p(t)$ and $q(t)$ are smooth positive functions on the real line and $q(t)$ is decreasing. Rewrite this equation as a linear system and show that the function

$$E(t, y, \dot{y}) = q(t)y^2 + (\dot{y})^2$$

is decreasing along the trajectories of the system.

Exercise 7.5 Consider, for $t > 0$, the equation

$$\ddot{x} + \frac{\lambda}{t^{2002}}x = 0.$$

(1) Prove that for any interval $(a, b) \subset (0, \infty)$ there exists λ such that if $\phi(t)$ is a solution of the above equation, then there $t_0 \in (a, b)$ satisfying $\phi(t_0) = 0$.
(2) Prove that if $\lambda < 0$, then any solution $\phi(t)$ which is not identically zero has no more than one root.

Exercise 7.6 Find eigenvalues and eigenfunctions of

$$y'' + \lambda y = 0, \ y(0) + y'(0) = 0, \ y(1) = 0.$$

Exercise 7.7 Show that the solution $v(x, a)$ of initial value problem (7.10) with $0 < a < \pi$ satisfies $\lim_{x \to \infty} v(x, a) = 0$.

Exercise 7.8 Consider the following boundary value problem

$$(P)_\alpha = \begin{cases} v''(x) = K^3 x \sin v, \ 0 \le x \le 1, K > 0, \\ v'(0) = 0, v(1) = \alpha, 0 \le \alpha \le \pi, \end{cases}$$

show that

(1) If $\alpha = 0$, then $(P)_0$ has a unique solution

$$v(x) \equiv 0, \ 0 \le x \le 1, \ \text{for any } K > 0;$$

(2) If $K^3 < \sqrt{45}$, then $(P)_\alpha$ has a unique solution for every $\alpha \in [0, \pi]$.

Exercise 7.9 Complete the proof of Example 7.3.3.

Exercise 7.10 Find solvability condition for

$$u'' + u = f(x), \ u(0) - u(2\pi) = \alpha, \ u'(0) - u'(2\pi) = \beta.$$

Exercise 7.11 Show that a Sturm-Liouville operator $Lu = (p(x)u')' + q(x)u$ with periodic boundary condition $u(a) = u(b), u'(a) = u'(b)$ on $[a, b]$ is self-adjoint if and only if $p(a) = p(b)$.

Exercise 7.12 Convert the following differential equations to Fredholm integral equations of the form $u = \lambda \int k(x, \xi)u(\xi) + g(x)$:

(1) $u'' + \lambda u = f(x), \ u(0) = \alpha, \ u(1) = \beta,$
(2) $u'' + \lambda u = f(x), \ u(0) = u'(0), \ u(1) = -u'(1).$

Exercise 7.13 The convolution of regular distribution is defined by

$$(f * g)(x) = \int_{-\infty}^{\infty} f(x - t)g(t)dt.$$

(1) Define the convolution of distributions,
(2) What is $\delta * \delta$?

Exercise 7.14 Find the derivative in the sense of distribution of the function

$$f(x) = \begin{cases} 1 & x \text{ irrational}, \ 0 < x < 1, \\ 0 & x \text{ rational}. \end{cases}$$

Exercise 7.15 Find the Green's function for $Lu = u'' + k^2u$, $k \neq 0$, subjected to homogeneous boundary condition $u(0) = u(1) = 0$.

Exercise 7.16 Find the adjoint operator L^*, its domain.

(1) $Lu = u'' + a(x)u' + b(x)u$, $u(0) = u'(1)$, $u(1) = u'(0)$.
(2) $Lu = -(p(x)u')' + q(x)u$, $u(0) = u(1)$, $u'(0) = u'(1)$.

Exercise 7.17 Find the solvability conditions for

(1) $u'' + u = f(x)$, $u(0) - u(2\pi) = \alpha$, $u'(0) - u'(2\pi) = \beta$.
(2) $u'' = f(x)$, $u(0) - u(1) = \alpha$, $u'(0) - u'(1) = \beta$.
(3) $(xu')' = f(x)$, $u'(1) = \alpha$, $xu'(x)|_{x=0} = \beta$.

Exercise 7.18

(1) Show that the eigenvalue problem

$$-u'' + 4\pi^2 \int_0^1 u(x)dx = \lambda u, \ 0 < x < 1,$$

$$u(0) = u(1), \ u'(0) = u'(1),$$

has $\lambda = 4\pi^2$ as an eigenvalue of multiplicity 3.
(2) Find the all eigenvalues and eigenfunctions of this problem.

Exercise 7.19 Show that a test function $\psi(x)$ is of the form $\psi = (x\varphi)'$ where φ is a test function if and only if

$$\int_{-\infty}^{\infty} \psi(x)dx = 0 \text{ and } \int_{0}^{\infty} \psi(x)dx = 0.$$

Hint: $\varphi(x) - \varphi(0) = x \int_{0}^{1} \varphi'(xt)dt$.

Exercise 7.20 Show that

$$1 + 2\sum_{n=1}^{\infty} \cos 2n\pi x = \sum_{k=-\infty}^{\infty} \delta(x-k).$$

Chapter 8

THE INDEX THEORY AND BROUWER DEGREE

8.1 Index Theory in the Plane

Given a simple closed curve C in $x - y$ plane and a vector field $\vec{f}(x, y) = (P(x, y), Q(x, y))$ such that $\vec{f}$ does not vanish on any point of C. Let C be parameterized with positive orientation as $x = x(s)$, $y = y(s)$, $a \leq s \leq b$, $x(a) = x(b)$, $y(a) = y(b)$. Let

$$\theta(s) = \tan^{-1} \frac{Q(x(s), y(s))}{P(x(s), y(s))}$$

be a continuous function on $[a, b]$ and the index

$$I_f(C) \overset{\text{def}}{=} \frac{1}{2\pi} (\theta(b) - \theta(a)) = \frac{1}{2\pi} \int_a^b \frac{d\theta}{ds} ds$$

is an integer (see Fig. 8.1)

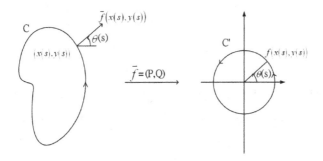

Fig. 8.1

215

$$I_f(C) = \text{ winding number}$$
$$= \text{ number of times that the "circle" } C'$$
$$\text{winds in the counterclockwise direction.}$$

Remark 8.1.1 $I_f(C)$ is independent of the choice of parameterization $\theta(s)$.

Example 8.1.1 Let C be the unit circle and

$$f(x,y) = (P(x,y), Q(x,y)) = \left(2x^2 - 1, 2xy\right).$$

Then $(P, Q) = (\cos 2\theta, \sin 2\theta)$, $0 \le \theta \le 2\pi$ (see Fig. 8.2).

$$I_f(C) = \frac{1}{2\pi} \cdot 4\pi = 2.$$

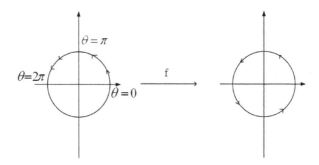

Fig. 8.2

Example 8.1.2 Let C be a circle of radius ρ and $f(z) = z^n$. If $z = \rho e^{i\varphi}$, then $z^n = \rho^n e^{in\varphi}$ and hence we have $I_f(C) = n$.

Since

$$\tan \theta(s) = \frac{Q(s)}{P(s)},$$

differentiating both sides of the above identity with respect to s yields

$$\frac{d\theta}{ds} \sec^2 \theta(s) = \frac{d\theta}{ds} \left(1 + \tan^2 \theta(s)\right) = \frac{P(s)\frac{dQ}{ds} - Q(s)\frac{dP}{ds}}{P^2(s)},$$

and we have

$$\frac{d\theta}{ds} = \frac{P\frac{dQ}{ds} - Q\frac{dP}{ds}}{P^2 + Q^2}.$$

Thus we have the formula

$$I_f(C) = \frac{1}{2\pi} \oint_C \frac{-Q}{P^2 + Q^2} dP + \frac{P}{P^2 + Q^2} dQ. \tag{8.1}$$

Lemma 8.1.1 (Homotopy invariance property)
Let f_t, $0 \le t \le 1$, be a continuous family of vector fields such that $f_t(C) \ne 0$ for all t. Then $I_{f_t}(C) \equiv$ constant.

Proof. From formula (8.1), obviously $I_{f_t}(C)$ is continuous in t. Hence $I_{f_t}(C) \equiv \sigma$ constant integer. $\quad\square$

Theorem 8.1.1 *Let C be a smooth Jordan curve and there is no equilibrium of f inside C. Then $I_f(C) = 0$.*

Proof. Let Ω be the region enclosed by C. Then

$$\begin{aligned}
I_f(C) &= \frac{1}{2\pi} \oint \frac{-Q}{P^2 + Q^2} dP + \frac{P}{P^2 + Q^2} dQ \\
&= \frac{1}{2\pi} \int\int_\Omega \frac{\partial}{\partial P}\left(\frac{P}{P^2 + Q^2}\right) + \frac{\partial}{\partial Q}\left(\frac{Q}{P^2 + Q^2}\right) dP dQ \\
&= 0.
\end{aligned}$$

$\quad\square$

Theorem 8.1.2 *Let C_1, C_2 be two Jordan curves with C_2 enclosed by C_1 and the vector field $\vec{f}$ does not vanish on the region between C_1 and C_2. Then $I_f(C_1) = I_f(C_2)$.*

Proof. Consider the Jordan curve $\Gamma : ABAA'B'A'A$ (see Fig. 8.3).
Then from Theorem 8.1.1, it follows that

$$\begin{aligned}
0 &= I_f(\Gamma) \\
&= \frac{1}{2\pi} \int_\Gamma \frac{d\theta}{ds} = \frac{1}{2\pi}\left[\int_{C_1} + \int_{AA'} - \int_{C_2} + \int_{A'A}\right]\frac{d\theta}{ds}ds.
\end{aligned}$$

Hence $I_f(C_1) = I_f(C_2)$. $\quad\square$

Theorem 8.1.3 [CL]
Let C be a continuously differentiable, simple, closed curve with nonzero tangent vector field f in xy plane (i.e. C is a periodic orbit of $x' = f(x)$). Then $I_f(C) = 1$.

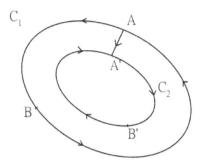

Fig. 8.3

Proof. If $u(P)$ is the unit tangent vector to C at P, then clearly $I_u(C) = I_f(C)$, and so it suffices to prove theorem for vector field u. Without loss of generality, assume that C lies entirely in the region $y \geq 0$ and that the point P of C are given by $P(t) : (\alpha_1(t), \alpha_2(t)), 0 \leq t \leq 1$. Thus $v(t) = (\alpha'_1(t), \alpha'_2(t))$, and it can be further assumed that the positive x axis is tangent to C at $P_0 = P(0)$, i.e., $v(0)$ has the same direction as the positive x axis (see Fig. 8.4).

We shall prove Theorem 8.1.3 by constructing an auxiliary vector V on the closed triangular region

$$T : 0 \leq s \leq 1, s \leq t \leq 1$$

in (s,t) plane (see Fig. 8.5) as follows:

Define $V(s,s) = u(s)$ for $0 \leq s \leq 1$, $V(0,1) = -u(0)$, and for all other (s,t) in T define $V(s,t)$ to be the unit vector in the direction from $P(s)$ to $P(t)$ on C (see Fig. 8.4).

Let $\theta(s,t)$ be the angle that $V(s,t)$ makes with positive x-axis. Clearly $\theta(0,0) = 0$ and since C remains in the region $y \geq 0$, $\theta(0,t)$ varies from zero to π as t varies from zero to one. Similarly $\theta(s,1)$ varies from π to 2π as s varies from zero to one. From the definition of V it is clear that V is continuous on T and $V \neq 0$ there. Hence by Theorem 8.1.1, applied to the boundary B of T, $I_v(B) = 0$. This means that the variation of $\theta(s,s)$ as s goes from zero to one is 2π. But this is precisely the variation in the angle that u makes with positive x axis as C is transversed once in the positive direction. Hence $I_u(C) = 1$ and we complete the proof of Theorem 8.1.3.   □

Fig. 8.4

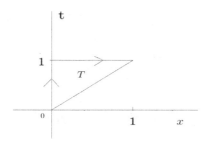

Fig. 8.5

Corollary 8.1.1 *Suppose the autonomous system $x' = f(x)$, $x \in \mathbb{R}^2$ has a periodic orbit C, then there exists an equilibrium of f inside C.*

Proof. If there exists no equilibrium of f inside C, then $I_f(C) = 0$ which is a contradiction to the result of Theorem 8.1.3, $I_f(C) = 1$. □

Definition 8.1.1 *Let z_0 be an* __isolated__ *zero of the vector field f. Assume $f \neq 0$ on a simple closed curve C enclosing z_0 with dist (z_0, C) sufficiently small. We define the index of f at z_0 to be*

$$I_f(z_0) = I_f(C).$$

Remark 8.1.2 From Theorem 8.1.2, Definition 8.1.1 is well-defined.

In the following, we shall compute the index of the saddle point, center, spiral, node of a nonlinear vector field. As a first step, we compute these indices for linear vector fields, then we do for nonlinear vector fields through __linearization__ and __homotopy invariance__.

Consider the linear vector field

$$\begin{pmatrix} x \\ y \end{pmatrix}' = \begin{pmatrix} a & b \\ c & d \end{pmatrix}\begin{pmatrix} x \\ y \end{pmatrix} = \begin{pmatrix} ax + by \\ cx + dy \end{pmatrix} = \begin{pmatrix} P(x,y) \\ Q(x,y) \end{pmatrix}. \tag{8.2}$$

Assume $(0,0)$ is a nondegenerate critical point of (8.2), i.e., the determinant $\Delta = ad - bc \neq 0$. The eigenvalues of $\begin{pmatrix} a & b \\ c & d \end{pmatrix}$ satisfies

$$\lambda^2 - (a + d)\lambda + (ad - bc) = 0.$$

If $\Delta > 0$, then $(0,0)$ is a source, a sink, or a center. On the other hand, if $\Delta < 0$, then $(0,0)$ is a saddle point. Let $C = \{(x,y) : x = \cos s, \ y = \sin s, \ 0 \leq s \leq 2\pi\}$ be the unit circle. Then the index for the linear vector field $f = (P, Q)$ is

$$I_f(C) = \frac{1}{2\pi} \int_0^{2\pi} \frac{d\theta}{ds} ds,$$

where

$$\theta(s) = \tan^{-1} \frac{Q(x(s), y(s))}{P(x(s), y(s))} = \tan^{-1} \frac{cx + dy}{ax + by}.$$

Since $\frac{dx}{ds} = -y$, $\frac{dy}{ds} = x$, we have

$$\frac{d\theta}{ds} = \frac{\frac{d}{ds}\left(\frac{cx+dy}{ax+by}\right)}{1 + \left(\frac{cx+dy}{ax+by}\right)^2}$$

$$= \frac{(ax + by)\left[c\frac{dx}{ds} + d\frac{dy}{ds}\right] - \left(a\frac{dx}{ds} + b\frac{dy}{ds}\right)(cx + dy)}{(ax + by)^2 + (cx + dy)^2}$$

$$= \frac{ad - bc}{(ax + by)^2 + (cx + dy)^2}.$$

Hence we have

$$I_f(C) = \frac{ad - bc}{2\pi} \int_0^{2\pi} \frac{dt}{(a\cos t + b\sin t)^2 + (c\cos t + d\sin t)^2}. \tag{8.3}$$

We note the R.H.S. of (8.3) is an integer and is continuous in a, b, c, d for $ad - bc \neq 0$. If $ad - bc > 0$, we have two cases:

(i) $ad > 0$:

Let $b, c \to 0$ and $d \to a$ then we have $I_f(C) = \frac{1}{2\pi} \int_0^{2\pi} d\theta = 1$ by homotopy invariance property.

(ii) $ad \le 0$:

Then $bc < 0$.

Let $a, d \to 0$ and $b \to -c$, it follows that $I_f(C) = 1$.

If $ad - bc < 0$, then the same arguments with $d \to -a$, $b, c \to 0$ yields $I_f(C) = -1$.

Theorem 8.1.4 Let f and g be vector fields which never have opposite directions at any point of Jordan curve C. Then $I_f(C) = I_g(C)$.

Proof. Define a homotopy between f and g,

$$h_s = (1 - s)f + sg, \quad 0 \le s \le 1.$$

Then $h_0 = f$ and $h_1 = g$. We claim that $h_s \ne 0$ at every point of C, for all $0 < s < 1$. If not, then, for some $0 < s < 1$, there exists a point p on C such that $h_s(p) = 0$, i.e., $f(p) = -\frac{s}{1-s}g(p)$ which contradicts the assumption. Hence the claim holds and by homotopy invariance property $I_f(C) = I_{h_s}(C) = I_g(C)$. □

Theorem 8.1.5 Let $ad - bc \ne 0$ and

$$
\begin{aligned}
f(x, y) &= \begin{pmatrix} f_1(x, y) \\ f_2(x, y) \end{pmatrix} = \begin{pmatrix} a & b \\ c & d \end{pmatrix} \begin{pmatrix} x \\ y \end{pmatrix} + \begin{pmatrix} g_1(x, y) \\ g_2(x, y) \end{pmatrix} \\
&= \begin{pmatrix} a & b \\ c & d \end{pmatrix} \begin{pmatrix} x \\ y \end{pmatrix} + g(x, y)
\end{aligned}
\tag{8.4}
$$

where $g_1(x, y) = O(\sqrt{x^2 + y^2})$, $g_2(x, y) = O(\sqrt{x^2 + y^2})$ as $(x, y) \to (0, 0)$. Then $I_f(0) = I_v(0)$ where v is the linear vector field $v(x, y) = D_x f(0, 0) \begin{pmatrix} x \\ y \end{pmatrix} = \begin{pmatrix} a & b \\ c & d \end{pmatrix} \begin{pmatrix} x \\ y \end{pmatrix}$.

Proof. Let $C_r = \{(x, y) : x = r \cos \theta, \ y = r \sin \theta\}$ be the circle with center $(0, 0)$ and radius r which is sufficiently small. We show the vector fields f and v never point in opposite direction on C_r. Then, from Theorem 8.1.4, it follows that $I_f(C_r) = I_v(C_r)$ and hence $I_f(0) = I_v(0)$.

Suppose, on the contrary, the vector fields f and v point in opposite direction at some point of C_r. Then

$$f(x_0, y_0) + sv(x_0, y_0) = 0 \quad \text{for some} \ s > 0$$

and for some $(x_0, y_0) \in C_r$. From (8.4), we have $(1 + s)v(x_0, y_0) = -g(x_0, y_0)$ and hence

$$(1 + s)^2 \parallel v(x_0, y_0) \parallel^2 = \parallel g(x_0, y_0) \parallel^2 . \tag{8.5}$$

Since $ad - bc \neq 0$, v vanishes only at $(0, 0)$. Let $m = \inf \{\|v(x, y)\| : x^2 + y^2 = 1\}$. Then $m > 0$. Since

$$v \left(\frac{x}{\sqrt{x^2 + y^2}}, \frac{y}{\sqrt{x^2 + y^2}} \right) = \frac{1}{\sqrt{x^2 + y^2}} v(x, y),$$

it follows that $\frac{1}{r} \parallel v(x, y) \parallel \geq m$ for $(x, y) \in C_r$. From (8.5), we have

$$m^2 (1 + s)^2 r^2 \leq \parallel g(x_0, y_0) \parallel^2,$$

or

$$m^2 < m^2 (1 + s)^2 \leq \frac{\parallel g(x_0, y_0) \parallel^2}{r^2} \to 0, \quad \text{as} \quad r \to 0.$$

This is a contradiction and we complete the proof. $\square$

From Theorems 8.1.1, 8.1.3, 8.1.5 and the indices of linear vector fields, we have the following:

Theorem 8.1.6

(i) *The index of a sink, source, center is* $+1$*;*
(ii) *The index of a saddle is* -1*;*
(iii) *The index of a closed* **orbit** *is* $+1$*;*
(iv) *The index of a closed* **curve** *not containing any fixed points is* 0*;*
(v) *The index of a closed curve is equal to the sum of indices of the fixed points within it.*

Corollary 8.1.2 *Inside any periodic orbit* Γ*, there exists at least one fixed point. If there is only one, then it must be a sink, a source or a center. If all fixed points within* Γ *are hyperbolic, there must be odd number* $2n + 1$ *of which* n *are saddle,* $n + 1$ *either sink, source.*

Now we shall prove two classical theorems using index theory. Let $z = x + iy$ and $f(z) = z^n$, n is a positive integer. Then from Example 8.1.2, $I_f(C) = n$ for $C = \{z : |z| = \rho\}$.

Theorem 8.1.7 (Fundamental Theorem of Algebra)
Let $p(z) = a_n z^n + \cdots + a_0$*,* $a_n \neq 0$*, be a polynomial in complex variable* $z = x + iy$*,* $a_i \in \mathbb{C}$*,* $i = 1, \cdots, n$*. Then* $p(z) = 0$ *has at least one zero.*

Proof. Assume $a_n = 1$. Define a homotopy of vector fields

$$f_t(z) = z^n + t\left(a_{n-1}z^{n-1} + \cdots + a_0\right).$$

Since

$$\frac{|a_{n-1}z^{n-1} + \cdots + a_0|}{|z^n|} \to 0 \text{ as } |z| \to \infty,$$

there exists $\rho > 0$ such that

$$f_t(z) \neq 0 \text{ for } |z| \geq \rho \text{ and } 0 \leq t \leq 1.$$

We note that $f_0(z) = z^n$, $f_1(z) = p(z)$. Let $C = \{z : |z| = \rho\}$. Then

$$n = I_{f_0}(C) = I_{f_1}(C) = I_p(C) \neq 0.$$

If $p(z)$ has no zeros inside C then $I_p(C) = 0$. This is a contradiction. Hence $p(z)$ has at least one zero inside C. $\qquad\square$

Theorem 8.1.8 (Brouwer Fixed Point Theorem for $\mathbb{R}^2$)
Let $D = \{z \in \mathbb{R}^2 : |z| \leq 1\}$ and $g : D \to D$ be continuous. Then g has a fixed point.

Proof. Let $C = \partial D = \{z : |z| = 1\}$ be the unit circle in $\mathbb{R}^2$. We shall prove by contradiction. Suppose $g(z) \neq z$ for all $z \in D$, i.e., there is no fixed point of g in D. Define a continuous family of vector fields,

$$f_t(z) = tg(z) - z.$$

Verify $f_t(z) \neq 0$ for all $z \in C$ and $0 \leq t \leq 1$. If not, for some $0 < t < 1$ and some $z \in C$ such that $f_t(z) = 0$ or $tg(z) = z$. Then $1 > t = \frac{|z|}{|g(z)|} = 1$. This is a contradiction. Hence $I_{f_1}(C) = I_{f_0}(C)$. Claim: $I_{f_0}(C) = 1$. Since $f_0(z) = -z$ is the linear vector field $f_0(z) = Az$ where $A = \begin{pmatrix} -1 & 0 \\ 0 & -1 \end{pmatrix}$ has $(0,0)$ as a sink. Thus $I_{f_0}(C) = 1$. Hence $I_{f_1}(C) = 1$. But $f_1(z) = g(z) - z \neq 0$ on D, which implies $I_{f_1}(C) = 0$. This is a contradiction. $\qquad\square$

8.2 Introduction to the Brouwer Degree in $\mathbb{R}^n$

Brouwer degree $d(f, p, D)$ is a useful tool to study the existence of solutions of the nonlinear equation $f(x) = p$, $f : D \subseteq \mathbb{R}^n \to \mathbb{R}^n$, where D is a bounded domain with boundary $\partial D = C$ and $f(x) \neq p$ for all $x \in C$. To motivate the definition of $d(f, p, D)$, we consider the special case $n = 2, p = 0$. From Theorem 8.1.6(v), (i), (ii), we have

$$I_f(C) = \sum_{i=1}^{N} I_f(z_i) = \sum_{i=1}^{N} I_{v_i}(0)$$

$$= \sum_{i=1}^{N} sgn(\det J_f(z_i))$$

where $J_f(x) = Df(x)$ is the Jacobian matrix of $v_i(x) = Df(x_i)x$ and $\{z_i\}_{i=1}^{N}$ is the set of all fixed points inside C.

We now define the Brouwer degree $d(f, p, D)$ as follows:
Let $f : D \subseteq \mathbb{R}^n \to \mathbb{R}^n, p \in \mathbb{R}^n$ and $f(x) \neq p$ for all $x \in \partial D$.

Step 1: Let $f \in C^1(\overline{D})$ and assume $\det J_f(x) \neq 0$ for all $x \in f^{-1}(p)$. We define $d(f, p, D) = \sum_{x \in f^{-1}(p)} sgn(\det J_f(x))$.

From the assumption $\det J_f(x) \neq 0$ for all $x \in f^{-1}(p)$ and the Inverse Function Theorem, it follows that $f^{-1}(p)$ is a discrete subset in the compact set $\overline{D}$. Thus $f^{-1}(p)$ is a finite set and $d(f, p, D)$ is well-defined.

Step 2: Let $f \in C^1(\overline{D})$ and assume that there exists $x \in f^{-1}(p)$ such that $\det J_f(x) = 0$. From Sard's Theorem (see Theorem 8.2.1) there exists a sequence $\{p_m\}$, $p_m \to p$ as $m \to \infty$ such that $\det J_f(x) \neq 0$ for all $x \in f^{-1}(p_m)$. Then we define $d(f, p, D) = \lim_{m \to \infty} d(f, p_m, D)$. We note that it can be shown that the definition is independent of the choices of $\{p_m\}$.

Step 3: Let $f \in C(\overline{D})$. Since $C^1(\overline{D})$ is dense in $C(\overline{D})$, there exists $\{f_m\} \subseteq C^1(\overline{D})$ such that $f_m \to f$ uniformly on $\overline{D}$. Then we define $d(f, p, D) = \lim_{m \to \infty} d(f_m, p, D)$. We note that it can be shown that the definition is independent of the choices of $\{f_m\}$.

In the following we state Sard's Theorem without proof and explain the theorem. The proof can be found in [BB].

Theorem 8.2.1 (Special case of Sard's Theorem)
Let $f : \overline{D} \subseteq \mathbb{R}^n \to \mathbb{R}^n$ be C^1 and $B = \{x \in D : \det J_f(x) = 0\}$. Then $f(B)$ has empty interior.

Remark 8.2.1 If $p \in f(B)$, then $p = f(x)$ for some $x \in B$. Hence $x \in f^{-1}(p)$ and $\det J_f(x) = 0$. Since p is not an interior point, there exists $\{p_m\}, p_m \to p, m \to \infty$ such that $p_m \notin f(B)$. Then for all $x \in f^{-1}(p_m)$, we have $\det J_f(x) \neq 0$.

Before we state Sard's Theorem, we introduce the following definition.

Definition 8.2.1 *Let* $f : D \subseteq \mathbb{R}^n \to \mathbb{R}^m$ *be* C^1, $m \leq n$. *We say that* $x \in \mathbb{R}^n$ *is a regular point if* $\mathrm{Rank} J_f(x) = m$. *(I.e., the* $m \times n$ *matrix* $J_f(x)$ *is of full rank.) We say that* $x \in \mathbb{R}^n$ *is a critical point if* x *is not a regular point. We say that* $p \in \mathbb{R}^n$ *is a regular value if for each* $x \in f^{-1}(p)$, x *is a regular point. We say that* $p \in \mathbb{R}^m$ *is a critical value if* p *is not a regular value.*

Theorem 8.2.2 (Sard's Theorem)
Let $f : D \subseteq \mathbb{R}^n \to \mathbb{R}^m$ *be* C^1 *and* $C = \{x \in D : \mathrm{Rank} J_f(x) < m\}$, *i.e.,* C *is the set of critical points. Then* $f(C)$ *is of measure zero.*

Remark 8.2.2 Sard's Theorem implies that for almost all $p \in \mathbb{R}^m$, p is a regular value. Let $E = f(C)$. Then E is of measure zero. Let $p \in \mathbb{R}^m \backslash E$, then we have $f^{-1}(p) \cap C$ is empty, i.e., for each $x \in f^{-1}(p)$, x is a regular point. Hence p is a regular value.

Remark 8.2.3 If $m = n$ then $C = \{x \in \mathbb{R}^n : \det J_f(x) = 0\}$. By Sard's Theorem, $f(C)$ is of measure zero and thus $f(C)$ has empty interior. Hence we complete the proof of Theorem 8.2.1.

Properties of $d(f, p, D)$
_____

(i) Homotopy invariance
Let $H : \overline{D} \times [0, 1] \to \mathbb{R}^n$, $\overline{D} \subseteq \mathbb{R}^n$ be continuous in x and t and $H(x, t) \neq p$ for all $x \in \partial D, 0 \leq t \leq 1$. Then $d(H(\cdot, t), p, D) = constant$.

Proof. Let $h(t) = d(H(\cdot, t), p, D)$. Then $h(t)$ is continuous in t and $h(t)$ is integer-valued. Hence $h(t) \equiv constant$ for all t. $\square$

Remark 8.2.4 Consider linear homotopy: $H(x, t) = tf(x) + (1 - t)g(x)$. Then $H(x, 0) = g(x), H(x, 1) = f(x)$ and $d(f, p, D) = d(g, p, D)$ provided $tg(x) + (1 - t)f(x) \neq 0$ for all $x \in \partial D$ and $0 \leq t \leq 1$.

(ii) $d(f, p, D)$ **is uniquely determined by** $f \mid_{\partial D}$.

Proof. Let $\widetilde{f}(x)$ satisfies $\widetilde{f}(x) = f(x)$ for $x \in \partial D$. Define $H(x,t) = t\widetilde{f}(x) + (1-t)f(x)$, $x \in \overline{D}$, $0 \le t \le 1$. Then for $x \in \partial D$, $H(x,t) = f(x) \ne p$. By (i) we have $deg(f, p, D) = deg(\widetilde{f}, p, D)$. □

(iii) $d(f, p, D)$ **is a continuous function of** f **and** p.

Proof. (iii) follows directly from the definition. □

(iv) (Poincaré-Bohl) Let $f(x) \ne p$, $g(x) \ne p$ **for all** $x \in \partial D$. **If** $f(x) - p$ **and** $g(x) - p$ **never point opposite on each** $x \in \partial D$ **then** $d(f, p, D) = d(g, p, D)$.

Proof. Define a homotopy

$$H(x,t) = t(f(x) - p) + (1-t)(g(x) - p).$$

We claim that $H(x,t) \ne 0$ for all $x \in \partial D, 0 \le t \le 1$. If not, then there exists t, $0 < t < 1$ and, $x \in \partial D$ such that $H(x,t) = 0$ and it follows that $f(x) - p = -\frac{1-t}{t}(g(x) - p)$. This contradicts that $f(x) - p$ and $g(x) - p$ never point opposite for each $x \in \partial D$. By (i), $deg(f, p, D) = deg(g, p, D)$. □

Corollary 8.2.1 *If* $f(x) \cdot x < 0$ *for all* $x \in \partial D$ *or* $f(x) \cdot x > 0$ *for all* $x \in \partial D$, *then there exists* $x_0 \in D$ *such that* $f(x_0) = 0$ *or there exists an equilibrium of* $x' = f(x)$.

Proof. Let $g(x) = x, p = 0$. Then $d(g, 0, D) = \sum_{x \in g^{-1}(0)} sgn(\det J_g(x)) = 1$. The assumption $f(x) \cdot x > 0$ for all $x \in \partial D$ says that $f(x)$ and $g(x)$ never point opposite on ∂D. From (iv) $d(f, 0, D) = d(g, 0, D) = 1 \ne 0$ and by the following (v) there exists $x_0 \in D$ such that $f(x_0) = 0$. If $f(x) \cdot x < 0$ for $x \in \partial D$, then we replace $f(x)$ by $-f(x)$. □

(v) If $d(f, p, D) \ne 0$ **then there exists** $x \in D$ **such that** $f(x) = p$.

Proof. We prove (v) by contradiction. Suppose $f(x) \neq p$ for all $x \in D$. Then by definition of $d(f, p, D)$, it follows that $d(f, p, D) = 0$. This contradicts the assumption $d(f, p, D) \neq 0$. $\square$

Next, we prove

Brouwer Fixed Point Theorem

Let $D^n = \{x \in \mathbb{R}^n : |x| \leq 1\}$ and $f : D^n \to D^n$ be continuous. Then f has a fixed point, i.e., there exists $x \in D^n$ such that $f(x) = x$.

Proof. We shall show that $f(x) - x = 0$ has a solution in D^n. Consider the following homotopy $H(x, t) = x - tf(x), x \in D^n, 0 \leq t \leq 1$. Then $H(x, 0) = x = g(x)$, $H(x, 1) = x - f(x)$. We claim that $H(x, t) \neq 0$ for all $x \in \partial D^n, 0 \leq t \leq 1$. Obviously $H(x, 0) = g(x) \neq 0$ on ∂D^n. If $H(x, t) = 0$ for some $0 < t < 1$ and some $x \in \partial D^n$, then $x = tf(x)$. Since $1 = |x| = t |f(x)| \leq t$, we obtain a contradiction. Hence $d(x - f(x), 0, D^n) = d(g, 0, D^n) = 1 \neq 0$ and it follows that there exists $x \in D^n$ such that $x = f(x)$. $\square$

Remark 8.2.5 Brouwer Fixed Point Theorem has many applications in Economics. In 1976, Kellogg, Li and Yorke gave a constructive proof of Brouwer Fixed Point Theorem. Interested readers may consult their work [KLY] in computing numerically Brouwer fixed points.

Generalized Brouwer Fixed Point Theorem

Let $K \subset \mathbb{R}^n$ be a compact convex set. If $f : K \to K$ is a continuous function then f admits a fixed point.

Proof. By translation we may assume Int $K \neq \phi$ and $0 \in$ Int K. Consider a map $J_K : \mathbb{R}^n \to \mathbb{R}^+$ defined by $J_K(x) = inf\{\lambda \geq 0 | x \in \lambda K\}$. Define $h(0) = 0, h(x) = J_K(x)\frac{x}{|x|}$ if $x \in K \backslash \{0\}$. Then $h : K \to D^n$ is a homeomorphism with $h(\partial K) \subset \partial D^n$. Let $g = h \circ f \circ h^{-1} : D^n \to D^n$. Then g is continuous, and by Brouwer Fixed Theorem, there exists $x_0 \in D^n$ such that $x_0 = g(x_0) = h \circ f \circ h^{-1}(x_0)$ or $h^{-1}(x_0) = f(h^{-1}(x_0))$. Hence $h^{-1}(x_0) \in K$ is a fixed point of f. $\square$

(vi) Domain decomposition

If $\{D_i\}_{i=1}^N$ is a finite disjoint open sets in D and $f(x) \neq p$ for all $x \in (\overline{D} - \cup_{i=1}^N D_i)$, then $d(f, p, D) = \sum_{i=1}^N d(f, p, D_i)$.

Proof. Assume $f \in C^1(\overline{D})$ and $det J_f(x) \neq 0$ for all $x \in f^{-1}(p)$. Then

$$d(f, p, D) = \sum_{x \in f^{-1}(p)} sgn(\det J_f(x))$$

$$= \sum_{i=1}^{N} \sum_{x \in D_i \cap f^{-1}(p)} sgn(\det J_f(x))$$

$$= \sum_{i=1}^{N} d(f, p, D_i).$$

$\square$

(vii) (Odd mapping) Let $D \subseteq \mathbb{R}^n$ symmetric domain about origin 0 and $f(-x) = -f(x)$ on D with $f : D \to \mathbb{R}^n, f(x) \neq 0$ on ∂D, then $d(f, 0, D)$ is an odd integer.

Proof. We may assume $f \in C^1$ and $\det(J_f(x)) \neq 0$ for all $x \in f^{-1}(0)$ in D. The solutions of $f(x) = 0$ can be group in the pair $(\bar{x}, -\bar{x})$ together with $\bar{x} = 0$. Since $f(x)$ is an odd mapping, we have $\det(J_f(\bar{x})) = (-1)^n \det(J_f(-\bar{x})) \neq 0$ and

$$d(f, p, D) = \sum_{\bar{x} \in f^{-1}(0)} sgn(\det J_f(\bar{x}))$$

$$= sgn \det J_f(0) + even\ integer$$

$$= odd\ integer.$$

$\square$

Borsuk-Ulam Theorem

Let $f : S^n \to \mathbb{R}^n$ be a continuous map where $S^n = \{x \in \mathbb{R}^{n+1} | \parallel x \parallel_2 = 1\}$. Then there exists $x_0 \in S^n$ with $f(x_0) = f(-x_0)$.

Proof. Define $\tilde{f} : S^n \to \mathbb{R}^{n+1}$, $\tilde{f} = (f_1(x), -f_1(x), f_2(x) - f_2(-x), \ldots, f_n(x) - f_n(-x), 0)$.

We prove by contradiction.

Assume for all $x \in S^n, f_i(x) \neq f_i(-x)$ for some $i \in \{1, 2, \ldots, n\}$.

Let $\sum = \{x \in \mathbb{R}^{n+1} | \parallel x \parallel_2 \leq 1\}$. Then $d(\tilde{f}, 0, \sum)$ is defined since $\tilde{f}(x) \neq 0$ for all $x \in \partial \sum = S^n$.

By (vii) and $\tilde{f}$ is an odd mapping, $d(\tilde{f}, 0, \sum) = odd \neq 0$.

Let $\tilde{g} : \sum \to \mathbb{R}^{n+1}$. Then $d(\tilde{g}, 0, \sum) = 0$.

Consider homotopy $H(x,t) = t\tilde{g}(x) + (1-t)\tilde{f}(x)$. Since $H(x,t) \neq 0$ on S^n for all $0 < t < 1$, then $d(\tilde{f}, 0, \sum) = d(H(\cdot, t); 0, \sum) = d(\tilde{g}, 0, \sum) = 0$. We obtain a contradiction. $\qquad\qquad\square$

Example 8.2.1 Let n be odd.

Let $f : S^{n-1} \to \mathbb{R}^n$ be continuous and $f(x) \neq 0$ for all $x \in S^{n-1}$. Then the direction of some normal vector is left unchanged under f, i.e., there exist $\lambda \neq 0$ and $x_0 \in S^{n-1}$ such that $f(x_0) = \lambda x_0$.

Proof. Let $\Sigma = \{x \in \mathbb{R}^n : |x| \leq 1\}$. Then $\partial\Sigma = S^{n-1}$ and by hypothesis, $d(f, 0, \Sigma)$ is defined. Consider the following homotopies,

$$H_1(x,t) = tf(x) + (1-t)x,$$
$$H_2(x,t) = tf(x) + (1-t)(-x).$$

Claim: One of the homotopies must vanish for some $t_0 \in (0,1]$, some $x_0 \in S^{n-1}$. Otherwise if both of $H_1(x,t)$, $H_2(x,t)$ do not vanish, then

$$d(f, 0, \Sigma) = d(x, 0, \Sigma) = 1$$
$$d(f, 0, \Sigma) = d(-x, 0, \Sigma) = -1 \qquad (\because n \, is \, odd).$$

We obtain a contradiction. Hence there exists $x_0 \in S^{n-1}$ such that $f(x_0) = \lambda x_0$ for some $\lambda \neq 0$. $\qquad\qquad\square$

Example 8.2.2 A nonvanishing tangent vector field exists on S^{n-1} if and only if n is even.

Proof. ($\Leftarrow$) If n is even, say $n = 2m, x = (x_1, x_2, \ldots, x_{2m-1}, x_{2m}) \in S^{n-1}$ set $V(x) = (x_1, x_2, \ldots, x_{2m-1}, x_{2m})$. Then $V(x) \cdot x = 0$ and $V(x) \neq 0$ for all $x \in S^{n-1}$. Hence $V(x)$ is a nonvanishing tangent vector field on S^{n-1}.

($\Rightarrow$) Suppose a tangent vector field $V(x)$ exists on S^{n-1}. We want to show n is even.
Consider

$$H(x,t) = x \cos \pi t - \frac{V(x)}{|V(x)|} \sin \pi t.$$

Then

$$H(x,0) = x$$
$$H(x,1) = -x.$$

Claim $|H(x,t)| = 1 \neq 0$ for all $x \in \partial\Sigma = S^{n-1} \subseteq \mathbb{R}^n$

$$
|H(x,t)|^2 = \sum_{i=1}^{n} \left(x_i \cos \pi t - \frac{(V(x))_i}{|V(x)|^2} \sin \pi t \right)^2
$$

$$
= \sum_{i=1}^{n} x_i^2 \cos^2 \pi t + \frac{(V(x))_i^2}{|V(x)|^2} \sin^2 \pi t - 2x_i \frac{(V(x))_i}{|V(x)|} \cos \pi t \sin \pi t
$$

$$
= \cos^2 \pi t + \sin^2 \pi t - x \cdot \frac{V(x)}{|V(x)|} \sum_{i=1}^{n} \sin 2\pi t
$$

$$
= 1.
$$

Then $d(f, 0, \Sigma) = d(x, 0, \Sigma) = 1$. Since $d(-x, 0, \Sigma) = (-1)^n$, we obtain that n is even. $\qquad\square$

Next we introduce the notion of index.

Definition 8.2.2 *Assume* $f(x) - p$ *has an isolated zero* x_0. *Then the index of* f *at* x_0 *is defined as*

$$
i(x_0, f(x) - p) = d(f(x) - p, 0, B_\epsilon)
$$
$$
= d(f, p, B_\epsilon),
$$

where B_ϵ *is the ball* $B(x_0, \varepsilon), 0 < \varepsilon \ll 1, B_\epsilon \subset D$.

As we did in Theorem 8.1.6(vi), we have the following result.

Theorem 8.2.3 $d(f, p, D) = \sum_{x_j \in f^{-1}(p)} i(x_j, f(x) - p)$.

Remark 8.2.6 For the infinite dimensional space, the Brouwer degree and Brouwer fixed point Theorem can be extended to Leray-Schauder degree and Schauder fixed point theorem, respectively. Interested reader may consult [BB].

8.3   Lienard Equation with Periodic Forcing

Consider the Lienard equation (see Section 6.2) with a periodic forcing term:

$$
x'' + f(x)x' + g(x) = p(t). \tag{8.6}
$$

Let $F(x) = \int_0^x f(s)\,ds$. Assume f, g, p satisfy

(A1) $f \in C(\mathbb{R}, \mathbb{R})$ is even, g, $p \in C^1(\mathbb{R}, \mathbb{R})$ and p is T-periodic.

(A2) $g(x)/x \to \infty$ as $|x| \to \infty$.

(A3) There exist b, $B > 0$ such that $|F(x) - bg(x)| \leq B|x|$ for any $x \in \mathbb{R}$.

Example 8.3.1 Given $\varepsilon > 0$. By the Levinson-Smith theorem (Theorem 6.2.1), the Lienard equation

$$x'' + \varepsilon(x^2 - 1)x' + x^3 = 0$$

has unique limit cycle, and it is globally orbital stable. Clearly $f(x) = \varepsilon(x^2 - 1)$ and $g(x) = x^3$ satisfy assumptions (A1), (A2), and (A3) with $b = \varepsilon/3$, $B = \varepsilon$. If a T-periodic forcing term p were added, where $T > 0$ is arbitrary,

$$x'' + \varepsilon(x^2 - 1)x' + x^3 = p(t),$$

it is not apparent whether T-periodic solutions always exist.

Theorem 8.3.1 *There exists a T-periodic solution for (8.6).*

Proof. [Lef] The equation (8.6) can be written

$$\begin{cases} x' = y - F(x), \\ y' = p(t) - g(x). \end{cases}$$

Assumption (A1) ensures existence and uniqueness of solution for (8.6) with arbitrary initial condition $(x(0), y(0)) = (x_0, y_0)$. Denote this solution by $\phi^t(x_0, y_0)$. The mapping $P : \mathbb{R}^2 \to \mathbb{R}^2$ defined by

$$P(x, y) = \phi^T(x, y)$$

is well-defined. If P has a fixed point (x_0, y_0), then clearly $\phi^t(x_0, y_0)$ is a T-periodic solution for (8.6). Thus, all we need to show is the existence of fixed point for P.

Consider the quadratic form on $\mathbb{R}^2$:

$$u(x, y) = \frac{x^2}{b} - xy + \frac{by^2}{2}.$$

It can be alternatively written

$$\left\langle A \begin{pmatrix} x \\ y \end{pmatrix}, \begin{pmatrix} x \\ y \end{pmatrix} \right\rangle, \quad A = \begin{pmatrix} \frac{1}{b} & -\frac{1}{2} \\ -\frac{1}{2} & \frac{b}{2} \end{pmatrix}.$$

The quadratic form u is positive definite since A is positive definite, so its level sets are ellipses.

Along any solution $(x(t), y(t))$, we have

$$u' = \left(\frac{2x}{b} - y\right) x' + (by - x)y'$$

$$= \left(\frac{2x}{b} - y\right) (y - F(x)) + (by - x)(p(t) - g(x))$$

$$= -y^2 - \left[\frac{2F(x)}{b} - g(x)\right] x + [F(x) - bg(x)] y + \frac{2}{b}xy + p(t)(by - x).$$

It follows from assumptions (A2) and (A3) that

$$\frac{g(x)}{x}, \quad \frac{F(x)}{x} \to \infty \quad \text{as } |x| \to \infty.$$

Since we have assumed that f is even, $F(x)/x$ is nonnegative for any $x \in \mathbb{R}$. Using this observation and assumption (A3), we can compare the two brackets $[\cdots]$ in the above equation for u':

$$\frac{\frac{2F(x)}{b} - g(x)}{x} \geq \frac{F(x) - bg(x)}{bx} \geq -\frac{B}{b} \quad \text{for any } x \in R.$$

The difference $F(x)/(bx)$ between the first and second terms is unbounded, so the first term tends to infinity as $|x| \to \infty$. By (A3), the second term in above is bounded as $|x| \to \infty$.

Our next goal is to show that $u' < 0$ outside a sufficiently large circular disc. We begin by choosing a cropped double cone C_R and then consider its complement. See Fig. 8.6.

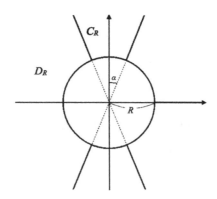

Fig. 8.6   Regions C_R and D_R.

Use polar coordinates $x = r\cos\theta$, $y = r\sin\theta$. Given $\alpha \in (0, \pi/2)$. Consider the double cone C given by $|\theta - \pi/2|$ or $|\theta - 3\pi/2| \leq \alpha$. Let

$$C_R = C \cap \{r \geq R\}.$$

In C_R,

$$\frac{u'}{r^2} = -\sin^2\theta - \left(\frac{\frac{2F(x)}{b} - g(x)}{x}\right)\cos^2\theta$$

$$+ \left(\frac{F(x) - bg(x)}{x} + \frac{2}{b}\right)\sin\theta\cos\theta + \frac{p(t)(b\sin\theta - \cos\theta)}{r}$$

$$\leq -\sin^2\theta + \frac{B}{b}\cos^2\theta + \left(B + \frac{2}{b}\right)|\sin\theta||\cos\theta| + \frac{|p(t)|(b+1)}{r}.$$

By choosing $\alpha > 0$ sufficiently small and $r_1 > 0$ sufficiently large, in C_{r_1} the last equation can be as close to -1 as possible. In particular, this implies the existence of $\alpha \in (0, \pi/2)$, $r_1 > 0$ such that $u' < 0$ on C_{r_1}.

With $\alpha \in (0, \pi/2)$ fixed as above. Consider (x, y) in the complement of C in $\mathbb{R}^2 \setminus \{r < R\}$, denoted by D_R; i.e.

$$D_R = C^c \cap \{r \geq R\}.$$

Since $|x|$ is away from zero and y/x is bounded for $(x, y) \in D_R$, the function h defined by

$$h(x, y, t) = \frac{p(t)(by - x)}{x^2}$$

is bounded on D_R, for any $R > 0$. Thus

$$u' = -y^2 - \left[\frac{2F(x)}{b} - g(x)\right]x + [F(x) - bg(x)]y + \frac{2}{b}xy + p(t)(by - x)$$

$$= -y^2 - \left\{\frac{\frac{2F(x)}{b} - g(x)}{x} - h(x, y, t)\right\}x^2 + \left[\frac{F(x) - bg(x)}{x} + \frac{2}{b}\right]xy.$$

As $|x| \to \infty$, the bracket $\{\cdots\}$ in the last line tends to ∞ but the bracket $[\cdots]$ in the last line is bounded. Thus we may choose $r_2 \geq r_1$ such that $u' < 0$ on D_r for $r \geq r_2$. Together with our discussions for u' on the cropped double cone C_{r_1}, we proved the existence of $r_2 > 0$ such that

$$u' < 0 \quad \text{whenever } r \geq r_2.$$

Choose $u_0 > \max_{r \leq r_2} u(x, y)$. Then on the ellipse $\{u = u_0\}$, we must have $r > r_2$, and so $u' < 0$ on this ellipse. This implies that the filled ellipse

$$E = \{(x, y) : u(x, y) \leq u_0\}$$

is positively invariant for the flow ϕ^t. Therefore, the mapping $P : \mathbb{R}^2 \to \mathbb{R}^2$ maps E into E. By the Brouwer fixed point theorem, there exists a fixed point of P in E. $\qquad\square$

8.4   Exercises

Exercise 8.1 Prove that the system of differential equations

$$\frac{dx}{dt} = x + P(x,y)$$

$$\frac{dy}{dt} = y + Q(x,y),$$

where $P(x,y)$ and $Q(x,y)$ are bounded functions on the entire plane, has at least one equilibrium.

Exercise 8.2 Let $f = Ax + b$ be a linear mapping of $\mathbb{R}^n$ into $\mathbb{R}^n$, where A is an $n \times n$ matrix.

Prove, by definition (ii) of degree, that if $f(x_0) = p$ then

$$d(f,p,B_\epsilon) = \begin{cases} 1 & \text{if } \det(A) > 0 \\ -1 & \text{if } \det(A) < 0 \end{cases}$$

where B_ϵ is a sphere of any radius ε centered about x_0 in $\mathbb{R}^n$. What happens if $\det(A) = 0$?

Exercise 8.3

(a) Let x_0 be an isolated zero of $f(x) = 0$ where $f : D \subseteq \mathbb{R}^n \to \mathbb{R}^n$. Suppose x_0 is a regular point, show that

$$i(x_0) = (-1)^\beta$$

where β is the number of real negative eigenvalues of the Jacobian $Df(x_0)$.

(b) Let $F(x_1, \cdots, x_n) = \sum a_{ij}x_i x_j$ be a real nonsingular quadratic form with $a_{ij} = a_{ji}$. Prove that $i(grad, F, 0) = (-1)^\lambda$ where λ is the number of negative eigenvalues of the symmetric matrix (a_{ij}).

Exercise 8.4 Let f be a continuous mapping of a bounded domain D of $\mathbb{R}^n$ into $\mathbb{R}^n$ with the property that, for all $x \in \partial D$, $f(x)$ never points in the direction $q(\neq 0)$, that is, $f(x) \neq kq$ for all real nonzero k. Then $d(f, 0, D) = 0$.

Exercise 8.5 Let $f(z) = u(x,y) + iv(x,y)$ be a complex analytic function defined on a bounded domain D and its closure $\overline{D}$ in $\mathbb{R}^2$, where $z = x + iy$ and $i^2 = -1$. Suppose $f(z) \neq p$ for $z \in \partial D$. Prove $d(f, p, D) \geq 0$, where f denote the mapping $(x, y) \to (u, v)$.

Exercise 8.6 (Frobenius Theorem)
Suppose A is an $n \times n$ matrix (a_{ij}) with $a_{ij} > 0$ for $i, j = 1, 2, \ldots, n$. Prove by using Brouwer Fixed Point Theorem, that A has a positive eigenvalue and a corresponding eigenvector $x = (x_1, \ldots, x_n)$, with all $x_i \geq 0$ and some $x_j > 0$.
(Hint: Let $|x| = |x_1| + \ldots + |x_n|$ and consider the mapping $\tilde{A}(x) = \frac{Ax}{|Ax|}$ on the closed convex set

$$\partial\Sigma_1^+ = \{x : |x| = 1, x = (x_1, \ldots, x_n), \ x_i \geq 0, \ i = 1, 2, \ldots, n\}.)$$

Exercise 8.7 Use the Poincaré-Bohl theorem to give an alternative proof for the Brouwer Fixed Point Theorem.

Exercise 8.8 Assume $f : \mathbb{R}^2 \to \mathbb{R}^2$ is continuous and

$$\lim_{\|x\| \to \infty} \frac{f(x) \cdot x}{\|x\|} = \infty.$$

Consider the system $\dot{x} = f(x)$. Show that there is an equilibrium point. Furthermore, if the equilibrium point is attractive, then either there is another equilibrium point, or there is periodic orbit. What happens if the space dimension is greater than or equal to 3?

Exercise 8.9 In the proof for Theorem 8.2.5 we assumed (A1)–(A3) hold. Show that in (A1), the assumption that f is even can be removed.

Chapter 9

PERTURBATION METHODS

9.1 Regular Perturbation Methods

Suppose we have a problem $P(\varepsilon)$ with small parameter $0 < \varepsilon \ll 1$ and the solution $x(\varepsilon)$, $\varepsilon \geq 0$. If $\lim_{\varepsilon \to 0} x(\varepsilon) = x(0)$, then we say that the problem $P(\varepsilon)$ is regular. In the following, we give several examples to show how to solve the regular perturbation problem $P(\varepsilon)$ by the perturbation methods. Then we explain how to verify that a problem is regular by the Implicit Function Theorem in Banach space.

Example 9.1.1 Solve the quadratic equation

$$P(\varepsilon) : x^2 - x + \varepsilon = 0, 0 < \varepsilon \ll 1. \tag{9.1}$$

When $\varepsilon = 0$, the problem $P(0)$ has two solutions $x = 0$ and $x = 1$. Let

$$x(\varepsilon) = \sum_{n=0}^{\infty} a_n \varepsilon^n = a_0 + a_1 \varepsilon + a_2 \varepsilon^2 + \cdots . \tag{9.2}$$

Substitute (9.2) into (9.1)

$$x^2(\varepsilon) - x(\varepsilon) + \varepsilon = 0 \tag{9.3}$$

or

$$(a_0 + a_1 \varepsilon + a_2 \varepsilon^2 + \cdots)^2 - (a_0 + a_1 \varepsilon + a_2 \varepsilon^2 + \cdots) + \varepsilon = 0. \tag{9.4}$$

The next step is to find the coefficients $a_0, a_1, a_2, a_3, \ldots, a_n, \cdots$ by comparing the coefficients ε^n in (9.4):

$O(1)$: Set $\varepsilon = 0$ into (9.3) or (9.4), we obtain

$x^2(0) - x(0) = 0$ or $a_0^2 - a_0 = 0$. Then $a_0 = 0$ or $a_0 = 1$.

$O(\varepsilon)$: Differentiating (9.3) with respect to ε yields

$$2x(\varepsilon)\frac{dx}{d\varepsilon}(\varepsilon) - \frac{dx}{d\varepsilon}(\varepsilon) + 1 = 0. \tag{9.5}$$

From (9.2), note that $x(0) = a_0$, $\frac{dx}{d\varepsilon}(0) = a_1$. If we set $\varepsilon = 0$ in (9.5), then

$$2a_0a_1 - a_1 + 1 = 0 \quad \text{or} \quad a_1 = \frac{-1}{2a_0 - 1}.$$

Thus $a_1 = 1$ if $a_0 = 0$ and $a_1 = -1$ if $a_0 = 1$.

$O(\varepsilon^2)$: Differentiating (9.5) with respect to ε yields

$$2x(\varepsilon)\frac{d^2x}{d\varepsilon^2}(\varepsilon) + 2\left(\frac{dx}{d\varepsilon}\right)^2 - \frac{d^2x}{d\varepsilon^2} = 0. \tag{9.6}$$

From (9.2), $\frac{d^2x}{d\varepsilon^2}(0) = 2a_2$. If we set $\varepsilon = 0$ in (9.6), then we have

$$4a_0a_2 + 2a_1^2 - 2a_2 = 0 \quad \text{or} \quad a_2 = \frac{-a_1^2}{2a_0 - 1}.$$

Thus $a_2 = -1$ if $a_0 = 0$, $a_1 = 1$ and $a_2 = -1$ if $a_0 = 1$, $a_1 = -1$.

For $O(\varepsilon^n)$, $n \geq 3$, we continue this process to find coefficients a_n, $n \geq 3$. Thus we obtain two solutions of $P(\varepsilon)$, namely,

$$x_1(\varepsilon) = \varepsilon + \varepsilon^2 + 2\varepsilon^3$$
$$x_2(\varepsilon) = 1 - \varepsilon - \varepsilon^2 - 2\varepsilon^3 + \dots.$$

Remark 9.1.1 From the above, we observe $(2a_0 - 1)a_k = f_k(a_0, a_1, \cdots, a_{k-1})$, $k \geq 1$.

To determine a_k, we need $L = \frac{d}{dx}(x^2 - x)\big|_{x=a_0} = 2a_0 - 1$ to be an invertible linear operator (see Theorem 9.1.5).

Example 9.1.2 Consider two-point boundary value problem

$$P(\varepsilon) : \begin{cases} u'' + \varepsilon u^2 = 0, & 0 < x < 1; \\ u(0) = 1, \quad u(1) = 1, & 0 < \varepsilon \ll 1. \end{cases}$$

Let

$$u(\varepsilon, x) = \sum_{n=0}^{\infty} \varepsilon^n u_n(x) = u_0(x) + \varepsilon u_1(x) + \varepsilon^2 u_2(x) + \cdots \tag{9.7}$$

be a solution of $P(\varepsilon)$. Substitute (9.7) into $P(\varepsilon)$. Then we have

$$u''(\varepsilon, x) + \varepsilon(u(\varepsilon, x))^2 = 0,$$
$$u(\varepsilon, 0) = 1, u(\varepsilon, 1) = 1, \tag{9.8}$$

or

$$u_0''(x) + \varepsilon u_1''(x) + \varepsilon^2 u_2''(x) + \ldots + \varepsilon(u_0(x) + \varepsilon u_1(x) + \cdots)^2 = 0,$$

$$u(\varepsilon, 0) = 1 = u_0(0) + \varepsilon u_1(0) + \varepsilon^2 u_2(0) + \cdots,$$
$$u(\varepsilon, 1) = 1 = u_0(1) + \varepsilon u_1(1) + \varepsilon^2 u_2(1) + \cdots. \tag{9.9}$$

In the following, we obtain $u_0(x)$, $u_1(x)$, $\cdots$ by comparing the coefficients of ε^n.

$O(1)$: Set $\varepsilon = 0$ in (9.8) and (9.9) we obtain

$$u_0''(x) = 0, \quad 0 < x < 1,$$
$$u_0(0) = u_0(1) = 1. \tag{9.10}$$

Hence $u_0(x) = 1$ for all x.

$O(\varepsilon)$: Differentiating (9.8) with respect to ε yields

$$\frac{d}{d\varepsilon} u''(x, \varepsilon) + (u(x, \varepsilon))^2 + \varepsilon \left(2u(x, \varepsilon) \frac{du}{d\varepsilon}(x, \varepsilon) \right) = 0. \tag{9.11}$$

Set $\varepsilon = 0$ in (9.11) and (9.9), we have

$$\begin{cases} u_1''(x) + (u_0(x))^2 = 0, \\ u_1(0) = 0, u_1(1) = 0. \end{cases} \tag{9.12}$$

From $u_0 \equiv 1$, we solve (9.12) and obtain

$$u_1(x) = -\frac{x^2}{2} + \frac{x}{2}. \tag{9.13}$$

$O(\varepsilon^2)$: Differentiating (9.11) with respect to ε and setting $\varepsilon = 0$ yield

$$\begin{cases} u_2''(x) + 2u_0(x)u_1(x) = 0, \\ u_2(0) = 0, u_2(1) = 0. \end{cases} \tag{9.14}$$

From $u_0 \equiv 1$ and (9.13), we solve (9.14) and obtain $u_2(x) = \frac{x^4}{12} - \frac{x^3}{6} + \frac{x}{12}$.

Continue this process in step $O(\varepsilon^n)$, $n \geq 3$, we obtain

$$u(\varepsilon, x) = 1 + \varepsilon \left(-\frac{x^2}{2} + \frac{x}{2} \right) + \varepsilon^2 \left(\frac{x^4}{12} - \frac{x^3}{6} + \frac{x}{12} \right) + O(\varepsilon^3).$$

The regular perturbation method works in Example 9.1.1 and Example 9.1.2 because of the following Implicit Function Theorem (IFT). Before we state the IFT, we introduce the following definition.

Definition 9.1.1 *Let X, Y be Banach spaces and $B(X, Y) = \{L | L : X \to Y$ is a bounded linear operator$\}$.*

We say $f : X \to Y$ is Fréchet-differentiable at $x_0 \in X$ if there exists a bounded linear operator $L \in B(X, Y)$ such that

$$\lim_{\|u\| \to 0} \frac{\|f(x_0 + u) - f(x_0) - Lu\|}{\|u\|} = 0.$$

Remark 9.1.2 It is easy to show that L is unique and we denote $L = f_x(x_0)$.

Definition 9.1.2 *We say $f \in C^1$ if $f_x : X \to B(X, Y)$ is continuous.*

Theorem 9.1.1 (Implicit Function Theorem)

Let X, Y, Z be Banach spaces, U be an open set in $X \times Y$, and $f : U \subseteq X \times Y \to Z$ be continuous. Assume f is Fréchet differentiable with respect to x and $f_x(x, y)$ is continuous in U. Suppose $(x_0, y_0) \in U$ and $f(x_0, y_0) = 0$. If $A = f_x(x_0, y_0) : X \to Z$ is an isomorphism of X <u>onto</u> Z, i.e., $f_x(x_0, y_0)$ is onto and has a bounded inverse $f_x^{-1}(x_0, y_0)$. Then

(i) *There exists a ball $B_r(y_0) \subseteq Y$ and a unique map $u : B_r(y_0) \to X$, satisfying $u(y_0) = x_0, f(u(y), y) = 0$;*

(ii) *If $f \in C^1$ then $u(y) \in C^1$ and $u_y(y) = -[f_x(u(y), y)]^{-1} \circ f_y(u(y), y)$.*

Proof. See the book [Kee1] or [Mur]. □

Now we want to explain why the perturbation method works in Example 9.1.2. First we decompose the problem $P(\varepsilon)$ into two parts

$$\begin{cases} -u'' = \varepsilon u^2, \\ u(0) = 0, u(1) = 0, \end{cases}$$

and

$$\begin{cases} -u'' = 0, \\ u(0) = 1, u(1) = 1. \end{cases}$$

Then as we did in §7.3, the solution $u(x)$ of $P(\epsilon)$ satisfies $u(x) = 1 + \varepsilon \int_0^1 g(x, \xi) u^2(\xi) d\xi$.

Define $F : C[0,1] \times \mathbb{R} \to C[0,1]$ by

$$F(u,\varepsilon) = u - 1 - \varepsilon \int_0^1 g(x,\xi) u^2(\xi) d\xi. \tag{9.15}$$

Obviously, $F(u_0, 0) = 0$, where $u_0(x) \equiv 1$.

We want to solve $F(u,\varepsilon)$ and express the solution $u = u(\varepsilon)$ satisfying $F(u(\varepsilon), \varepsilon) = 0$. In order to apply Implicit Function Theorem, we need to verify $F_u(u_0, 0) : C[0,1] \to C[0,1]$ is an isomorphism. We note that for $f : X \to Y$, where X, Y are Banach spaces, we may use Gateux derivative (directional derivative) to evaluate linear operator $f'(x_0) : X \to Y$, which satisfies

$$f'(x_0)v = \left(\frac{d}{dt} f(x_0 + tv) \right)\bigg|_{t=0} = f'(x_0 + tv) \cdot v \,|_{t=0} \,.$$

Now we compute $F_u(u_0, 0)v$, $v \in C[0,1]$, from

$$\frac{d}{dt} F(u_0 + tv, \varepsilon) = \frac{d}{dt} \left[u_0 + tv - 1 - \varepsilon \int_0^1 g(x,\xi)(u_0(\xi) + tv(\xi))^2 d\xi \right]$$

$$= v - 2\epsilon \int_0^1 g(x,\xi) \left(u_0(\xi) + tv(\xi) \right) v(\xi) d\xi.$$

Set $t = 0$ in the above identity, we have for $\varepsilon = 0$, $F_u(u_0, 0)v = v$.

Obviously $F_u(u_0, 0)$ is an isomorphism. Differentiating $F(u(\varepsilon), \varepsilon) = 0$ with respect to ε, we have

$$F_u(u(\varepsilon), \varepsilon) \frac{du}{d\varepsilon} + \frac{\partial F}{\partial \varepsilon}(u(\varepsilon), \varepsilon) = 0. \tag{9.16}$$

Let $\varepsilon = 0$ in (9.16), then $F_u(u_0, 0)u_1 - \int_0^1 g(x,\xi)(u_0(\xi))^2 d\xi = 0$, where $u_1(x) = \int_0^1 g(x,\xi)(u_0(\xi))^2 d\xi$, i.e., $u_1(x)$ satisfies

$$\begin{cases} -u_1'' = (u_0(x))^2, \\ u_1(0) = 0, u_1(1) = 0. \end{cases}$$

Example 9.1.3 Find a periodic solution of the forced logistic equation $u' = au(1 + \varepsilon \cos t - u)$.

Let the periodic solution $u(t, \epsilon)$ be of the form

$$u(t,\varepsilon) = u_0(t) + \varepsilon u_1(t) + \varepsilon^2 u_2(t) + \cdots.$$

Then
$$u_0'(t) + \varepsilon u_1'(t) + \varepsilon^2 u_2'(t) + \cdots$$
$$= a(u_0(t) + \varepsilon u_1(t) + \varepsilon^2 u_2(t) + \cdots) \qquad (9.17)$$
$$\cdot (1 + \varepsilon \cos t - u_0(t) - \varepsilon u_1(t) - \varepsilon^2 u_2(t) - \cdots).$$

$O(1)$: Set $\varepsilon = 0$ in (9.17), we obtain $u_0'(t) = au_0(t)(1 - u_0(t))$.

We choose $u_0(t) \equiv 1$.

$O(\varepsilon)$: Comparing coefficient of ε, we obtain
$$u_1'(t) = a[u_0(t)(\cos t - u_1(t)) + u_1(t)(1 - u_0(t))],$$
or
$$u_1'(t) + au_1(t) = a \cos t. \qquad (9.18)$$
Solving (9.18) by integration factor, we have $(e^{at}u_1)' = ae^{at} \cos t$ and the periodic solution $u_1(t) = \frac{1}{\sqrt{a^2+1}} \cos(t - \varphi)$ for some φ.

Continuing this process, we have the periodic solution
$$u(t) = 1 + \frac{\varepsilon}{\sqrt{a^2 + 1}} \cos(t - \varphi) + O(\varepsilon^2).$$
To explain why the perturbation method works, we define
$$F : C^1 \times \mathbb{R} \to C,$$
$$F(u, \varepsilon) = u' - au(1 + \varepsilon \cos t - u),$$
where C^1, C are the Banach spaces of 2π-periodic differentiable and continuous functions, respectively. Obviously, $F(u_0, 0) = 0$ where $u_0(x) \equiv 1$.

In order to apply Implicit Function Theorem we need to verify that $F_u(u_0, 0) : C^1 \to C$ is an isomorphism. It is easy to show that $Lv = F_u(u_0, 0)v = v' + av$.

To show L has bounded inverse, we need to solve the equation $Lv = f$ where $f \in C$. Then
$$v(t) = v(0)e^{-at} + e^{-at} \int_0^t e^{a\xi} f(\xi) d\xi.$$
In order to have that $v(t)$ is 2π-periodic, we need $v(0) = v(2\pi)$. We choose
$$v(0) = \frac{e^{-2\pi a} \int_0^{2\pi} e^{a\xi} f(\xi) d\xi}{1 - e^{-2\pi a}}.$$
It is easy to verify that $\|v\|_{C^1} = \|L^{-1}f\|_{C^1} \le M \|f\|_C$ where $\|v\|_{C^1} = \|v\|_C + \|v'\|_C$.

Example 9.1.4 Perturbation of eigenvalues:
  Let $A \in \mathbb{R}^{n \times n}$ and $Ax = \lambda x$ have eigenpair (λ_0, x_0).
  Consider a perturbed eigenproblem $(A + \varepsilon B)x = \lambda x$.

Let

$$\lambda(\varepsilon) = \lambda = \lambda_0 + \varepsilon \lambda_1 + \varepsilon^2 \lambda_2 + \cdots,$$
$$x(\varepsilon) = x = x_0 + \varepsilon x_1 + \varepsilon^2 x_2 + \cdots.$$

Then we have

$$(A + \varepsilon B)x(\varepsilon) = \lambda(\varepsilon)x(\varepsilon) \qquad (9.19)$$

$O(1)$: Set $\varepsilon = 0$, we have $Ax_0 = \lambda_0 x_0$.

$O(\varepsilon)$: Differentiate (9.19) with respect to ε, then·set $\varepsilon = 0$. We obtain

$$Ax_1 + Bx_0 = \lambda_1 x_0 + \lambda_0 x_1,$$

or

$$(A - \lambda_0 I)x_1 = \lambda_1 x_0 - Bx_0. \qquad (9.20)$$

In order to find λ_1, x_1 from Fredholm Alternative Theorem, (9.20) is solvable iff

$$\lambda_1 x_0 - Bx_0 \perp N(A^* - \overline{\lambda_0} I).$$

If we assume $N(A^* - \overline{\lambda_0} I) = \langle y_0 \rangle$ is one-dimensional, then we are able to find λ_1 by

$$\langle \lambda_1 x_0 - Bx_0, y_0 \rangle = 0, \text{ or } \lambda_1 = \frac{\langle Bx_0, y_0 \rangle}{\langle x_0, y_0 \rangle}, \quad \text{if } \langle x_0, y_0 \rangle \neq 0.$$

Then we solve x_1 by $(A - \lambda_0 I)x_1 = \lambda_1 x_0 - Bx_0$.

Remark 9.1.3 Regular perturbation is very useful in theory of bifurcation. Interested reader can consult [Kee1].

9.2 Singular Perturbation: Boundary Value Problem

We say a problem $P(\varepsilon)$ is singular if its solution $x(\varepsilon) \not\to x(0)$ as $\varepsilon \to 0$. In the following example, we introduce the notions of outer solution, inner solution, boundary layer and matching procedure.

Example 9.2.1 Consider boundary value problem

$$\begin{cases} \varepsilon u'' + u' = 0, 0 < x < 1, \\ u(0) = u_0, u(1) = u_1, 0 < \varepsilon \ll 1 \text{ is a small parameter.} \end{cases} \tag{9.21}$$

The exact solution of (9.21) is

$$u(x, \varepsilon) = \frac{u_0 \left[e^{-\frac{x}{\varepsilon}} - e^{-\frac{1}{\varepsilon}} \right]}{1 - e^{-\frac{1}{\varepsilon}}} + \frac{u_1 \left[1 - e^{-\frac{x}{\varepsilon}} \right]}{1 - e^{-\frac{1}{\varepsilon}}}.$$

Then, for $\varepsilon > 0$ sufficiently small, $u(x, \varepsilon) = u_0 e^{-\frac{x}{\varepsilon}} + u_1 (1 - e^{-\frac{x}{\varepsilon}})$.

In the following Fig. 9.1, we sketch the graph of $u(x, \epsilon)$:

Fig. 9.1

In the region near $x = 0$ with thickness $O(\varepsilon)$, the solution $u(x, \varepsilon)$ changes rapidly, i.e., $\varepsilon u''(x)$ is not small in this region. The region is called a boundary layer since it is near the boundary $x = 0$. If the region is in the interior, then we call it the interior layer. Given a singular perturbation problem, it is nontrivial to find where is the boundary layer or interior layer. It needs intuition, experiences and trials and errors. Now for the problem (9.21), we pretend we don't know anything about the solution $u(x, \varepsilon)$. In the following steps, we shall use perturbation method to obtain the approximate solution to true solution $u(x, \varepsilon)$.

Step 1: Blindly use regular perturbation
Let

$$u(x, \varepsilon) = u_0(x) + \varepsilon u_1(x) + \varepsilon^2 u_2(x) + \cdots \tag{9.22}$$

and substitute it into (9.21), we obtain

$$\varepsilon(u_0'' + \varepsilon u_1'' + \varepsilon^2 u_2'' + \cdots) + (u_0' + \varepsilon u_1' + \varepsilon^2 u_2' + \cdots) = 0,$$
$$u(0) = u_0 = u_0(0) + \varepsilon u_1(0) + \varepsilon^2 u_2(0) + \cdots,$$
$$u(1) = u_1 = u_0(1) + \varepsilon u_1(1) + \varepsilon^2 u_2(1) + \cdots.$$

Then we compare coefficients of $\varepsilon^n (n \geq 0)$

$O(1)$: $\begin{cases} u_0' = 0 \\ u_0(0) = u_0 \ , \ u_0(1) = u_1 \end{cases}$, $u_0(x) = constant.$

$O(\varepsilon)$: $\begin{cases} u_0'' + u_1' = 0 \\ u_1(0) = 0 \ , \ u_1(1) = 0 \end{cases}$, $u_1(x) = constant.$

Continue this process, we obtain $u_n(x) = const.$ for $n \geq 0$.

Thus for $u_0 \neq u_1$ we recognize that this is a singular problem, the regular perturbation method doesn't work for the solution $u(x, \varepsilon)$ of (9.21). The expansion (9.22) is not valid for all $0 \leq x \leq 1$.

Step 2: We guess the location of boundary layer or interior layer. Here we assume that the boundary layer is at $x = 0$. (If we assume the boundary layer is at $x = 1$, we shall find the following outer solution and inner solution will not be matched in the matching expansion.)

(1) Outer expansion:
Since we assume the boundary layer is at $x = 0$, we obtain outer solution $U(x, \varepsilon)$ by regular perturbation method under the condition $U(1, \varepsilon) = u_1$. Let

$$U(x, \varepsilon) = u_0(x) + \varepsilon u_1(x) + \varepsilon^2 u_2(x) + \cdots .$$

Then regular perturbation method yields

$$u_0(x) = u_1 \ , \ u_1(x) = 0 \ , \ u_2(x) = 0 \cdots .$$

(2) Inner expansion:
Assume the inner solution in the boundary layer is

$$W(\tau, \varepsilon) = w_0(\tau) + \varepsilon w_1(\tau) + \varepsilon^2 w_2(\tau) + \cdots ,$$

where $\tau = \frac{x}{\varepsilon^\alpha}$, where $\alpha > 0$ is to be determined and ε^α is the thickness of the boundary layer. The introduction of new variable τ means that we stretch the boundary layer.

(3) Matching:
Let the solution

$$u(x, \varepsilon) = U(x, \varepsilon) + W(\tau \varepsilon). \tag{9.23}$$

Substitute (9.23) into equation (9.21), we obtain

$$\varepsilon \left[U''(x,\varepsilon) + \varepsilon^{-2\alpha} \frac{d^2 W}{d\tau^2}(\tau,\varepsilon) \right] + \left[U'(x,\varepsilon) + \varepsilon^{-\alpha} \frac{dW}{d\tau}(\tau,\varepsilon) \right] = 0,$$

or

$$\begin{aligned} &\varepsilon(u_0'' + \varepsilon u_1'' + \varepsilon^2 u_2'' + \cdots) + \varepsilon^{1-2\alpha}(\tfrac{d^2 w_0}{d\tau^2} + \varepsilon \tfrac{d^2 w_1}{d\tau^2} + \cdots) \\ &+ (u_0' + \varepsilon u_1' + \varepsilon^2 u_2' + \cdots) + \varepsilon^{-\alpha}(\tfrac{dw_0}{d\tau} + \varepsilon \tfrac{dw_1}{d\tau} + \cdots) = 0. \end{aligned} \tag{9.24}$$

For the boundary conditions in (9.21), it follows that

$$\begin{aligned} u(0) = u_0 &= U(0,\varepsilon) + W(0,\varepsilon) \\ &= (u_0(0) + \varepsilon u_1(0) + \cdots) + (w_0(0) + \varepsilon w_1(0) + \cdots). \end{aligned}$$

Here we note that $\tau = 0$ if $x = 0$. Since $u_0(0) = u_1$, we have $w_0(0) = u_0 - u_1$.

As x lies in the boundary layer region, we stretch the boundary layer and consider $W(\tau,\varepsilon)$. Here we need $\varepsilon^{1-2\alpha} = \varepsilon^{-\alpha}$ or $\alpha = 1$. Multiplying ε on both sides of (9.24) yields

$$\begin{aligned} &\varepsilon^2(u_0'' + \varepsilon u_1'' + \varepsilon^2 u_2'' + \cdots) + (\tfrac{d^2 w_0}{d\tau^2} + \varepsilon \tfrac{d^2 w_1}{d\tau^2} + \cdots) \\ &+ \varepsilon(u_0' + \varepsilon u_1' + \varepsilon^2 u_2' + \cdots) + (\tfrac{dw_0}{d\tau} + \varepsilon \tfrac{dw_1}{d\tau} + \cdots) = 0. \end{aligned} \tag{9.25}$$

Comparing the coefficients of ε^n, $n \geq 0$, we obtain

$$O(1): \begin{cases} \frac{d^2 w_0}{d\tau^2} + \frac{dw_0}{d\tau} = 0, \\ w_0(0) = u_0 - u_1, \ \lim\limits_{\tau \to \infty} w_0(\tau) = 0, \ \lim\limits_{\tau \to \infty} \frac{dw_0}{d\tau}(\tau) = 0. \end{cases} \tag{9.26}$$

Integrating (9.26) from τ to ∞ yields $\frac{dw_0}{d\tau} + w_0(\tau) = 0$, $w_0(0) = u_0 - u_1$ and hence $w_0(\tau) = (u_0 - u_1)e^{-\tau}$.

$$O(\varepsilon): \begin{cases} \frac{d^2 w_1}{d\tau^2} + \frac{dw_1}{d\tau} = 0, \\ w_1(0) = 0, \ w_1(\infty) = 0, \ \frac{dw_1}{d\tau}(\infty) = 0. \end{cases}$$

Then $w_1(\tau) \equiv 0$.

If we compute the solution $u(x,\varepsilon)$ up to $O(\varepsilon)$, then

$$u(x,\varepsilon) = u_1 + (u_0 - u_1)e^{-x/\varepsilon} + O(\varepsilon^2).$$

Remark 9.2.1 If we assume the boundary layer is near $x = 1$, then we assume

$$u(x,\varepsilon) = U(x,\varepsilon) + W(\eta,\varepsilon), \ \eta = \frac{1-x}{\varepsilon}, \tag{9.27}$$

$$\begin{aligned} U(x,\varepsilon) &= u_0(x) + \varepsilon u_1(x) + \cdots, \\ W(\eta,\varepsilon) &= w_0(\eta) + \varepsilon w_1(\eta) + \cdots. \end{aligned} \tag{9.28}$$

Substitute (9.27) into (9.21), we have

$$\varepsilon^2 U''(x,\varepsilon) + \frac{d^2 W}{d\eta^2} + \varepsilon U'(x,\varepsilon) - \frac{dW}{d\eta} = 0. \tag{9.29}$$

Then, from (9.28) and (9.29), it follows that

$$\varepsilon^2 (u_0'' + \varepsilon u_1'' + \varepsilon^2 u_2'' + \cdots) + \left(\frac{d^2 w_0}{d\eta^2} + \varepsilon \frac{d^2 w_1}{d\eta^2} + \cdots \right)$$
$$+ \varepsilon (u_0' + \varepsilon u_1' + \varepsilon^2 u_2' + \cdots) - \left(\frac{dw_0}{d\eta} + \varepsilon \frac{dw_1}{d\eta} + \cdots \right) = 0.$$

$$O(1): \begin{cases} \frac{d^2 w_0}{d\eta^2} - \frac{dw_0}{d\eta} = 0, \\ w_0(0) = u_1 - u_0, \ w_0(\infty) = 0, \ \frac{dw_0}{d\eta}(\infty) = 0. \end{cases} \tag{9.30}$$

Integrating (9.30) from η to ∞ yields

$$\frac{dw_0}{d\eta} = w_0(\eta),$$
$$w_0(0) = u_1 - u_0.$$

Then $w_0(\eta) = (u_1 - u_0)e^\eta \to \infty$ as $\eta \to \infty$. This contradicts (9.30).

Thus the assumption that the boundary layer is near $x = 1$, is not valid.

Nonlinear boundary value problem:

$$\varepsilon y'' + f(x,y)y' + g(x,y) = 0,$$
$$y(0) = y_0, \ y(1) = y_1. \tag{9.31}$$

Assume the boundary layer is at $x = 0$. Let

$$y(x,\varepsilon) = U(x,\varepsilon) + W(\tau,\varepsilon), \ \tau = \frac{x}{\varepsilon}. \tag{9.32}$$

(1) Find outer solution. Let

$$U(x,\varepsilon) = u_0(x) + \varepsilon u_1(x) + \cdots,$$
$$U(1,\varepsilon) = y_1.$$

Substituting $U(x,\epsilon)$ into (9.31), we obtain

$$F(x,\varepsilon) = \varepsilon U''(x,\varepsilon) + f(x,U(x,\varepsilon))U'(x,\varepsilon) + g(x,U(x,\varepsilon)) = 0. \tag{9.33}$$

$O(1)$: Set $\varepsilon = 0$ in (9.33), then we have

$$\begin{cases} f(x,u_0(x))u_0'(x) + g(x,u_0(x)) = 0, \\ u_0(1) = y_1. \end{cases} \tag{9.34}$$

Thus, from (9.34), we obtain the solution $u_0(x)$.

$O(\varepsilon)$: Differentiate (9.33) with respect to ε and then set $\varepsilon = 0$, we obtain

$$\begin{cases} u_0''(x) + f(x, u_0(x))u_1'(x) + \frac{\partial f}{\partial y}(x, u_0(x))u_1(x)u_0'(x) + \frac{\partial g}{\partial y}(x, u_0(x))u_1(x) = 0, \\ u_1(1) = 0. \end{cases}$$

(9.35)

Then from (9.35), we obtain the solution $u_1(x)$.

(2) Find inner solution. Let

$$W(\tau, \varepsilon) = w_0(\tau) + \varepsilon w_1(\tau) + \varepsilon^2 w_2(\tau) + \cdots .$$

Substituting (9.32) into (9.31), we have

$$\varepsilon[U''(x, \varepsilon) + \varepsilon^{-2}\frac{d^2 W}{d\tau^2}(\tau, \varepsilon)] + f(x, U(x, \varepsilon) + W(\tau, \varepsilon))(U'(x, \varepsilon) + \varepsilon^{-1}\frac{dW}{d\tau}(\tau, \varepsilon))$$
$$+ g(x, U(x, \varepsilon) + W(\tau, \varepsilon)) = 0.$$

(9.36)

Multiply (9.36) by ε, we get

$$G(x, \tau, \varepsilon) = \varepsilon^2 U''(x, \varepsilon) + \frac{d^2 W}{d\tau^2}(\tau, \varepsilon) + f(x, U(x, \varepsilon) + W(\tau, \varepsilon))$$
$$\cdot \left(\varepsilon U'(x, \varepsilon) + \frac{dW}{d\tau}(\tau, \varepsilon) \right) + \varepsilon g(x, U(x, \varepsilon) + W(\tau, \varepsilon))$$
$$= 0,$$

(9.37)

$$y(0, \varepsilon) = y_0 = U(0, \varepsilon) + W(0, \varepsilon).$$

Comparing the coefficients of ε^n, $n \geq 1$, we have:

$O(1)$: Setting $\varepsilon = 0$ in (9.37), we obtain

$$\begin{aligned} &\frac{d^2 w_0}{d\tau^2} + f(0, u_0(0) + w_0(\tau))\frac{dw_0}{d\tau} = 0, \\ &w_0(0) = y_0 - u_0(0), \\ &w_0(\infty) = 0 , \ \frac{dw_0}{d\tau}(\infty) = 0. \end{aligned}$$

(9.38)

Case 1: If $y_0 < u_0(0)$ then $w_0(0) < 0$.

Integrating (9.38) from τ to ∞, we have

$$-\frac{dw_0}{d\tau}(\tau) + \int_\tau^\infty f(0, u_0(0) + w_0(\tau))\frac{dw_0}{d\tau} \, d\tau = 0.$$

Then

$$\frac{dw_0}{d\tau}(\tau) = \int_{w_0(\tau)+u_0(0)}^{u_0(0)} f(0,t)\, dt.$$

We need

$$\int_{y+u_0(0)}^{u_0(0)} f(0,t)\, dt > 0 \quad \text{for} \quad y_0 - u_0(0) < y < 0.$$

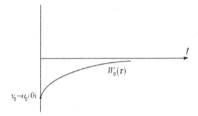

Fig. 9.2

Case 2: If $y_0 > u_0(0)$.

Then we need $\int_{y+u_0(0)}^{u_0(0)} f(0,t)dt < 0$ for $0 < y < y_0 - u_0(0)$.

9.3 Singular Perturbation: Initial Value Problem

We first consider two weakly nonlinear oscillators in Examples 9.3.1 and 9.3.2 and present the method of two timing (or method of multiple time scales).

Example 9.3.1 [Str]
Consider the weakly damped linear oscillator

$$\frac{d^2x}{dt^2} + 2\epsilon\frac{dx}{dt} + x = 0, \tag{9.39}$$

$$x(0) = 0, \quad \frac{dx}{dt}(0) = 1. \tag{9.40}$$

The exact solution of (9.39), (9.40) is

$$x(t,\epsilon) = \left(1-\epsilon^2\right)^{-\frac{1}{2}} e^{-\epsilon t} \sin\left(\left(1-\epsilon^2\right)^{\frac{1}{2}} t\right). \tag{9.41}$$

Now let's solve the problem (9.39), (9.40) by regular perturbation method. Suppose that

$$x(t, \epsilon) = \sum_{n=0}^{\infty} \epsilon^n x_n(t).$$

Substitute it into (9.39), (9.40) and then require that the coefficients of all powers of ϵ vanish, and we find that

$$O(1) : x_0'' + x_0 = 0, \ x_0(0) = 0, \ x_0'(0) = 1, \ \left(' = \frac{d}{dt}\right), \qquad (9.42)$$

$$O(\epsilon) : x_1'' + 2x_0' + x_1 = 0, \ x_1(0) = 0, \ x_1'(0) = 0. \qquad (9.43)$$

Then the solution of (9.42) is $x_0(t) = \sin t$. Plugging this solution to (9.43) gives

$$x_1'' + x_1 = -2\cos t. \qquad (9.44)$$

We find that the right-hand side of (9.44) is a resonant forcing. The solution of (9.44) subjected to $x_1(0) = 0$, $x_1'(0) = 0$ is

$$x_1(t) = -t \sin t \qquad (9.45)$$

which is a secular term, i.e., a term that grows without bound as $t \to \infty$. Hence by regular perturbation method the solution of (9.39), (9.40) is

$$x(t, \epsilon) = \sin t - \epsilon t \sin t + O(\epsilon^2). \qquad (9.46)$$

How does it compare with the exact solution (9.41)? It is easy to verify that the first two terms in (9.46) are exactly the first two terms of (9.41) expanded as power series in ϵ. In fact, (9.46) is the beginning of a convergent series expansion for the true solution. Hence for fixed t and small ϵ, (9.46) is a good approximation for true solution (9.41). The following Fig. 9.3 shows the difference between true solution (9.41) and approximation solution (9.46) when $\epsilon = 0.1$.

The reasons why regular perturbation method fails are that:

1. The true solution (9.41) exhibits two time scales: a fast time $t \sim O(1)$ for the sinusoidal oscillations, $\sin\left((1 - \epsilon^2)^{\frac{1}{2}}t\right)$, and a slow time $t \sim O\left\{\frac{1}{\epsilon}\right\}$ over which the amplitude $(1 - \epsilon^2)^{-\frac{1}{2}}e^{-\epsilon t}$ decays. The approximation solution (9.46) completely misrepresents the slow time scale behavior. In particular, because of the secular term $t\sin t$, (9.46) falsely suggests that the solution grows with time where we know from (9.41) that the amplitude $A = (1 - \epsilon^2)^{-\frac{1}{2}}e^{-\epsilon t}$ decay exponentially as $t \to \infty$.

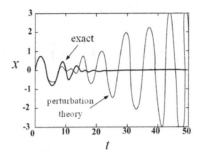

Fig. 9.3

2. The frequency of the oscillation in (9.41) is $\omega = (1 - \epsilon^2)^{\frac{1}{2}} \approx 1 - \frac{1}{2}\epsilon^2$ which is shifted slightly from the "resonance" frequency $\omega = 1$ from (9.46). After a very long time $t \sim O\left(\frac{1}{\epsilon^2}\right)$, this frequency error will have a significant cumulative effect.

Two-Timing Method

We guess that x is a function of two variables, a "fast" variable, $s = t$ and a "slow" variable, $\tau = \epsilon t$. Suppose $x = x(s, \tau, \epsilon)$ and seek a power series solution of the form

$$x(t) = x(s, \tau, \epsilon) = \sum_{n=0}^{\infty} x_n(s, \tau)\epsilon^n. \tag{9.47}$$

First we treat s and τ as two new independent variables, then

$$\frac{dx}{dt} = \frac{\partial x}{\partial s}\frac{ds}{dt} + \frac{\partial x}{\partial \tau}\frac{d\tau}{dt} = \frac{\partial x}{\partial s} + \epsilon\frac{\partial x}{\partial \tau}, \tag{9.48}$$

$$\frac{d^2x}{dt^2} = \frac{\partial^2 x}{\partial s^2} + 2\epsilon \cdot \frac{\partial^2 x}{\partial s \partial \tau} + \epsilon^2\frac{\partial^2 x}{\partial \tau^2}. \tag{9.49}$$

Now we substitute (9.47)–(9.49) into the governing equation (9.39), (9.40) and obtain

$$\left(\frac{\partial^2 x}{\partial s^2} + 2\epsilon\frac{\partial^2 x}{\partial s \partial \tau} + \epsilon^2\frac{\partial^2 x}{\partial \tau^2}\right) + 2\epsilon\left(\frac{\partial x}{\partial s} + \epsilon\frac{\partial x}{\partial \tau}\right) + x(s, \tau, \epsilon) = 0, \tag{9.50}$$

$$x(0, 0, \epsilon) = 0, \quad \frac{\partial x}{\partial s}(0, 0, \epsilon) + \epsilon\frac{\partial x}{\partial \tau}(0, 0, \epsilon) = 1. \tag{9.51}$$

Set $\epsilon = 0$ in (9.50), (9.51) and we obtain

$$O(1): \frac{\partial^2 x_0}{\partial s^2} + x_0(s, \tau) = 0,$$
$$x_0(0, 0) = 0, \quad \frac{\partial x_0}{\partial s}(0, 0) = 1. \tag{9.52}$$

Differentiate (9.50), (9.51) with respect to ϵ then set $\epsilon = 0$, and we obtain

$$O(\epsilon): \frac{\partial^2 x_1}{\partial s^2} + 2\frac{\partial^2 x_0}{\partial s \partial \tau} + 2\frac{\partial x_0}{\partial s} + x_1(s, \tau) = 0,$$
$$x_1(0,0) = 0, \quad \frac{\partial x_1}{\partial s}(0,0) + \frac{\partial x_0}{\partial t}(0,0) = 0. \tag{9.53}$$

To solve (9.52), we have

$$x_0(s, \tau) = A(\tau) \sin s + B(\tau) \cos s \tag{9.54}$$

and from the initial conditions in (9.52) we obtain

$$A(0) = 1, \quad B(0) = 0. \tag{9.55}$$

Next we solve (9.53). By (9.54), we rewrite (9.53) as

$$\frac{\partial^2 x_1}{\partial s^2} + x_1(s, \tau) = -2 \cos s \left(A(\tau) + A'(\tau) \right)$$
$$+ 2 \sin s \left(B(\tau) + B'(\tau) \right). \tag{9.56}$$

The right-hand side of (9.56) is a resonant forcing that will produce secular terms like $s \cos s$, $s \sin s$. Hence we set the coefficients of the resonant terms to zero. Here it yields

$$A'(\tau) = -A(\tau)$$
$$B'(\tau) = -B(\tau).$$

With initial conditions in (9.55) we obtain

$$A(\tau) = e^{-\tau} \text{ and } B(\tau) \equiv 0.$$

Hence

$$x(t) = e^{-\tau} \sin s + O(\epsilon)$$
$$= e^{-\epsilon t} \sin t + O(\epsilon) \tag{9.57}$$

is the approximate solution predicted by two-timing. The following figure compares the two-timing solution (9.57) and exact solution (9.41) for $\epsilon = 0.1$. The two curves are almost indistinguishable.

Example 9.3.2 [Str]
We shall use two-timing method to show that the weakly nonlinear van der Pol oscillator

$$x'' + x + \epsilon(x^2 - 1)x' = 0 \tag{9.58}$$

has a stable limit cycle that is nearly circular, with radius $2 + O(\epsilon)$ and frequency $1 + O(\epsilon^2)$. Substitute (9.48), (9.49) into (9.58), we have

$$\left(\frac{\partial^2 x}{\partial s^2} + 2\epsilon \frac{\partial^2 x}{\partial s \partial \tau} + \epsilon^2 \frac{\partial^2 x}{\partial \tau^2} \right) + \epsilon \left(x^2(s, \tau, \epsilon) - 1 \right) \left(\frac{\partial x}{\partial s} + \epsilon \frac{\partial x}{\partial \tau} \right)$$
$$+ x(s, \tau, \epsilon) = 0. \tag{9.59}$$

Fig. 9.4

Set $\epsilon = 0$ in (9.59), we obtain

$$O(1): \frac{\partial^2 x_0}{\partial s^2} + x_0(s,\tau) = 0. \tag{9.60}$$

Differentiating (9.59) with respect to ϵ and then set $\epsilon = 0$, we obtain

$$\frac{\partial^2 x_1}{\partial s^2} + 2\frac{\partial^2 x_0}{\partial s \partial \tau} + \left(x_0^2(s,\tau) - 1\right)\frac{\partial x_0}{\partial s} + x_1(s,\tau) = 0. \tag{9.61}$$

The general solution of (9.60) is

$$x_0(s,\tau) = r(\tau)\cos(s + \phi(\tau)) \tag{9.62}$$

where $r(\tau)$ and $\phi(\tau)$ are the slowly-varying amplitude and phase of $x_0(s,\tau)$. To find the equations governing $r(\tau)$ and $\phi(\tau)$, we insert (9.62) into (9.61). This yields

$$\begin{aligned}
\frac{\partial^2 x_1}{\partial s^2} + x_1(s,\tau) &= -2\frac{\partial^2 x_0}{\partial s \partial \tau} - \left(x_0^2(s,\tau) - 1\right)\frac{\partial x_0}{\partial s}\\
&= 2\left(r'(\tau)\sin(s + \phi(\tau)) + r(\tau)\phi'(\tau)\cos(s + \phi(\tau))\right)\\
&\quad + r(\tau)\sin(s + \phi(\tau))\left(r^2(\tau)\cos^2(s + \phi(\tau)) - 1\right).
\end{aligned} \tag{9.63}$$

Use the trigonometric identity

$$\sin(s + \phi(\tau))\cos^2(s + \phi(\tau)) = \frac{1}{4}\left[\sin(s + \phi(\tau)) + \sin 3(s + \phi(\tau))\right].$$

Substituting it into (9.63) yields

$$\begin{aligned}
\frac{\partial^2 x_1}{\partial s^2} + x_1(s,\tau) &= \left[2r'(\tau) - r(\tau) + \frac{1}{4}r^3(\tau)\right]\sin(s + \phi(\tau))\\
&\quad + 2r(\tau)\phi'(\tau)\cos(s + \phi(\tau))\\
&\quad + \frac{1}{4}r^3(\tau)\sin 3(s + \phi(\tau)).
\end{aligned}$$

To avoid secular terms, we require

$$2r'(\tau) - r(\tau) + \frac{1}{4}r^3(\tau) = 0, \tag{9.64}$$

$$2r(\tau)\phi'(\tau) = 0. \tag{9.65}$$

First we consider (9.64), it can be rewritten as

$$r'(\tau) = \frac{1}{8}r(4 - r^2). \tag{9.66}$$

Obviously, from (9.66), $r(\tau) \to 2$ as $\tau \to \infty$. Secondly, from (9.65), we have $\phi'(\tau) \equiv 0$ or $\phi(\tau) \equiv \phi_0$ for some ϕ_0. Hence, from (9.62), $x_0(s, \tau) \to 2\cos(s + \phi_0)$ as $\tau \to \infty$ and therefore

$$x(t) \to 2\cos(t + \phi_0) + O(\epsilon) \text{ as } t \to \infty. \tag{9.67}$$

Thus $(x(t), x'(t))$ approaches a stable limit cycle of radius $2 + O(\epsilon)$. To find the frequency implied by (43), let $\theta = t + \varphi(\tau)$ denote the argument of the cosine. Then the angular frequency ω is given by

$$\omega = \frac{d\theta}{dt} = 1 + \frac{d\phi}{d\tau}\frac{d\tau}{dt} = 1 + \epsilon\varphi' \equiv 1.$$

Hence $\omega = 1 + O(\epsilon^2)$.

Next we consider the singular perturbation for the following first order system

$$\frac{dx}{dt} = f(x, y),$$

$$\epsilon\frac{dy}{dt} = g(x, y),$$

$$x(0, \epsilon) = x_0, \quad y(0, \epsilon) = y_0.$$

Example 9.3.3 Consider enzymatic reactions proposed by Michaelis-Menten [Mur] involving a substrate (molecule) S reacting with an enzyme E to form a complex SE which in turn is converted into a product P. Schematically we have

$$S + E \underset{k_{-1}}{\overset{k_1}{\rightleftharpoons}} SE, \quad SE \overset{k_2}{\to} P + E.$$

Let

$$s = [S], e = [E], c = [SE], p = [P]$$

where [ ] denotes the concentration. By law of mass action, we have the system of nonlinear equations

$$\begin{cases} \frac{ds}{dt} = -k_1 es + k_{-1} c, \\ \frac{de}{dt} = -k_1 es + (k_{-1} + k_2) c, \\ \frac{dc}{dt} = k_1 es - (k_{-1} + k_2) c, \\ \frac{dp}{dt} = k_2 c, \\ s(0) = s_0, \ e(0) = e_0, \ c(0) = 0, \ p(0) = p_0. \end{cases} \tag{9.68}$$

From (9.68), we have

$$\frac{de}{dt} + \frac{dc}{dt} = 0 \qquad \text{or} \qquad e(t) + c(t) \equiv e_0. \tag{9.69}$$

By (9.69), we substitute $e(t) = e_0 - c(t)$ into (9.68) and obtain

$$\begin{cases} \frac{ds}{dt} = -k_1 e_0 s + (k_1 s + k_{-1}) c, \\ \frac{dc}{dt} = k_1 e_0 s - (k_1 s + k_{-1} + k_2) c, \\ s(0) = s_0, \quad c(0) = 0. \end{cases} \tag{9.70}$$

With the following scaling

$$\begin{aligned} \tau = k_1 e_0 t, \ u(\tau) = \frac{s(t)}{s_0}, \ v(\tau) = \frac{c(t)}{e_0}, \\ \lambda = \frac{k_2}{k_1 s_0}, \ K = \frac{k_{-1} + k_2}{k_1 s_0}, \ \varepsilon = \frac{e_0}{s_0}, \end{aligned} \tag{9.71}$$

the system (9.70) is in the following non-dimensional form:

$$\begin{cases} \frac{du}{d\tau} = -u + (u + K - \lambda) v, \\ \varepsilon \frac{dv}{d\tau} = u - (u + K) v, \\ u(0) = 1, \quad v(0) = 0, \end{cases} \tag{9.72}$$

where $0 < \varepsilon \ll 1$ and, from (9.71), $K > \lambda$.

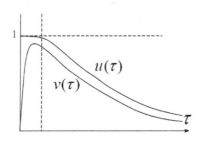

Fig. 9.5

From Fig. 9.5, the solution $v(\tau)$ changes rapidly in dimensionless time $\tau = O(\epsilon)$. After that $v(\tau)$ is essentially in a steady state, or $\epsilon \frac{dv}{d\tau} \approx 0$, i.e., the v-reaction is so fast that it is more or less in equilibrium at all times. This is so-called "Michaelis and Menten's pseudo-steady state hypothesis". In the following, we introduce method of singular perturbation for the system (9.72).

Singular Perturbation: Initial Value Problem

Consider the following system

$$
\begin{aligned}
\frac{dx}{dt} &= f(x, y), \\
\epsilon \frac{dy}{dt} &= g(x, y), \quad 0 < |\varepsilon| \ll 1, \\
x(0, \epsilon) &= x_0, \; y(0, \epsilon) = y_0.
\end{aligned}
\tag{9.73}
$$

If we set $\epsilon = 0$ in (9.73), then

$$
\begin{aligned}
\frac{dx}{dt} &= f(x, y), \; x(0) = x_0, \\
0 &= g(x, y).
\end{aligned}
\tag{9.74}
$$

Assume $g(x, y) = 0$ can be solved as

$$
y = \varphi(x).
\tag{9.75}
$$

Substitute (9.75) into (9.74), then we have

$$
\begin{aligned}
\frac{dx}{dt} &= f\left(x, \varphi(x)\right) \\
x(0) &= x_0.
\end{aligned}
\tag{9.76}
$$

Let $X_0(t)$, $0 \le t \le 1$ be the unique solution of (9.76) and $Y_0(t) = \varphi\left(X_0(t)\right)$. In general $Y_0(0) \ne y_0$. Assume the following hypothesis:

There exists $K > 0$ such that for $0 \le t \le 1$

$$
\left. \frac{\partial g}{\partial y} \right|_{\substack{x = X_0(t) \\ y = Y_0(t)}} \le -K
$$

$$
\tag{H}
$$

and

$$
\left. \frac{\partial g}{\partial y} \right|_{\substack{x = X_0(t) \\ y = \lambda}} \le -K
$$

for all λ lying between $Y(0)$ and y_0. We shall prove that

$$
\lim_{\epsilon \downarrow 0} x(t, \epsilon) = X_0(t), \quad \lim_{\epsilon \downarrow 0} y(t, \epsilon) = Y_0(t)
$$

uniformly on $0 < t \le 1$. Since $Y_0(0) \ne y_0$, we expect $Y_0(t)$ to be non-uniformly valid near $t = 0$. Introduce a new variable, the stretch variable $\xi = t/\epsilon$, and write

$$x(t, \epsilon) = X(t, \epsilon) + u(\xi, \epsilon),$$
$$y(t, \epsilon) = Y(t, \epsilon) + v(\xi, \epsilon), \qquad (9.77)$$

where $X(t, \epsilon)$, $Y(t, \epsilon)$ are called "outer solutions" and $u(\xi, \epsilon)$, $v(\xi, \epsilon)$ are called "inner solutions". There is a matching condition between inner and outer solutions, (see Fig. 9.6)

$$\lim_{\xi \uparrow \infty} u(\xi, \epsilon) = 0, \quad \lim_{\xi \uparrow \infty} v(\xi, \epsilon) = y_0 - Y(0). \qquad (9.78)$$

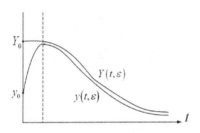

Fig. 9.6

Step 1: Find the outer solutions $X(t, \epsilon)$ and $Y(t, \epsilon)$. Let

$$X(t, \epsilon) = \sum_{n=0}^{\infty} \epsilon^n X_n(t), \quad Y(t, \epsilon) = \sum_{n=0}^{\infty} \epsilon^n Y_n(t). \qquad (9.79)$$

Do the regular perturbation for system (9.73), i.e., substitute (9.79) into (9.73). Then $X(t, \epsilon)$, $Y(t, \epsilon)$ satisfy

$$\frac{dX}{dt} = f(X, Y)$$
$$\epsilon \frac{dY}{dt} = g(X, Y) \qquad (9.80)$$
$$X(0, \epsilon) = x_0, \quad Y(0, \epsilon) = y_0.$$

Compare ϵ^n-term for $n = 0, 1, 2, \ldots$. Compute

$$f(X(t, \epsilon), Y(t, \epsilon)) = f\left(\sum \epsilon^n X_n, \sum \epsilon^n Y_n\right)$$
$$= f(X_0, Y_0) + \epsilon \left[\left(\frac{\partial f}{\partial x}\right)_{X_0, Y_0} X_1 + \left(\frac{\partial f}{\partial y}\right)_{X_0, Y_0} Y_1\right] + O(\epsilon^2),$$

and

$$g(X(t, \epsilon), Y(t, \epsilon)) = g\left(\sum \epsilon^n X_n, \sum \epsilon^n Y_n\right)$$
$$= g(X_0, Y_0) + \epsilon \left[\left(\frac{\partial g}{\partial x}\right)_{X_0, Y_0} X_1 + \left(\frac{\partial g}{\partial y}\right)_{X_0, Y_0} Y_1\right] + O(\epsilon^2).$$

The comparison of ϵ^n term by substituting (9.79) into (9.80) yields

O(1):

$$\begin{aligned} \frac{dX_0}{dt} &= f(X_0, Y_0), \\ 0 &= g(X_0, Y_0), \end{aligned} \tag{9.81}$$

O(ϵ):

$$\begin{aligned} \frac{dX_1}{dt} &= \left(\frac{\partial f}{\partial x}\right)_{X_0,Y_0} X_1 + \left(\frac{\partial f}{\partial y}\right)_{X_0,Y_0} Y_1, \\ \frac{dY_0}{dt} &= \left(\frac{\partial g}{\partial x}\right)_{X_0,Y_0} X_1 + \left(\frac{\partial g}{\partial y}\right)_{X_0,Y_0} Y_1. \end{aligned} \tag{9.82}$$

In (9.81) $X_0(t), Y_0(t)$ satisfy

$$\begin{aligned} Y_0(t) &= \varphi\left(X_0(t)\right), \\ \frac{dX_0}{dt} &= f\left(X_0, \varphi(X_0)\right), \\ X_0(0) &= x_0. \end{aligned} \tag{9.83}$$

From (9.82), we obtain

$$Y_1(t) = \left[\frac{dY_0}{dt} - \left(\frac{\partial g}{\partial x}\right)_{X_0,Y_0} X_1\right] \Big/ \left(\frac{\partial g}{\partial y}\right)_{X_0,Y_0}. \tag{9.84}$$

From (9.82) and (9.84), $X_1(t)$ satisfies

$$\begin{aligned} \frac{dX_1}{dt} &= \psi_1(t)X_1 + \mu_1(t), \\ X_1(0) &= 0, \end{aligned} \tag{9.85}$$

where

$$\psi_1(t) = \left(\frac{\partial f}{\partial x}\right)_{X_0,Y_0} - \left[\frac{\left(\frac{\partial f}{\partial y}\right)\left(\frac{\partial g}{\partial x}\right)}{\left(\frac{\partial g}{\partial y}\right)}\right]_{X_0,Y_0},$$

$$\mu_1(t) = \frac{\left(\frac{\partial f}{\partial y}\right)_{X_0,Y_0} \frac{dY_0}{dt}}{\left(\frac{\partial g}{\partial y}\right)_{X_0,Y_0}}.$$

Inductively we shall have, for $i = 2, 3, \ldots,$

$$\begin{cases} Y_i(t) = \alpha_i(t) + \beta_i(t)X_i(t), \\ \frac{dX_i}{dt} = \psi_i(t)X_i + \mu_i(t), \\ X_i(0) = 0. \end{cases} \tag{9.86}$$

For $x(0, \epsilon) = X(0, \epsilon) = x_0 = \sum_{i=0}^{\infty} X_i(0)\epsilon^n$, it follows that $X_0(0) = x_0$ and $X_i(0) = 0$ for $i = 1, 2, \ldots.$

Step 2: Inner expansion at singular layer near $t = 0$.

From (9.73), (9.77) and (9.80), $\xi = t/\epsilon$, we have

$$\frac{du}{d\xi} = \frac{d}{d\xi}\left(x(\epsilon\xi,\epsilon) - X(\epsilon\xi,\epsilon)\right) = \epsilon\frac{dx}{dt}(\epsilon\xi,\epsilon) - \epsilon\frac{dX}{dt}(\epsilon\xi,\epsilon)$$

$$= \epsilon f\left(X(\xi\epsilon,\epsilon) + u(\xi,\epsilon), Y(\xi\epsilon,\epsilon) + v(\xi,\epsilon)\right)$$

$$-\epsilon f\left(X(\xi\epsilon,\epsilon), Y(\xi\epsilon,\epsilon)\right),$$

$$\frac{dv}{d\xi} = g\left(X(\xi\epsilon,\epsilon) + u(\xi,\epsilon), Y(\xi\epsilon,\epsilon) + v(\xi,\epsilon)\right) \qquad (9.87)$$

$$-g\left(X(\xi\epsilon,\epsilon), Y(\xi\epsilon,\epsilon)\right),$$

$$u(0,\epsilon) = x(0,\epsilon) - X(0,\epsilon) = x_0 - x_0 = 0,$$

$$v(0,\epsilon) = y_0 - Y(0,\epsilon) = y_0 - Y(0) \neq 0.$$

Let

$$u(\xi,\epsilon) = \sum_{n=0}^{\infty} u_n(\xi)\epsilon^n, \quad v(\xi,\epsilon) = \sum_{n=0}^{\infty} v_n(\xi)\epsilon^n. \qquad (9.88)$$

Expand (9.87) in power series in ϵ by (9.88) and compare the coefficients on both sides of (9.87). Set $\epsilon = 0$, and we obtain

O(1):

$$\begin{cases} \frac{du_0}{d\xi} = 0 \\ \\ u_0(0) = 0 \end{cases} \Rightarrow u_0(\xi) \equiv 0. \qquad (9.89)$$

By mean value theorem, we have

$$\begin{cases} \frac{dv_0}{d\xi} = g\left(X_0(0), Y_0(0) + v_0(\xi)\right) - g\left(X_0(0), Y_0(0)\right) \\ \qquad \equiv^{M.V.T} v_0(\xi)G\left(v_0(\xi)\right), \\ v_0(0) = y_0 - Y_0(0). \text{ (Boundary layer jump)} \end{cases} \qquad (9.90)$$

From hypothesis (H), $G\left(v_0(\xi)\right) \leq -K < 0$, $|v_0(\xi)|$ initially decreases and $G(v_0(\xi))$ will remain negative and $|v_0(\xi)|$ will decrease monotonically to zero as $\xi \to \infty$ such that $|v_0(\xi)| \leq |v_0(0)|e^{-K\xi}$.

O(ϵ):

$$\begin{cases} \frac{du_1}{d\xi} = f\left(X_0(0), Y_0(0) + v_0(\xi)\right) - f\left(X_0(0), Y_0(0)\right) \\ \qquad \equiv v_0(\xi)F\left(v_0(\xi)\right). \end{cases} \qquad (9.91)$$

Once $v_0(\xi)$ is solved by (9.90), we solve (9.91) and obtain

$$u_1(\xi) = \int_{\infty}^{\xi} v_0(s)F\left(v_0(s)\right)ds$$

by the matching condition (9.78) $u_1(\infty) = 0$.

Hence

$$x(t, \epsilon) \sim X_0(t) + \epsilon \left[X_1(t) + u_1 \left(t/\epsilon \right) \right] + O(\epsilon^2),$$
$$y(t, \epsilon) \sim Y_0(t) + v_0 \left(t/\epsilon \right) + O(\epsilon).$$

Now we go back to the Michaelis-Menten kinetics. Consider (9.72),

$$\begin{aligned}
&\frac{dx}{dt} = f(x, y) = -x + (x + K - \lambda) y, \quad K > 0, \lambda > 0, \\
&\epsilon \frac{dy}{dt} = g(x, y) = x - (x + K) y, \\
&x(0) = 1, \quad y(0) = 0.
\end{aligned} \tag{9.92}$$

Let

$$x(t, \epsilon) = X(t, \epsilon) + u(\xi, \epsilon) = \sum_{n=0}^{\infty} \epsilon^n X_n(t) + \sum_{n=0}^{\infty} \epsilon^n u_n(\xi),$$

$$y(t, \epsilon) = Y(t, \epsilon) + v(\xi, \epsilon) = \sum_{n=0}^{\infty} \epsilon^n Y_n(t) + \sum_{n=0}^{\infty} \epsilon^n v_n(\xi).$$

Then, from (9.83),

$$Y_0(t) = \varphi \left(X_0(t) \right) = \frac{X_0(t)}{X_0(t) + K}, \tag{9.93}$$

where $X_0(t)$ satisfies

$$\begin{cases}
\frac{dx}{dt} = -x + (x + K - \lambda) \frac{x}{x+K} = \frac{-\lambda x}{x+K}, \\
x(0) = x_0 = 1.
\end{cases}$$

Then $X_0(t)$ satisfies

$$X_0(t) + K \ln X_0(t) = 1 - \lambda t, \quad t \geq 0. \tag{9.94}$$

From (9.90) we obtain

$$\begin{aligned}
\frac{dv_0}{d\xi} &= \left[x_0 - (x_0 + K) \left(Y_0(0) + v_0(\xi) \right) \right] - \left[x_0 - (x_0 + K) Y_0(0) \right] \\
&= -(x_0 + K) v_0(\xi), \\
v_0(0) &= y_0 - Y_0(0).
\end{aligned}$$

Since $x_0 = 1$, $y_0 = 0$ and $Y_0(0) = \frac{1}{1+K}$, it follows that

$$\begin{aligned}
v_0(\xi) &= \left(y_0 - \frac{x_0}{x_0 + K} \right) e^{-(x_0+K)\xi} \\
&= \left(\frac{-1}{1 + K} \right) e^{-(1+K)\xi}.
\end{aligned}$$

Hence

$$y(t, \epsilon) \sim \frac{X_0(t)}{X_0(t) + K} + \left(\frac{-1}{1 + K} \right) e^{-(1+K)(t/\epsilon)}.$$

From (9.91) we have

$$\frac{du_1}{d\xi} = f(X_0(0), Y_0(0) + v_0(\xi)) - f(X_0(0), Y_0(0))$$

$$= (1 + K - \lambda) v_0(\xi) = \frac{\lambda - (1 + K)}{1 + K} e^{-(1+K)\xi},$$

$$u_1(\infty) = 0,$$

and it follows that $u_1(\xi) = ((1 + K) - \lambda) e^{-(1+K)\xi}$. Hence

$$x(t, \epsilon) \sim X_0(t) + \epsilon ((1 + K) - \lambda) e^{-(1+K)(t/\epsilon)}.$$

9.4   Exercises

Exercise 9.1 Approximate all roots of the equation

$$x^3 + x^2 + x + \varepsilon = 0, \ 0 < \varepsilon \ll 1.$$

Exercise 9.2 Approximate all roots of $\varepsilon x^3 + x - 1 = 0$.
(Hint: Set $x = \frac{1}{y}$ and find y as a function of ε.)

Exercise 9.3 Estimate the eigenvalues and eigenvectors of the matrix

$$\begin{pmatrix} 1 & 1 - \varepsilon \\ \varepsilon & 1 \end{pmatrix}.$$

Exercise 9.4 Approximate the eigenvalues and eigenvectors of

$$y'' + \lambda(1 + \varepsilon x)y = 0,$$
$$y(0) = y(\pi) = 0,$$

for all $\varepsilon \neq 0$ small.

Exercise 9.5 Assume

$$\begin{cases} \varepsilon y'' + y' + y = 0, \\ y(0) = \alpha_0, \ y(1) = \beta_0, \end{cases}$$

has boundary layer at $x = 0$. Find the solution $u(x, \varepsilon)$ by singular perturbation.

Exercise 9.6 Consider the linear system

$$x' = ax + by$$
$$\epsilon y' = cx + dy$$

where $d < 0$. Rewrite the linear system as $z' = A(\epsilon)z$ for $z = \begin{pmatrix} x \\ y \end{pmatrix}$ and

$$A(\epsilon) = \begin{pmatrix} a & b \\ \frac{c}{\epsilon} & \frac{d}{\epsilon} \end{pmatrix}.$$

(1) Determine two linearly independent solutions of the $z' = A(\epsilon)z$ in the form $z(t) = e^{\lambda(\epsilon)t}p(\epsilon)$.
(2) Show that the solution $z(t)$ obtained in (i) implies an asymptotic solution of the form

$$x(t, \epsilon) = X_0(t) + O(\epsilon)$$
$$y(t, \epsilon) = Y_0(t) + \eta_0(\tau) + O(\epsilon)$$

on any bounded interval $[0, \tau]$ where $\begin{pmatrix} X_0 \\ Y_0 \end{pmatrix}$ satisfies the reduced problem and $\eta_0 \to 0$ as $\tau \equiv \frac{t}{\epsilon} \to \infty$.

Exercise 9.7 Use the multi-scale technique to find solutions of the initial value problem

$$u'' + \varepsilon(\cos t)(u')^2 = 0,$$
$$u(0) = 0, \ u'(0) = -1.$$

Exercise 9.8 Analyze the following singular perturbation problem similar to Michaelis-Menten kinetics

$$\frac{du}{dt} = -u + (u - a_3 u + a_1)v_1 + (a_4 + u)v_2 = f(u, v_1, v_2),$$

$$\varepsilon \frac{dv_1}{dt} = u - (u + a_3 u + a_1 + a_2)v_1 + (a_4 + a_5 - u)v_2 = g_1(u, v_1, v_2),$$

$$\varepsilon \frac{dv_2}{dt} = a_3 u v_1 - (a_4 + a_5)v_2 = g_2(u, v_1, v_2),$$

$$u(0) = 1, \ v_1(0) = v_2(0) = 0.$$

Chapter 10

INTRODUCTION TO MONOTONE DYNAMICAL SYSTEMS

10.1 Monotone Dynamical System with Applications to Cooperative Systems and Competitive Systems

The theory of monotone systems was discovered by Morris Hirscch in 1980 [Hir1; Hir2]. During the past thirty years, it was successfully applied to analyzing n-species's cooperative systems, two species and three species competitive systems and their variants. The nice part of the theory is that it is a non-Lyapunov approach for the global behavior of the solutions of the above mentioned systems. It also has discrete-version monotone dynamical systems [HS2; Z]. In this chapter we state without proof the main theorems of monotone dynamical system. The proofs are deferred to the book [Smi]. We shall emphasize on the application of theory to mathematical models in population biology.

We shall use the same notations in Section 5.1 of Chapter 5. Let $\pi(x, t)$ denote the dynamical system generated by the autonomous system of differential equations

$$\frac{dx}{dt} = f(x) \tag{10.1}$$

where f is continuously differentiable on an open subset $D \subseteq R^n$.

We recall that $\pi(x, t)$ is a solution of (10.1) which starts at position x at time $t = 0$. $\pi(x, t)$ satisfies the following properties:

(i) $\pi(x, 0) = x$
(ii) $\pi(x, t + s) = \pi(\pi(x, s), t)$
(iii) π is continuous in t and x

where (i) states that the trajectory $\pi(x, t)$ starting at position x at initial time $t = 0$, (ii) is the semi-group property and (iii) is the property of continuous dependence on initial data.

Definition 10.1.1 *Let X be a Banach Space. We say that set $K \subseteq X$ is a cone if the following conditions are satisfied:*

(i) *the set K is closed;*
(ii) *if $x, y \in K$ then $\alpha x + \beta y \in K$ for all $\alpha, \beta \geq 0$;*
(iii) *for each pair of vectors $x, -x$ at least one does not belong to K provided $x \neq 0$, where 0 is the origin of X.*

Definition 10.1.2 *Let $x, y \in X$. We say that*

(i) *$x \leq_K y$ if and only if $y - x \in K$,*
(ii) *$x \ll_K y$ if and only if $y - x \in Int(K)$,*
(iii) *$x <_K y$ if and only if $x \leq_K y, x \neq y$.*

Example 10.1.1

(i) Let $X = \mathbb{R}^n$, $K = \mathbb{R}^n_+ = \{x = (x_1, \cdots, x_n) \in \mathbb{R}^n : x_i \geq 0, i = 1, 2, \cdots, n\}$ Then

$$x \leq_K y \text{ iff } x_i \leq y_i, \ i = 1, 2, \cdots, n$$

$$x \ll_K y \text{ iff } x_i < y_i, \ i = 1, 2, \cdots, n$$

We call that $\leq_K$ is an ordinary order.

(ii) Let $X = \mathbb{R}^n \times \mathbb{R}^n$, $K = \mathbb{R}^n_+ \times (-\mathbb{R}^n_+)$. Then $x = (x_1, x_2) \leq_K y = (y_1, y_2)$ iff $x_1 \leq x_2, y_1 \geq y_2$.
We call that $\leq_K$ is a competitive order.

Definition 10.1.3 *A dynamical system $\pi(t, x)$ is called a monotone dynamical system with respect to $\leq_K$ if it has the property that $\pi(x, t) \leq_K \pi(y, t)$ for all $t \geq 0$ whenever $x \leq_K y$. The dynamical system will be called strongly monotone with respect to $\leq_K$ if $x \leq_K y$ and $x \neq y$ implies that $\pi(x, t) \ll_K \pi(y, t)$ for all $t > 0$.*

Definition 10.1.4 *Let $\frac{dx}{dt} = f(x)$, $x \in D \subseteq \mathbb{R}^n$, where D is a convex, open subset of $\mathbb{R}^n$, $f = (f_1, \cdots, f_n) : D \to \mathbb{R}^n$ be C^1. We say that (1.1) is a* *cooperative system if $\frac{\partial f_i}{\partial x_j} \geq 0$ for $i \neq j$. We say that (1.1) is irreducible if the Jacobian $Df(x) = \left(\frac{\partial f_i}{\partial x_j}(x)\right)$ is irreducible for all $x \in D$, i.e. the matrix $Df(x)$ cannot be put into the form $\begin{bmatrix} A & B \\ 0 & C \end{bmatrix}$ where A, C are square matrices by reordering $(x_1, ..., x_n)$.*

In the following, we may use the language of graph theory to verify the irreducibility of a $n \times n$ matrix A [SW].

Definition 10.1.5 *Let $A = (a_{ij})$ be an $n \times n$ matrix. Let $P_1, P_2, \cdots, P_n$ be n vertices. If $a_{ij} \neq 0$, we draw a directed line segment $P_i P_j$ connecting P_i to P_j. The resulting graph is said to be strongly connected if, for each pair (P_i, P_j), there is a directed path $P_i P_{k_1}, P_{k_1} P_{k_2}, \cdots, P_{k_r} P_j$. A square matrix is irreducible if and only if its directed graph is strongly connected.*

In the following, a sufficient condition is given for the system (10.1) to generate a monotone (strongly monotone) dynamical system.

Theorem 10.1.1 *If the system (10.1) is cooperative in D then π is a monotone dynamical system with respect to $\leq_K$ in D, where $K = \mathbb{R}_+^n$. If (10.1) is cooperative and irreducible in D then π is a strongly monotone system with respect to $\ll_K$.*

Proof. From Theorem 2.6.3 [Smi], π is a monotone dynamical system. The proof of strongly monotonicity follows from Theorem 4.1.1 in [Smi]. □

Theorem 10.1.2 (Convergence Criterion)
Let $\gamma^+(x)$ be an orbit of the monotone dynamical system (10.1) which has compact closure in D. If $x \ll_K \pi(x, T)$ or $\pi(x, T) \ll_K x$ for some $T > 0$ then $\lim_{t \to \infty} \pi(x, t) = e$ for some equilibrium point e.

Proof. If $x \ll_K \pi(x, T)$ then monotonicity implies that for $n = 1, 2, ...,$

$$\pi(x, nT) \leq_K \pi(x, (n+1)T)$$

then from assumption that $\gamma^+(x)$ has compact closure in D it follows that $\lim_{t \to \infty} \pi(x, nT) = e$ for some e.

From continuity of π,

$$\pi(e, T) = \pi \left(\lim_{t \to \infty} \pi(x, nT), T \right) = \lim_{t \to \infty} \pi(x, (n+1)T) = e.$$

It follows that the omega limit set of $x, \omega(x)$ is a T-periodic orbit $\{\pi(e, t) : t \in \mathbb{R}\}$. However, T may not be the minimal period of $\pi(e, t)$. Let $P = \{\tau : \pi(e, t + \tau) = \pi(e, t)\}$ be the set of all periods of the solution $\pi(e, t)$. It is easy to verify that P is a closed set which is closed under addition and substraction and which contains nT for every positive integer n. Since the assumption $x \ll_K \pi(x, T)$ holds, $x \ll_K \pi(x, T + s)$ for all s satisfying $|s| < \varepsilon$ for some $\varepsilon > 0$ small. Repeating the same argument as above, we have that $\omega(x)$ is a periodic orbit of the point e with period $T + s$. Hence P contains the interval $(T - \varepsilon, T + \varepsilon)$. Since P is closed under addition, it must contain an open interval of length 2ε centered on each of its points. It follows that $P = \mathbb{R}$ which implies that e is a rest point and $\lim_{t \to \infty} \pi(x, t) = p$. ☐

Theorem 10.1.3 *A monotone dynamical system cannot have a nontrivial attracting periodic orbit.*

Proof. If there were an attracting periodic orbit, then there exists a point x in its domain of attraction such that $x \ll_K p$ for some point p on the periodic orbit. As p is the limit point of the positive point $\gamma^+(x)$, there exists $T > 0$ such that $x \ll_K \pi(x, T)$. Then $\pi(x, t)$ converges to a rest point by Theorem 10.1.2, contradicting to the fact that it converges to a nontrivial periodic orbit. ☐

Remark 10.1.1 From the [Sel], the author considered a simple model of positive feedback control in a biochemical pathway.

$$\begin{aligned} x_1' &= f(x_5) - \alpha x_1 \\ x_i' &= x_{i-1} - \alpha x_i, 2 \le i \le 5 \end{aligned} \tag{10.2}$$

where $f(x) = \frac{x^p}{(1+x^p)}$ with $p > 0$ and $\alpha > 0$.

The author derives a necessary condition for a Hopf bifurcation to a periodic solution: If $p > \cos^{-5}\left(\frac{2\pi}{5}\right) \cong 355$, then a Hopf bifurcation occurs at $\alpha = \left(\frac{f(c)}{c}\right)^{\frac{1}{5}}$ where $c = \left(p \cos^5\left(\frac{2\pi}{5}\right) - 1\right)^{\frac{1}{p}}$. The periodic orbits that arise at the Hopf bifurcation point are necessarily unstable since from Theorem 10.1.3 and that (10.2) is a cooperative system.

Let $x(t)$ be a solution of the monotone dynamical system (10.1) on an interval I. A subinterval $[a, b]$ of I is called a rising interval if $x(a) \leq_K x(b)$ and $x(a) \neq x(b)$; it is called a falling interval $x(b) \leq_K x(a)$ and $x(a) \neq x(b)$.

Lemma 10.1.1 *A solution $x(t)$ cannot have a rising interval and falling interval that are disjoint.*

Proof. See [SW] p. 272.   □

Theorem 10.1.4 *A compact limit set of a monotone dynamical system cannot contain two points related by $\ll_K$. If the system is strongly monotone then the limit set is unordered.*

Proof. First we consider the case that the limit set L is the omega limit set of $\gamma^+(x_0)$. Suppose that L contains distinct points x_1 and x_2 satisfying $x_1 \ll_K x_2$. Then there exists $t_1 > 0$ such that $\pi(x_0, t_1) \ll_K x_2$. Similarly, there exists $t_2 > t_1$ such that $\pi(x_0, t_1) \ll_K \pi(x_0, t_2) = \pi(\pi(x_0, t_1), t_2 - t_1)$. By convergence criterion, (Theorem 10.1.2) L is a rest point. We obtain a contradiction. Consider the case that L is the alpha limit set of $\gamma^-(x_0)$. Let $x(t) = \pi(x_0, t), t \leq 0$. Arguing as before, there exists $t_1 < 0$ such that $x_1 <_K x(t_1)$ and $t_2 < t_1$ such that $x(t_2) \ll_K x(t_1)$. Continue by choosing $t_3 < t_2$ such that $x(t_3) \ll_K x_2$ and $t_4 < t_3$ such that $x(t_3) \ll_K x(t_4)$. Therefore the interval $I = [t_4, t_1]$ contains the falling interval $[t_4, t_3]$ and rising interval $[t_2, t_1]$ and these two intervals are disjoint. This contradicts to Lemma 10.1.1. Thus we complete the proof of the Theorem.   □

The following result of Hirsch exploits the strong restriction on how a limit set of monotone dynamical system is imbedded in space.

Theorem 10.1.5 *A compact limit set of a monotone dynamical system in $\mathbb{R}^n$ can be deformed by a Lipschitz homeomorphism (with a Lipschitz) inverse to a compact invariant set of a Lipschitz system in $\mathbb{R}^{n-1}$.*

Proof. Let L be the compact limit set of $\mathbb{R}^n$. Let v be a unit vector with $0 \ll_K v$ and let H_v be the hyperplane orthogonal to v, $H_v = \{x \in \mathbb{R}^n : x \cdot v = 0\}$. Let Q be the orthogonal projection onto H_v, that is, $Qx = x - (x \cdot v)v$. From Theorem 10.1.5, L is unordered and we have that Q is one-to-one on L. Therefore Q_L, the restriction of Q to L, is a Lipschitz homeomorphism of L onto a compact subset of H_v. It is easy to show that $|Q_L x_1 - Q_L x_2| \geq m |x_1 - x_2|$ for some $m > 0$ whenever $x_1 \neq x_2$ are points in L. Hence Q_L^{-1} is Lipschitz on $Q(L)$. We can construct a dynamical system on $Q(L) \subset H_v$. Let $y \in Q(L)$ then $y = Q_L(x)$ for a unique $x \in L$. Define flow $\Phi(y, t) \equiv Q_L(\pi(x, t))$. It is easy to verify that $\Phi(y, t)$ is a dynamical system generated by the vector field

$$F'(y) = Q_L(f(Q_L^{-1}(y)))$$

on $Q(L)$ and Q_L send the trajectories of $\dot{x} = f(x)$ on L onto the trajectories of $\dot{y} = F(y)$ on $Q(L)$ (see Fig. 10.1). $\qquad\square$

$$\mathfrak{R}^n \xrightarrow{\;\pi\;} \mathfrak{R}^n$$

$$Q \downarrow \qquad\qquad \downarrow Q$$

$$Q(L) \xrightarrow[\Phi]{\;\;} Q(L)$$

Fig. 10.1

Remark 10.1.2 Let $f : D \subseteq \mathbb{R}^n \to \mathbb{R}^n$ and

$$\frac{dx}{dt} = f(x), \quad x(0) = x_0 \tag{10.3}$$

be a competitive system, i.e., $\frac{\partial f_i}{\partial x_j} \leq 0, i \neq j$. Then with reverse time scaling $\tau = -t$ we obtain a cooperative system.

$$\frac{dx}{d\tau} = -f(x), \quad x(0) = x_0. \tag{10.4}$$

Thus the omega limit set of $\gamma^+(x_0)$ of system (10.3) is exactly the alpha limit set of $\gamma^-(x_0)$ of the system (10.4). Similarly, the alpha limit set of $\gamma^-(x_0)$ of (10.3) is the omega limit set $\gamma^+(x_0)$ of (10.4).

In the following, we shall consider a special case of the system (10.3) with $n = 3$. Let (10.5) be an irreducible competitive system in $\mathbb{R}^3$.

$$
\begin{aligned}
x_1' &= f_1(x_1, x_2, x_3) \\
x_2' &= f_2(x_1, x_2, x_3) \\
x_3' &= f_3(x_1, x_2, x_3)
\end{aligned}
\tag{10.5}
$$

where $\frac{\partial f_i}{\partial x_j} \leq 0$, $i \neq j$, $i, j = 1, 2, 3$.

Theorem 10.1.6 (Poincaré-Bendixson like theorem for 3-dimensional competitive systems)

(i) *A compact limit set L of the system (1.5) which contains no equilibrium is a periodic orbit.*

(ii) *Let γ be a periodic orbit of system (1.5). Then there exists at least one equilibrium in the "interior" of γ.*

Proof. From Theorem 10.1.5, the limit set L can be deformed to a compact invariant set A of a planar vector field. By the Poincaré-Bendixson Theorem, A either contains equilibrium points or A contains periodic orbit. Thus we complete the proof of (i).

The proof of (ii) can be found in [Smi] where the author applies Brouwer Fixed Point Theorem to show that there is at least one equilibrium in the "interior" of γ. $\qquad\square$

Remark 10.1.3 The term interior needs to be interpreted. It is the bounded component of the set $J = (\gamma + \mathbb{R}_+^3)^c \cap (\gamma - \mathbb{R}_+^3)^c$, where $\mathbb{R}_+^3$ is the positive cone in $\mathbb{R}^3$ and the subscript c denotes complement. See [Smi] for more details.

10.2 Uniform Persistence

Consider a population model of n-species interaction which takes the form

$$
\begin{aligned}
x_i' &= x_i f_i(x_1, \cdots, x_n) \\
x_i(0) &= x_{i0} \geq 0, \ i = 1, 2, \cdots, n.
\end{aligned}
\tag{10.6}
$$

Definition 10.2.1 *The system* (10.6) *is said to be persistent if*

$$\liminf_{t\to\infty} x_i(t) > 0, \; i = 1, 2, \cdots, n$$

for every trajectory with positive initial conditions. The system (10.6) *is said to be uniformly persistent if there exists a positive number δ such that*

$$\liminf_{t\to\infty} x_i(t) \geq \delta, \; i = 1, 2, \cdots, n$$

for every trajectory with positive initial condition.

To prove uniform persistence of (10.6), we need the following lemma.

Theorem 10.2.1 (Butler-McGehee Lemma [SW] p. 12)
Let P be a hyperbolic equilibrium of the system (10.6). *Suppose $P \in \omega(x)$, $\{P\} \neq \omega(x)$, the omega limit set of $\gamma^+(x), x \in \mathbb{R}_+^n$. Then there exists points $q \in W^s(P) \cap \omega(x)$ and $\hat{q} \in W^u(P) \cap \omega(x)$, where $W^s(P)$, $W^u(P)$ are stable and unstable manifold of equilibrium P, respectively.*

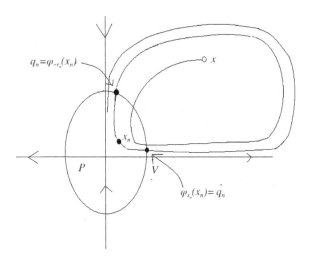

Fig. 10.2

Proof. Since P is a hyperbolic equilibrium, from the definition of stable and unstable manifold, there exists a bounded open set $U \subseteq \mathbb{R}^n$ containing P, but not x with the property that if $\varphi_t(y) \in U$ for all $t > 0$ $(t < 0)$, then $y \in W^s(P)$ $(W^u(P))$. By taking a smaller open set $V, P \in V \subseteq \bar{V} \subset U$, we have that $\varphi_t(y) \in \bar{V}$ for all $t > 0$ $(t < 0)$ implies $y \in W^s(P)$ $(W^u(P))$.

Since $P \in \omega(x)$, there exists a sequence $\{t_n\}$, $\lim_{n\to\infty} t_n = \infty$, such that $\lim_{n\to\infty} x_n = \lim_{n\to\infty} \varphi_{t_n}(x) = P$. It follows that $x_n \in V$ for all large n. Since $x \notin W^s(P)$ and $\omega(x) \neq \{P\}$, from the property of the neighborhood V, there exists $r_n, s_n > 0$ such that $r_n < t_n, \varphi_t(x_n) \in V$ for $-r_n < t < s_n$ and $q_n = \varphi_{-r_n}(x_n), \hat{q}_n = \varphi_{s_n}(x_n) \in \partial V$ (see Fig. 10.2). By continuity of $\varphi_t(x)$, solutions that start near P must remain near P; hence it follows that $\lim_{n\to\infty} r_n = \lim_{n\to\infty} s_n = \infty$. However, $\bar{V}$ is compact, so (passing to a subsequence if necessary) we have that $\lim_{n\to\infty} \varphi_{-r_n}(x_n) = q \in \bar{V}$ and $\lim_{n\to\infty} \varphi_{s_n}(x_n) = \hat{q} \in \bar{V}$. We continue the proof for q; the other case for $\hat{q}$ is similar. Claim that $\varphi_t(q) \in \bar{V}$ for all $t > 0$. Recall that $\lim_{n\to\infty} q_n = q$, where $q_n = \varphi_{-r_n}(x_n)$. Fix $t > 0$. By the continuity of $\varphi_t(x)$, $\lim_{n\to\infty} \varphi_t(q_n) = \varphi_t(q)$. Since $-r_n < t - r_n < 0$ for all large n, $\varphi_t(q_n) = \varphi_{t-r_n}(x_n) \in V$ for all large n. It follows that $\varphi_t(q) \in \bar{V}$. Since $t > 0$ was arbitrary, the claim is established.

Since $\varphi_t(q) \in \bar{V}$ for all $t > 0$, we have $q \in W^s(P)$ by the property that $P \in V \subseteq \bar{V} \subset U$. However, $q \in \overline{\gamma^+(x)} = \gamma^+(x) \cup \omega(x)$. Since $q \in W^s(P)$, $q \notin \gamma^+(x)$ and hence $q \in \omega(x)$, which established. $\qquad\square$

Example 10.2.1 Consider the predator-prey model in Example 4.2.3.

$$\frac{dx}{dt} = rx\left(1 - \frac{x}{K}\right) - \frac{mx}{a+x}y$$
$$\frac{dy}{dt} = \left(\frac{mx}{a+x} - d\right)y \tag{10.7}$$
$$x(0) = x_0 > 0, y(0) = y_0 > 0.$$

If $0 < \lambda < \frac{K-a}{2}$, then the positive equilibrium (x^*, y^*) is an unstable focus. By Poincaré-Bendixson Theorem, if the omega limit set $\omega(x_0, y_0)$ of the trajectory $\gamma^+(x_0, y_0)$ does not contain any equilibrium then $\omega(x_0, y_0)$ contains a periodic orbit. Thus the existence of a periodic solution follows. Since (x^*, y^*) is an unstable focus, then $(x^*, y^*) \notin \omega(x_0, y_0)$.

Claim that $E_1 = (K, 0), E_0 = (0, 0) \notin \omega(x_0, y_0)$. If $E_1 \in \omega(x_0, y_0)$, then from Butler-McGehee Lemma, there exists $(\hat{x}, 0) \in \omega(x_0, y_0)$, $\hat{x} > 0$ since the stable manifold of E_1 is positive x-axis.

If $\hat{x} > K$, then by invariance of omega limit set $\gamma^-(\hat{x}, 0) \subseteq \omega(x_0, y_0)$, where $\gamma^-(\hat{x}, 0) = \{(x, 0) : x > \hat{x}\}$. It contradicts to the fact that the solution $(x(t), y(t))$ is bounded.

If $\hat{x} < K$, then $(0,0) \in \gamma^-(\hat{x},0) \subseteq \omega(x_0,y_0)$. Apply Butler-McGehee Lemma again to $P = (0,0)$, then there exists $\hat{y} > 0$ such that $(0,\hat{y}) \in W^s(0,0)$. $\gamma^-(0,\hat{y}) \subseteq \omega(x_0,y_0)$, where $\gamma^-(0,\hat{y}) = \{(0,y) : y > \hat{y}\}$. It contradicts to the boundedness of solution $(x(t),y(t))$.

Hence we prove the existence of periodic solutions for the system (10.7).

Next we state without proof a theorem of uniform persistence for the system (10.6).

Definition 10.2.2

(i) *The flow $\mathcal{F}$, $\mathcal{F} = \varphi_t(x)$, generated by (10.6) is called dissipative if for each $x \in \mathbb{R}^n_+$, $\omega(x) \neq \emptyset$ and the invariant set $\Omega(\mathcal{F}) = \bigcup_{x \in \mathbb{R}^n_+} \omega(x)$ has compact closure.*

(ii) *We said that $M \subseteq \mathbb{R}^n_+$ is an isolated invariant set for the flow $\mathcal{F}$ generated by (10.6) if M is an nonempty invariant set which is maximal invariant set in some neighborhood of itself. Note that if M is a compact, isolated invariant set. One may always choose a compact, isolated neighborhood. A typical examples of isolated invariant are equilibrium and periodic orbit.*

(iii) *The stable set $W^s(M)$ of an isolated invariant set M is defined to be*

$$\{x \in \mathbb{R}^n_+ : \omega(x) \neq \emptyset, \ \omega(x) \subset M\}$$

and the unstable set $W^u(M)$ is similarly defined in terms of alpha limit set $\alpha(x)$.

(iv) *If M, N are isolated invariant sets for the flow $\varphi_t(x)$, we shall say that M is chained to N and write this as $M \to N$ if there exists $x \notin M \cup N$ such that $x \in W^u(M) \cap W^s(N)$.*

(v) *A chain of isolated invariant set is a finite sequence $M_1, M_2, \cdots, M_k$ with $M_1 \to M_2 \to M_3 \to \cdots \to M_k$ ($M_1 \to M_1$ if k=1). The chain is called a cycle if $M_k = M_1$.*

(vi) *Let $\partial\mathcal{F} = \varphi_t(x)\big|_{\partial(\mathbb{R}^n_+)} : \partial(\mathbb{R}^n_+) \to \partial(\mathbb{R}^n_+)$, the boundary flow of (10.6). $\partial\mathcal{F}$ will be called acyclic if for some isolated covering M of $\Omega(\partial\mathcal{F})$, $M = \bigcup_{i=1}^k M_i$, no subset of $\{M_i\}$ forms a cycle.*

Theorem 10.2.2 (Uniform Persistence [SW] p. 280)
Let $\mathcal{F} = \varphi_t(x) : \mathbb{R}^n_+ \to \mathbb{R}^n_+$ be the flow generated by (10.6) and $\partial\mathcal{F} : \partial(\mathbb{R}^n_+) \to \partial(\mathbb{R}^n_+)$. Assume that

(i) $\mathcal{F}$ is dissipative.

(ii) The boundary flow $\partial\mathcal{F}$ is acyclic with acyclic covering $M = \{M_1, M_2, \cdots, M_k\}$.

Then $\mathcal{F}$ is uniformly persistent if and only if

$$W^s(M_i) \cap Int(\mathbb{R}^n_+) = \emptyset \text{ for each } M_i \in M. \tag{10.8}$$

Furthermore, if (10.8) holds then there is an equilibrium in $Int(\mathbb{R}^n_+)$.

Example 10.2.2 Food Web Model [FW]

$$
\begin{aligned}
x' &= xg(x) - yp(x) \\
y' &= y(-d_1 + c_1 p(x)) - zq(y) \\
z' &= z(-d_2 + c_2 q(y))z \\
x(0) &= x_0 > 0, y(0) = y_0 \geq 0, z(0) = z_0 \geq 0
\end{aligned}
\tag{10.9}
$$

where

$$g(x) = r\left(1 - \frac{x}{K}\right), p(x) = \frac{m_1 x}{a_1 + x}, q(y) = \frac{a_2 y}{a_2 + y}.$$

The equilibrium $E^* = (x^*, y^*, 0)$ in the xy-plane exists if there is a point $x^* < K$ such that $p(x^*) = \frac{d_1}{c_1}$, in which case $y^* = \frac{x^* g(x^*)}{p(x^*)}$. The origin $E_0 = (0, 0, 0)$ and the equilibrium point $E_1 = (K, 0, 0)$ are each unstable with two dimensional unstable manifold. E^* may be stable or unstable in the plane.

Assume E^* is globally stable in the $x - y$ plane. From Theorem 10.2.2, the system (10.9) is uniformly persistent if and only if $-d_2 + c_2 q(y^*) > 0$.

Example 10.2.3 [ML; CHW] Rock-Scissor-Paper Model
Consider the system three competing species

$$
\begin{aligned}
x_1' &= x_1(1 - x_1 - \alpha_1 x_2 - \beta_2 x_3) \\
x_2' &= x_2(1 - \beta_2 x_1 - x_2 - \alpha_2 x_3) \\
x_3' &= x_3(1 - \alpha_3 x_1 - \beta_3 x_2 - x_3)
\end{aligned}
\tag{10.10}
$$

$x_1(0) > 0, x_2(0) > 0, x_3(0) > 0$, subject to

$$0 < \alpha_i < 1 < \beta_i, \alpha_i + \beta_i > 2, \ i = 1, 2, 3. \tag{10.11}$$

The condition (10.11) says that x_1 out-competes x_2 ($x_3 = 0$, orbits in $x_1 - x_2$ plane tend to a point a point $x_2 = 0$), x_2 out-competes x_3, and x_3 out-competes x_2. In [CHW], the authors proved that if $A_1 A_2 A_3 \neq B_1 B_2 B_3$

where $A_i = 1 - \alpha_i > 0$, $B_i = \beta_i - 1 > 0$, $i = 1, 2, 3$, then there is no periodic orbits in the interior of $\mathbb{R}_+^3$ by applying Stoke's Theorem. Thus from Theorem 10.1.7, the w-limit set of the orbit contains equilibrium points.

If $A_1 A_2 A_3 > B_1 B_2 B_3$, then the interior equilibrium $P = (x^*, y^*, z^*)$ exists and is globally asymptotically stable with respect to the interior of $\mathbb{R}_+^3$. If $A_1 A_2 A_3 < B_1 B_2 B_3$, then $P = (x^*, y^*, z^*)$ is saddle point with one-dimensional stable manifold Γ. If $(x_0, y_0, z_0) \notin \Gamma$, then the w-limit set $\omega(x_0, y_0, z_0)$ of $\gamma^+(x_0, y_0, z_0)$ is $\omega(x_0, y_0, z_0) = O_1 \cup O_2 \cup O_3$, where O_1 is the orbit on the $x_2 x_3$ plane connecting equilibrium $e_3 = (0, 0, 1)$ to the equilibrium $e_2 = (0, 1, 0)$; O_2 is the orbit on the $x_1 x_3$ plane connecting equilibrium $e_1 = (1, 1, 0)$ to e_3; O_3 is the orbit on $x_1 x_2$ plane connecting e_2 to e_1. In this case, the system (10.10) is not persistent and the solutions of (10.10) exhibit aperiodic oscillation.

Proof of nonexistence of periodic orbits of (10.10) by using Stoke's Theorem:

We shall prove that if $A_1 A_2 A_3 \neq B_1 B_2 B_3$, then the system (10.10) has no nontrivial periodic solutions.

Consider the system (10.10) with the assumptions (10.11),

$$
\begin{aligned}
\dot{x}_1 &= f_1(x_1, x_2, x_3) = x_1(1 - x_1 - \alpha_1 x_2 - \beta_1 x_3), \\
\dot{x}_2 &= f_2(x_1, x_2, x_3) = x_2(1 - \beta_2 x_1 - x_2 - \alpha_2 x_3), \\
\dot{x}_3 &= f_3(x_1, x_2, x_3) = x_3(1 - \alpha_3 x_1 - \beta_3 x_2 - x_3), \\
x_i(0) &> 0, i = 1, 2, 3.
\end{aligned}
\tag{10.12}
$$

Define a new vector field

$$
(M_1, M_2, M_3) = (x_1, x_2, x_3) \times (f_1, f_2, f_3).
$$

Then the routine computations yield

$$
\begin{aligned}
M_1 &= x_2 x_3 [(\beta_2 - \alpha_3) x_1 + (1 - \beta_3) x_2 + (\alpha_2 - 1) x_3], \\
M_2 &= x_1 x_3 [(\alpha_3 - 1) x_1 + (\beta_3 - \alpha_1) x_2 + (1 - \beta_1) x_3], \\
M_3 &= x_1 x_2 [(1 - \beta_2) x_1 + (\alpha_1 - 1) x_2 + (\beta_1 - \alpha_2) x_3],
\end{aligned}
\tag{10.13}
$$

and

$$
\begin{aligned}
&curl(M_1, M_2, M_3) \\
&= \left(\frac{\partial M_3}{\partial x_2} - \frac{\partial M_2}{\partial x_3}, \frac{\partial M_1}{\partial x_3} - \frac{\partial M_3}{\partial x_1}, \frac{\partial M_2}{\partial x_1} - \frac{\partial M_1}{\partial x_2} \right) \\
&= \begin{pmatrix} x_1[(A_3 - B_2)x_1 - (3A_1 + B_3)x_2 + (3B_1 + A_2)x_3] \\ x_2[(3B_2 + A_3)x_1 + (A_1 - B_3)x_2 - (3A_2 + B_1)x_3] \\ x_3[-(3A_3 + B_2)x_1 + (A_1 + 3B_3)x_2 + (A_2 - B_1)x_3] \end{pmatrix}.
\end{aligned}
\tag{10.14}
$$

Let

$$\Gamma = \{(p_1 t, p_2 t, p_3 t) |\, t > 0\}. \qquad (10.15)$$

Lemma 10.2.1 Γ *is a positive invariant set under* (10.12), *and the solution* $\psi(t)$ *of* (10.12) *with initial condition in* Γ *satisfies*

$$\lim_{t \to \infty} \psi(t) = P.$$

Proof. If $x(0) \in \Gamma$, then $x(0) = (p_1 \xi, p_2 \xi, p_3 \xi)$ for some $\xi > 0$. Let $\phi(t)$ satisfy $\phi'(t) = \phi(t)(1 - \phi(t)), \phi(0) = \xi$. Then it is easy to verify that $\psi(t) = (p_1 \phi(t), p_2 \phi(t), p_3 \phi(t))$ satisfies (10.12). Hence Γ is positively invariant and $\lim_{t \to \infty} \psi(t) = P$. $\qquad\square$

Lemma 10.2.2 *Let* $(x_1, x_2, x_3) \in \mathbb{R}_+^3$ *and* $x_i > 0, i = 1, 2, 3$. *If* $(x_1, x_2, x_3) \notin \Gamma$, *then* $(M_1, M_2, M_3) \neq 0$ *at* (x_1, x_2, x_3).

Proof. Since $(M_1, M_2, M_3) = (x_1, x_2, x_3) \times (f_1, f_2, f_3)$, if $(M_1, M_2, M_3) = 0$, then either $(f_1, f_2, f_3) = 0$ or $(f_1, f_2, f_3) = (x_1, x_2, x_3)t$ for some $t \in \mathbb{R}$. If $(f_1, f_2, f_3) = 0$, then $(x_1, x_2, x_3) = P$. If $(f_1, f_2, f_3) = (x_1, x_2, x_3)t$, then

$$(1 - x_1 - \alpha_1 x_2 - \beta_1 x_3) = (1 - \beta_2 x_1 - x_2 - \alpha_2 x_3)$$
$$= (1 - \alpha_3 x_1 - \beta_3 x_2 - x_3) = t.$$

It follows that $(x_1, x_2, x_3) = (1 - t)(p_1, p_2, p_3) \in \Gamma$.

Hence either of the above two cases leads to a contradiction to the assumption $(x_1, x_2, x_3) \notin \Gamma$. $\qquad\square$

Lemma 10.2.3 *The solution of* (10.12) *are positive and bounded, and furthermore, for any* $\varepsilon > 0$, *there exists* $T \geq 0$ *such that for each* $i = 1, 2, 3, x_i(t) < 1 + \varepsilon$ *for all* $t \geq T$.

We omit the proof of Lemma 10.2.3 because it is trivial.

Theorem 10.2.3 *If* $A_1 A_2 A_3 \neq B_1 B_2 B_3$, *then the system* (2.7) *has no periodic solutions in the interior of* $\mathbb{R}_+^3$.

Proof. Suppose there exists a periodic solution $x(t) = (x_1(t), x_2(t), x_3(t))$, with period w, in the interior of $\mathbb{R}^3_+$. Let

$$C = \{(x_1(t), x_2(t), x_3(t)) \mid 0 \le t \le T\}.$$

We claim that the periodic orbit C is disjoint from the set Γ. From Lemma 10.2.1, it follows that if $C \cap \Gamma \ne \emptyset$, then $x(t) \to P$ as $t \to \infty$. This contradicts the fact that $x(t)$ is a periodic solution. Next, we construct the following conical surface S:

$$S = \{\lambda\,(x_1(t), x_2(t), x_3(t)) \mid \lambda \in [0,1] \text{ and } t \in [0, w]\}.$$

Since (10.12) is a competitive system, from Theorem 10.1.4, for any two points $x, y \in C, x \ne y, x, y$ are unrelated; i.e. $x - y \notin Int(\mathbb{R}^3_+)$ or $y - x \notin Int(\mathbb{R}^3_+)$. Hence the surface S does not cross itself.

Given a point $(x_1(t_0), x_2(t_0)), x_3(t_0)) \in C$, consider the segment from 0 to $x(t_0)$. Then from Lemma 10.2.2

$$\vec{N} = (x_1(t_0), x_2(t_0)), x_3(t_0)) \times (f_1, f_2, f_3)|_{x=x(t_0)}$$
$$= (M_1, M_2, M_3)|_{x=x(t_0)} \ne 0$$

is a normal vector of the surface S at each point of the segment $\overline{(0, x(t_0))}$. Normalize the vector $\vec{N}$. Then we have the unit normal vector,

$$\vec{n} = \frac{1}{K_1}\,(M_1, M_2, M_3)|_{x=x(t_0)},$$

where $K_1 = \left|\vec{N}\right| \ne 0$. For each point on the segment $\overline{(0, x(t_0))}$, we compute $curl(M_1, M_2, M_3) \cdot \vec{n}$ at the point $x = s(x_1(t_0), x_2(t_0)), x_3(t_0)), s \in [0, 1]$. Then from (10.13) and (10.14), it follows that

$$curl(M_1, M_2, M_3) \cdot \vec{n}$$
$$= s^2 curl\,(M_1, M_2, M_3)|_{x=x(t_0)} \cdot \frac{1}{K_1}\,(M_1, M_2, M_3)|_{x=x(t_0)}$$
$$= s^2 \frac{1}{K_1} x_1 x_2 x_3\,(G(x_1, x_2, x_3))|_{x=x(t_0)}$$

where

$$G(x_1, x_2, x_3)$$

$$= (x_1, x_2, x_3) \left[\begin{pmatrix} B_2 + A_3 \\ -B_3 \\ -A_2 \end{pmatrix} (A_3 - B_2, 3A_1 - B_3, 3B_1 + A_2) \right.$$

$$+ \begin{pmatrix} -A_3 \\ B_3 + A_1 \\ -B_1 \end{pmatrix} (3B_2 + A_3, A_1 - B_3, -3A_2 - B_1)$$

$$\left. + \begin{pmatrix} -B_2 \\ -A_1 \\ B_1 + A_2 \end{pmatrix} (-3A_3 - B_2, 3B_3 + A_1, A_2 - B_1) \right] \begin{pmatrix} x_1 \\ x_2 \\ x_3 \end{pmatrix}.$$

A routine computation shows $G(x_1, x_2, x_3) = 0$. Hence

$$curl(M_1, M_2, M_3) \cdot \vec{n} = 0 \text{ on segment } \overline{(0, x(t_0))} \text{ for all } t_0 \in [0, w]$$

and

$$curl(M_1, M_2, M_3) \cdot \vec{n} = 0 \text{ on the surface } S. \tag{10.16}$$

Let the surface $C' = \{(x_1, x_2, x_3) | x_1^{\delta_1} x_2^{\delta_2} x_3^{\delta_3} = c\}$, where the positive numbers $\delta_1, \delta_2, \delta_3$ will be selected and $c > 0$ is sufficiently small such that C' is disjoint from the periodic orbit C. Let Y be the intersection of the surface C' and the cone (bounded by S). Then C' divides the surface S into two parts S_1 and S_2 such that $C \subset S_1$ and $(0, 0, 0) \in S_2$.

Let $S' = Y \cup S_1$. Then S' is the surface with $\partial S' = C$. On the surface Y, the outward normal vector $\vec{N} = -\nabla(x_1^{\delta_1} x_2^{\delta_2} x_3^{\delta_3}) = -c(\frac{\delta_1}{x_1}, \frac{\delta_2}{x_2}, \frac{\delta_3}{x_3})$. Thus the outward unit normal vector $\vec{n}$ on Y is $\vec{n} = -\frac{c}{K_2}(\frac{\delta_1}{x_1}, \frac{\delta_2}{x_2}, \frac{\delta_3}{x_3})$, where $K_2 = |\vec{N}|$. From (10.14), it follows that on the surface Y, we have

$$curl(M_1, M_2, M_3) \cdot \vec{n}$$

$$= -\frac{c}{K_2}[x_1((\delta_1 + \delta_2 - 3\delta_3)A_3 - (\delta_1 - 3\delta_2 + \delta_3)B_2)$$

$$+ x_2(-(3\delta_1 - \delta_2 - \delta_3)A_1 - (\delta_1 + \delta_2 - 3\delta_3)B_3)$$

$$+ x_3((\delta_1 - 3\delta_2 + \delta_3)A_2 + (3\delta_1 - \delta_2 - \delta_3)B_1)].$$

Choose $\delta_1, \delta_2, \delta_3$ satisfying

$$\delta_1 + \delta_2 - 3\delta_3 = -A_1 B_2$$

$$\delta_1 - 3\delta_2 + \delta_3 = -A_1 A_3$$

$$3\delta_1 - \delta_2 - \delta_3 = B_2 B_3$$

or

$$\delta_1 = \frac{1}{4}(A_1B_2 + A_1A_3 + 2B_2B_3) > 0,$$

$$\delta_2 = \frac{1}{4}(A_1B_2 + 2A_1A_3 + B_2B_3) > 0,$$

$$\delta_3 = \frac{1}{4}(2A_1B_2 + B_2B_3 + A_1A_3) > 0.$$

Then we have

$$curl(M_1, M_2, M_3) \cdot \vec{n} = -\frac{c}{K_2}x_3(B_1B_2B_3 - A_1A_2A_3) < 0 \text{ or } > 0 \text{ for all } x \in Y.$$
(10.17)

Now we are in a position to prove nonexistence of periodic solution by Stoke's Theorem. Since S_1 and Y are smooth enough for the application of Stoke's Theorem,

$$\oint_C M_1dx_1 + M_2dx_2 + M_3dx_3 = \int\int_{S_1 \cup Y} curl(M_1, M_2, M_3) \cdot \vec{n}dA.$$
(10.18)

From the fact that $(M_1, M_2, M_3) = (x_1, x_2, x_3) \times (f_1, f_2, f_3)$, it follows that

$$\oint_C M_1dx_1 + M_2dx_2 + M_3dx_3 = \int_0^w (M_1f_1 + M_2f_2 + M_3f_3)dt = 0.$$
(10.19)

From (10.16) and (10.17)

$$\int\int_{S_1 \cup Y} curl(M_1, M_2, M_3) \cdot \vec{n}dA$$

$$= \int\int_{S_1} curl(M_1, M_2, M_3) \cdot \vec{n}dA + \int\int_Y curl(M_1, M_2, M_3) \cdot \vec{n}dA \quad (10.20)$$

$$= 0 - \frac{c}{K_2}\int\int_Y x_3(B_1B_2B_3 - A_1A_2A_3)dA \neq 0.$$

Thus (10.18), (10.19), (10.20) lead to a desired contradiction.   □

10.3 Application: Competition of Two Species in a Chemostat with Inhibition

Let $S(t)$ denote the nutrient concentration at time t in the culture vessel; $x_1(t)$, $x_2(t)$, the concentration of the competitors; and $p(t)$, the concentration of the inhibitor (or toxicant or pollutant). The equations of the model take the form [HW]

$$S' = (S^{(0)} - S)D - \frac{m_1 x_1 S}{a_1 + S}f(p) - \frac{m_2 x_2 S}{a_2 + S},$$

$$x_1' = x_1 \left(\frac{m_1 S}{a_1 + S}f(p) - D \right),$$

$$x_2' = x_2 \left(\frac{m_2 S}{a_2 + S} - D \right), \tag{10.21}$$

$$p' = (p^{(0)} - p)D - \frac{\delta x_2 p}{K + p},$$

$$S(0) \geqq 0, \quad x_i(0) > 0, \quad p(0) \geqq 0, \quad i = 1,\, 2.$$

$S^{(0)}$ is the input concentration of the nutrient, and $p^{(0)}$ is the input concentration of the inhibitor, both of which are assumed to be constant. D is the dilution rate of the chemostat. $S^{(0)}, p^{(0)}$, and D are under the control of the experimenter. $m_i, a_i, i = 1,\, 2$ are the maximal growth rates of the competitors (without an inhibitor) and the Michaelis-Menten (or half saturation) constants, respectively. These parameters, inherent properties of the organism, are measurable in the laboratory. δ and K play similar roles for the pollutant, δ being the uptake by x_2, and K being a half saturation parameter. The function $f(p)$ represents the degree of inhibition of p on the growth rate (or uptake rate) of x_1. By suitable scaling, for example $S \to \frac{S}{S^{(0)}}, p \to \frac{p}{p^{(0)}}, t \to Dt$, we may assume $S^{(0)} = 1, D = 1, p(0) = 1$ and system (10.21) takes the form

$$S' = 1 - S - \frac{m_1 x_1 S}{a_1 + S}f(p) - \frac{m_2 x_2 S}{a_2 + S},$$

$$x_1' = x_1 \left(\frac{m_1 S}{a_1 + S}f(p) - 1 \right),$$

$$x_2' = x_2 \left(\frac{m_2 S}{a_2 + S} - 1 \right), \tag{10.22}$$

$$p' = 1 - p - \frac{\delta x_2 p}{K + p},$$

$$S(0) \geqq 0, \quad x_i(0) > 0, \quad i = 1,\, 2, \quad p(0) \geqq 0.$$

Concerning the function $f(p)$, we assume that

(i) $f(p) \geqq 0, f(0) = 1,$

(ii) $f'(p) < 0, p > 0.$ \hfill (10.23)

The function $f(p) = e^{-\eta p}, \eta > 0$ has these properties.

Let $\Sigma = 1 - S - x_1 - x_2$. Then $\Sigma' = -S' - x_1' - x_2' = -1 + S + x_1 + x_2 = -\Sigma$. System (10.22) may then be replaced by

$$\Sigma' = -\Sigma,$$

$$x_1' = x_1 \left(\frac{m_1(1 - \Sigma - x_1 - x_2)}{a_1 + 1 - \Sigma - x_1 - x_2} f(p) - 1 \right),$$

$$x_2' = x_2 \left(\frac{m_2(1 - \Sigma - x_1 - x_2)}{a_2 + 1 - \Sigma - x_1 - x_2} - 1 \right),$$

$$p' = 1 - p - \delta \frac{x_2 p}{K + p}.$$

(10.24)

Clearly, $\lim_{t\to\infty} \Sigma(t) = 0$. Hence the solutions in the omega limit set of (10.24) must satisfy

$$x_1' = x_1 \left(\frac{m_1(1 - x_1 - x_2)}{1 + a_1 - x_1 - x_2} f(p) - 1 \right),$$

$$x_2' = x_2 \left(\frac{m_2(1 - x_1 - x_2)}{1 + a_2 - x_1 - x_2} - 1 \right),$$

$$p' = 1 - p - \delta \frac{x_2 p}{K + p},$$

(10.25)

$$x_i(0) > 0, \quad i = 1, 2, \quad p(0) \geqq 0, \quad x_1(0) + x_2(0) < 1.$$

More directly, we could also apply the theory of asymptotically autonomous systems (see [SW] p. 294). System (10.25) is competitive. Let

$$\lambda_1 = \frac{a_1}{m_1 - 1}, \qquad \lambda_2 = \frac{a_2}{m_2 - 1}.$$

(10.26)

These are the break-even concentration for species 1 and 2 respectively for the chemostat and would determine the outcome if the inhibitor p were not present. The form of (10.25) guarantees that if $x_i(0) > 0$, $i = 1, 2$, then $x_i(t) > 0$ for $t > 0$. Moreover, $p'|_{p=0} = 1 > 0$; so, if $p(0) \geqq 0$, $p(t) > 0$ for $t > 0$. $x_1(t)$ and $x_2(t)$ satisfy

$$x_1' \leqq x_1 \left(\frac{m_1(1 - x_1 - x_2)}{1 + a_1 - x_1 - x_2} - 1 \right),$$

$$x_2' \leqq x_2 \left(\frac{m_2(1 - x_1 - x_2)}{1 + a_2 - x_1 - x_2} - 1 \right),$$

(10.27)

so that an application of Kamke's theorem, (Theorem 2.6.3) and the elementary knowledge of the behavior of trajectories of (10.27) with equalities, establishes the following proposition.

Proposition 10.3.1 *If $m_i \leqq 1$ or if $\lambda_i \geqq 1$, $\lim_{t\to\infty} x_i(t) = 0$, $i = 1$ or 2.*

This simply states the biologically intuitive fact that if one of the competitors could not survive in the simple chemostat, that competitor will not survive in the chemostat with an inhibitor. Thus we may assume that $m_i > 1$ and $0 < \lambda_i < 1$, $i = 1, 2$.

Lemma 10.3.1 *There exists a number $\gamma > 0$ such that $p(t) \geqq \gamma$ for t sufficiently large.*

Proof. Suppose $\liminf_{t \to \infty} p(t) = 0$. If $p(t)$ decreased to zero monotonically, then there would be a point t_0 such that for $t > t_0$, $p(t) + \delta p(t)/(K + p(t)) < 1$. For such values, $p'(t) > 0$, which contradicts $p(t)$ decreasing. Hence there exists a set of points t_n, $t_n \to \infty$, such that $p'(t_n) = 0$ and $p(t_n) \to 0$ as $t_n \to \infty$. For such values of t_n,

$$0 = 1 - p(t_n) - \frac{\delta p(t_n) x_2(t_n)}{K + p(t_n)}$$

$$> 1 - p(t_n) - \frac{\delta p(t_n)}{K + p(t_n)}$$

$$> 0$$

for n large. This establishes the lemma. □

Theorem 10.3.1 *If $0 < \lambda_2 \leqq \lambda_1 < 1$, then*

$$\lim_{t \to \infty} x_1(t) = 0, \ \lim_{t \to \infty} x_2(t) = 1 - \lambda_2 = x_2^*, \ \lim_{t \to \infty} p(t) = p_2^* < 1,$$

where p_2^* is the positive root of the quadratic

$$(1 - p)(K + p) - \delta(1 - \lambda_2)p = 0. \tag{10.28}$$

Proof. The reason for labeling it p_2^* will become clear below. $p_2^* < 1$ follows from the fact that $p(t)$ satisfies

$$p' < 1 - p \tag{10.29}$$

and the basic comparison theorem for differential inequalities. In view of Lemma 10.3.1 inequalities (10.27) can be replaced by

$$x_1' \leqq x_1 \left(\frac{m_1(1 - x_1 - x_2)f(\gamma)}{1 + a_1 - x_1 - x_2} - 1 \right),$$

$$x_2' \leqq x_2 \left(\frac{m_2(1 - x_1 - x_2)}{1 + a_2 - x_1 - x_2} - 1 \right),$$

for t sufficiently large. This system of inequalities can be compared to the equations for the chemostat with λ_2 and λ_γ as parameters, where

$$\lambda_2 \lneqq \lambda_1 < \frac{a_1}{m_1 f(\gamma) - 1} = \lambda_\gamma,$$

so that the first component of the comparison system tends to zero as t tends to infinity. Hence so does $x_1(t)$. $\square$

Thus, for the remainder of this section, we may assume that

$$m_i > 1, \quad i = 1, 2; \quad 0 < \lambda_1 < \lambda_2 < 1 \tag{10.30}$$

to make the problem interesting. Note also that this provides the boundedness of solutions. The results below provide conditions for one or both of the competitors to wash out of the chemostat. To avoid "unlikely" cases, we tacitly assume that all rest points and periodic orbits are hyperbolic, i.e., that their stability is determined by their linearization.

The rest point set. As noted above, system (10.25) is competitive system. From Theorem 10.1.6, a type of Poincaré-Bendixson theory holds. The only possible omega limit sets are those of a two-dimensional system, specifically a rest point, a periodic orbit, or a finite set of rest points connected by trajectories. Moreover, if there is a periodic orbit, it must have a rest point "inside," where "inside" is defined in terms of an order. This has the consequence that when there is no interior rest point, there cannot be a periodic orbit in the open positive octant, and hence the limit is on the boundary. Thus the existence of an interior rest point is crucial for coexistence.

There are three potential rest points on the boundary, which we label $E_0 = (0, 0, 1)$, $E_1 = (x_1^*, 0, 1)$, and $E_2 = (0, x_2^*, p_2^*)$. These correspond to one or both competitors becoming extinct.

E_0 always exists. E_2 exists with $x_2^* = 1 - \lambda_2$ and p_2^* the root of (10.25) if $0 < \lambda_2 < 1$, which is contained in our basic assumption (10.30). The existence of E_1 is a bit more delicate. In keeping with the definitions in (10.26), define $\lambda_0 = a_1/(m_1 f(1) - 1)$. The inequality $0 < \lambda_0 < 1$ corresponds to the survivability of the first population in a chemostat under maximal levels of the inhibitor. Easy computations show that $E_1 = (1 - \lambda_0, 0, 1)$ will exist if $\lambda_0 > 0$ and will have positive coordinates and be asymptotically stable in the x_1-p plane if $0 < \lambda_0 < 1$. If $1 - \lambda_0$ is negative, E_1 is not meaningful, nor is it accessible from the given initial conditions since the x_2-p plane is an invariant set. The stability of either E_1 or E_2 will depend on comparisons between the subscripted λ's. The local stability of each rest point depends on the eigenvalues of the linearization around those points. The Jacobian matrix for the linearization of (10.25) takes the form

$$J = \begin{pmatrix} m_{11} & m_{12} & m_{13} \\ m_{21} & m_{22} & 0 \\ 0 & m_{32} & m_{33} \end{pmatrix}. \tag{10.31}$$

At E_0,

$$J = \begin{pmatrix} \frac{m_1 f(1)}{1+a_1} - 1 & 0 & 0 \\ 0 & \frac{m_2}{1+a_2} - 1 & 0 \\ 0 & -\frac{\delta}{1+K} & -1 \end{pmatrix}.$$

The eigenvalues are the diagonal elements. One eigenvalue is -1, and the eigenvector lies along the p axis. This corresponds to the growth of the inhibitor to its limiting value in the absence of a consumer. The set $\{(0,0,p)\,|_{p>0}\}$ is invariant and is part of the stable manifold of E_0. $m_{22} = (m_2/(1 + a_2)) - 1$ is positive since $\lambda_2 < 1$. Similarly, the remaining diagonal term m_{11} is positive if $0 < \lambda_0 < 1$, and negative otherwise. When this eigenvalue is negative, the stable manifold of E_0 is the entire $(x_1 - p)$ plane.

Remark 10.3.1 When $\lambda_0 > 0$, no trajectory of (10.25) has E_0 as an omega limit point.

At E_1, $m_{21} = 0$; since $m_{23} = m_{31} = 0$, the eigenvalues are just the diagonal elements of J. Thus

$$\mu_1 = -\frac{m_1 a_1 (1 - \lambda_0)}{(a_1 + \lambda_0)^2} f(1), \quad \mu_2 = \frac{(m_2 - 1)(\lambda_0 - \lambda_2)}{a_2 + \lambda_0}, \quad \mu_3 = -1.$$

If $0 < \lambda_0 < \lambda_2 < 1$, then E_1 is asymptotically stable. This reflects the fact that x_1, in the presence of the maximal inhibitor concentration, is still a better competition than x_2. If $\lambda_0 > \lambda_2$, E_1 is unstable and, of course, if $\lambda_0 > 1$, E_1 does not exist.

Lemma 10.3.2 *If* $\lambda_0 > \lambda_2$, *then any solution of* (10.25) *satisfies* $\liminf_{t \to \infty} x_2(t) > 0$.

Table 10.1

	Exists	Stability
E_0	always	1- or 2-dimensional stable manifold
E_1	$0 < \lambda_0 < 1$	asymptotically stable if $0 < \lambda_0 < \lambda_2$
E_2	$\lambda_2 < 1$	asymptotically stable if $0 < \lambda_2 < \lambda^*$

Proof. Suppose that Lemma 10.3.2 is not true. Then some trajectory Γ has an omega limit point in the $(x_1$-$p)$ plane. Moreover, the initial conditions preclude that Γ is on the stable manifold of E_1. Thus, by the Butler-McGehee lemma (Theorem 10.2.1), the omega limit set of Γ must contain a point of the stable manifold of E_1 and hence the entire trajectory through that point. To remain bounded, such a trajectory must connect to E_0. We have already noted in Remark 10.3.1 that this is not possible.   □

At E_2, $m_{12} = m_{13} = m_{23} = 0$, so again the eigenvalues are just the following diagonal elements:

$$\mu_1 = \frac{m_1 \lambda_2 f(p_2^*)}{a_1 + \lambda_2} - 1,$$

$$\mu_2 = -\frac{m_2 a_2 (1 - \lambda_2)}{(a_2 + \lambda_2)^2},$$

$$\mu_3 = -1 - \frac{\delta K (1 - \lambda_2)}{(K + p_2^*)^2}.$$

Clearly, μ_2 and μ_3 are negative, so E_2 always has a two-dimensional stable manifold. $\mu_1 < 0$ is equivalent to

$$\lambda_2 < \frac{a_1}{m_1 f(p_2^*) - 1} = \lambda^*.$$

The local behavior of the rest point set on the boundary is summarized in Table 10.1, where $0 < \lambda_1 < \lambda_2 < 1$ is assumed.

The more interesting case is that of an interior rest point. As noted above, the competitiveness of the system and a type of Poincaré-Bendixson theorem require its existence for coexistence to be possible. Let $E_c = (x_{1c}^*, x_{2c}^*, p_c^*)$ denote the coordinates of a possible interior rest point. First, it must be the case that

$$1 - x_{1c} - x_{2c} = \lambda_2 \tag{10.32}$$

for this is the only nontrivial zero of the derivative of x_2. Using this, we set the derivative of x_1 equal to zero to find that

$$\frac{m_1 \lambda_2}{a_1 + \lambda_2} f(p) = 1$$

or that we need $(a_1 + \lambda_2)/m_1\lambda_2$ to be in the range of $f(p)$. It is, then

$$p_c^* = f^{-1}\left(\frac{a_1 + \lambda_2}{m_1\lambda_2}\right). \tag{10.33}$$

Since f is monotone, this number is unique. Given p_c^*, then x_{2c}^* can be determined from setting $p'(t)$ equal to zero, yielding

$$1 - p_c^* - \frac{\delta x_{2c}^* p_c^*}{K + p_c^*} = 0$$

or

$$x_{2c}^* = \frac{(1 - p_c^*)(K + p_c^*)}{\delta p_c^*}. \tag{10.34}$$

This number is unique since p_c^* is unique. If $x_{2c}^* < 1 - \lambda_2$, then x_{1c}^* is uniquely determined from (10.32) as

$$x_{1c}^* = 1 - x_{2c}^* - \lambda_2. \tag{10.35}$$

Since $1 - \lambda_2 = x_2^*$, it follows that if x_{2c}^* exists, then $x_{2c}^* < x_2^*$. This is the biologically expected statement that x_2 will do less well in the coexistent steady state than in the steady state where it is the sole survivor. This is true if and only if

$$x_{2c}^* = \frac{(1 - p_c^*)(K + p_c^*)}{\delta p_c^*} < \frac{(1 - p_2^*)(K + p_2^*)}{\delta p_2^*} = x_2^*$$

and hence, in view of the monotonicity of the expression in p, if and only if $p_2^* < p_c^*$. From (10.23) we have that this is equivalent to

$$f(p_2^*) > \frac{a_1 + \lambda_2}{m_1\lambda_2} \tag{10.36}$$

or to the instability of (E_2). See Table 10.1, where the value of λ^* has been substituted to obtain (10.36). Thus we have the following result.

Proposition 10.3.2 *If $(a_1 + \lambda_2)/m_1\lambda_2$ is in the range of $f(p)$, then a necessary condition for the existence of an interior equilibrium for (3.2) is that E_2 exist and be unstable.*

We see below that the interior equilibrium may exist even if E_1 does not. Before considering the stability of E_c, it remains to investigate whether (10.33) is feasible, to investigate whether $(a_1 + \lambda_2)/m_1\lambda_2$ is in the range of $f(p)$.

If $0 < \lambda_0 < \lambda_2$, then x_1 is a better competitor than x_2 even at the maximal level of the inhibitor. A simple consequence of the definition of

λ_0 is that $f(1) > (a_1 + \lambda_2)/m_1\lambda_2$ in this case or that there is no value of p_c^*, $0 \leqq p_c^* \leqq 1$, which satisfies (10.33). Hence $\lambda_0 \geqq \lambda_2$ is a necessary condition for $(a_1 + \lambda_2)/m_1\lambda_2$ to be in the range of $f(p)$, $0 \leqq p \leqq 1$. It is also sufficient since $f(0) = 1$. Hence that Proposition 10.3.2 can be improved to $\lambda_0 \geqq \lambda_2 > \lambda^*$ is necessary and sufficient for the existence of E_c.

There remains the question of the stability of E_c. The matrix J in (10.31) takes the form

$$
J = \begin{bmatrix} -\frac{m_1 a_1}{(a_1+\lambda_2)^2} f(p_c^*) x_{1c}^* & -\frac{m_1 a_1}{(a_1+\lambda_2)^2} f(p_c^*) x_{1c}^* & \frac{m_1\lambda_2}{a_1+\lambda_2} x_{1c}^* f'(p_c^*) \\ -\frac{m_2 a_2}{(a_2+\lambda_2)^2} x_{2c}^* & -\frac{m_2 a_2}{(a_2+\lambda_2)^2} x_{2c}^* & 0 \\ 0 & -\frac{\delta p_c^*}{K+p_c^*} & -1 - \frac{\delta K x_{2c}^*}{(K+p_c^*)^2} \end{bmatrix}.
$$

By expanding the determinant of J in the last row, we see that it is negative or that the dimension of the stable manifold is one or three.

If $\delta < 1$, the Gersgorin theory immediately gives two roots with negative real parts — and hence three such roots — so we easily have that E_c is asymptotically stable if $\delta < 1$.

The characteristic roots of J satisfy

$$
\mu^3 + \mu^2 \left(1 + \frac{\delta K x_{2c}^*}{(K+p_c^*)^2} + \frac{a_1 x_{1c}^*}{(a_1+\lambda_2)\lambda_2} + \frac{a_2 x_{2c}^*}{(a_2+\lambda_2)\lambda_2} \right)
$$
$$
+ \mu \left(1 + \frac{\delta K x_{2c}^*}{(K+p_c^*)^2} \right) \left(\frac{a_1 x_{1c}^*}{(a_1+\lambda_2)\lambda_2} + \frac{a_2 x_{2c}^*}{(a_2+\lambda_2)\lambda_2} \right) \qquad (10.37)
$$
$$
- \frac{f'(p_c^*)}{f(p_c^*)} \frac{a_2}{(a_2+\lambda_2)\lambda_2} \frac{\delta p_c^*}{K+p_c^*} x_{1c}^* x_{2c}^* = 0.
$$

Since $f'(p) < 0$, the constant term is positive, so the Routh-Hurwitz criterion states that E_c will be asymptotically stable if and only if

$$
\left(1 + \frac{\delta K x_{2c}^*}{(K+p_c^*)^2} + \frac{a_1 x_{1c}^*}{(a_1+\lambda_2)\lambda_2} + \frac{a_2 x_{2c}^*}{(a_2+\lambda_2)\lambda_2} \right) \left(1 + \frac{\delta K x_{2c}^*}{(K+p_c^*)^2} \right)
$$
$$
\cdot \left(\frac{a_1 x_{1c}^*}{(a_1+\lambda_2)\lambda_2} + \frac{a_2 x_{2c}^*}{(a_2+\lambda_2)\lambda_2} \right) > -\frac{f'(p_c^*)}{f(p_c^*)} \frac{a_2}{(a_2+\lambda_2)\lambda_2} \frac{\delta p_c^*}{K+p_c^*} x_{1c}^* x_{2c}^*.
$$
$$
\tag{10.38}
$$

Dynamics without an interior rest point. If E_c is not to exist, the inequality

$$
\lambda_0 \geqq \lambda_2 > \lambda^* \tag{10.39}
$$

must be violated. Recall that we are assuming that $0 < \lambda_1 < \lambda_2 < 1$ and, moreover, that $\lambda_0 > \lambda^*$ holds by definition of these quantities and the monotonicity of $f(p)$. There are two possible outcomes depending on the way inequality (10.39) is violated.

Since system (10.25) is competitive, the possible dynamics are limited. Two results are of interest here.

Remark 10.3.2 If E_c does not exist, from Theorem 10.1.6(ii) all omega limit sets lie on the boundary of R^3_+.

Theorem 10.3.2 *If* $0 < \lambda_1 < \lambda_2 < \lambda^*$, *then*

$$\lim_{t\to\infty} x_1(t) = 0, \ \lim_{t\to\infty} x_2(t) = x_2^*, \ \lim_{t\to\infty} p(t) = p_2^*.$$

Proof. $\lambda_0 > 0$ is implied by $\lambda^* > 0$. If $\lambda_0 > 1$, then the only viable equilibrium is E_2 (Remark 10.3.1, Table 10.1), and it is locally asymptotically stable. If $\lambda_0 < 1$, then E_1 exists but is unstable and is not an omega limit point of a trajectory of (10.25). In either case, Remark 10.3.2 completes the proof. □

Theorem 10.3.3 *If* $0 < \lambda_0 < \lambda_2$ *and* $\lambda^* < \lambda_2$, *then*

$$\lim_{t\to\infty} x_1(t) = x_1^*, \ \lim_{t\to\infty} x_2(t) = 0, \ \lim_{t\to\infty} p(t) = 1.$$

Proof. E_1 is locally asymptotically stable and E_2 is unstable. Remarks 10.3.1 and 10.3.2 complete the proof. □

Since $\lambda^* > \lambda_0$ always holds, the above two theorems complete the asymptotic description of the dynamics under the basic hypothesis $0 < \lambda_1 < \lambda_2 < 1$ when there is no interior rest point.

Dynamics with an interior rest point. It was shown that a necessary condition for the existence of the interior equilibrium point E_c was that E_2 be unstable. Ostensibly, there are three cases depending on E_1:

(i) E_1 exists and is asymptotically stable,
(ii) E_1 exists and is unstable,
(iii) E_1 does not exist.

Case (i) does not occur, however, since (see Table 10.1) E_1 being asymptotically stable requires that $\lambda_0 > \lambda_2$, and E_2 being unstable requires $\lambda_2 \leqq \lambda^*$. However, from (10.33),

$$f(p_c^*) = \frac{a_1 + \lambda_2}{m_1 \lambda_2}$$

or

$$\lambda_2 = \frac{a_1}{mf(p_c^*) - 1} < \frac{a_1}{mf(1) - 1} = \lambda_0,$$

since $0 < p_2^* < p_c^* < 1$ and f is decreasing. This contradicts $\lambda_2 > \lambda_0$. Hence, we need to only consider cases (ii) and (iii).

Theorem 10.3.4 *If case* (ii) *or* (iii) *holds, then there exists a* $\gamma > 0$ *such that every solution of* (10.25) *satisfies*

$$\liminf_{t \to \infty} x_1(t) \geqq \gamma, \ \liminf_{t \to \infty} x_2(t) \geqq \gamma.$$

Proof. To use the Theorem 10.2.2 on uniform persistence, first, it is necessary to have a system of equations defined on an open region with a boundary. We use, instead of (10.25),

$$\begin{aligned}
x_1' &= x_1(g_1(x_1 + x_2)f(p) - 1), \\
x_2' &= x_2(g_2(x_1 + x_2) - 1), \\
p' &= 1 - p - \frac{\delta x_2 p}{K + |p|},
\end{aligned} \tag{10.40}$$

where

$$g_i(u) = \begin{cases} \frac{m_i(1-u)}{1 + a_i - u}, & 0 \leqq u \leqq 1, \\ 0, & u > 1, \end{cases} \quad i = 1, 2.$$

Since the only initial conditions of interest are with $p(0) \geqq 0$, $x_i(0) > 0$, $x_1(0) + x_2(0) < 1$, solutions of these initial conditions are, for $t \geqq 0$, solutions of (10.40). The open region is the wedge $x_i > 0$, $i = 1, 2$ whose boundaries are the x_1-p and x_2-p planes in R^3, given by $x_2 = 0$, $x_1 \geqq 0$ and $x_1 = 0$, $x_2 \geqq 0$. The system is dissipative since in the extended region $x_1 + x_2 > 1$, $(x_1(t) + x_2(t))' = -(x_1(t) + x_2(t))$ and in the extended region $p < 0$, $p'(t) \geqq 1$. In case (ii) or (iii), no portion of the stable manifolds of E_0, E_1, and E_2 intersect the interior of the wedge.

Moreover, since there is only one equilibrium point interior to each of the x_1-p and x_2-p faces and an unstable rest point on the p axis whose stable manifold is the p axis, there are no connecting orbits to form a cycle. Hence, by Theorem 10.2.2 (10.40) is uniformly persistent, and the theorem is established. □

Theorem 10.3.4 guarantees the coexistence of both the x_1 and x_2 populations. However, it does not give the global asymptotic behavior. The further analysis of the system is complicated by the possibility of multiple limit cycles. Since this is a common difficulty in general two-dimension systems, it is not surprising that this presents difficulties in the analysis of three-dimension competitive systems.

Theorem 10.3.5 *Suppose that system* (10.25) *has no limit cycles. Then* E_c *is globally asymptotically stable.*

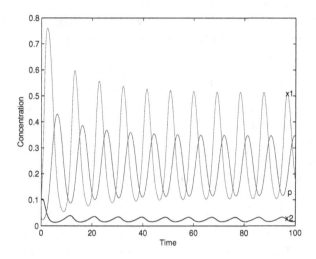

Fig. 10.3

Proof. In view of Theorem 10.3.4, the omega limit set of any trajectory cannot be on the boundary $x_1 = 0$ or $x_2 = 0$. Away from the boundary, the system is irreducible. Since there are no limit cycles, all trajectories must tend to E_c by Theorem 10.1.6(i). □

Conjecture *In case (ii), system* (10.25) *has no limit cycles.*

We note that since we are assuming hyperbolicity, there must be at least two limit cycles for the conjecture to fail. Because of the assumed stability of E_c, there must be an unstable limit cycle with E_c in its "interior". However, since the system is dissipative, there must be an asymptotically stable limit cycle as well.

Theorem 10.3.6 *Let $f(p) = e^{-\eta p}$ in (10.25) and let case* (iii) *hold. Then for η sufficiently large, there exists a $\delta_0 > 0$ and a K_0, such that for $\delta > \delta_0$ and $K < K_0$, (10.25) has an attracting limit cycle.*

Figure 10.3 shows the time course for an example of this type of behavior. Figure 10.4 shows the limit cycle plotted in phase space.

Proof. The theorem follows if E_c is unstable. To show this, we must show that (10.38) is violated. In the case under consideration, $-f'(p)/f(p) = \eta$.

Define c by $c = \ln(m_1\lambda_2/(a_1 + \lambda_2))$, and note that $c = \eta p_c^*$. It follows that

$$1 - p_c^* = 1 - \frac{c}{\eta} = \frac{(\eta - c)}{\eta}.$$

Note that when η is fixed, p_c^* is fixed. From the definition of x_{2c}^* (see (10.34)), it follows that

$$\frac{\delta K x_{2c}^*}{(K + p_c^*)^2} = \frac{K}{K + p_c^*} \frac{1 - p_c^*}{p_c^*} = \left(\frac{K}{K + p_c^*}\right)\left(\frac{\eta}{c} - 1\right) \tag{10.41}$$

for any choice of η and the corresponding p_c^*. Fix η satisfying

$$\eta > c + 2\left(3 + \frac{a_1(1 - \lambda_2)}{(a_1 + \lambda_2)\lambda_2}\right)\left(\frac{a_1(1 - \lambda_2)}{(a_1 + \lambda_2)\lambda_2} + 1\right)\left(\frac{2(a_2 + \lambda_2)\lambda_2}{(1 - \lambda_2)a_2}\right),$$

and $(m_1/(a_1+1))e^{-\eta} < 1$. Let K_0 be so small that the expression in (10.41) is less than 1. To show that (10.38) is violated, we estimate both sides. Note that, since $x_{1c}^* = 1 - \lambda_2 - x_{2c}^*$, $\lim_{\delta \to \infty} x_{2c}^* = 0$ and $\lim_{\delta \to \infty} x_{1c}^* = 1 - \lambda_2$. Hence, for δ sufficiently large, $a_2 x_{2c}^*/(a_2 + \lambda_2)\lambda_2 < 1$ and $1 - \lambda_2 > x_{1c}^* > (1 - \lambda_2)/2$. The right-hand side of (10.38) is bounded below by

$$\frac{a_2}{2(a_2 + \lambda_2)\lambda_2}(1 - \lambda_2)(\eta - c).$$

The left-hand side of (10.38) has three factors, which we denote by F_1, F_2, F_3, respectively. By the discussion above, for δ sufficiently large,

$$F_1 < \left(3 + \frac{a_1(1 - \lambda_2)}{(a_1 + \lambda_2)\lambda_2}\right), \quad F_2 < 2, \quad F_3 < \frac{a_1(1 - \lambda_2)}{(a_1 + \lambda_2)\lambda_2} + 1.$$

It follows that

$$F_1 F_2 F_3 < \frac{(\eta - c)a_2}{(a_2 + \lambda_2)\lambda_2}\frac{(1 - \lambda_2)}{2},$$

which contradicts (10.38). Hence E_c is unstable (with a two-dimensional unstable manifold). Choose a trajectory not on the stable manifold of E_c. By Theorem 10.1.6(i), its omega limit set must be a periodic orbit. By hyperbolicity, there must be an attracting orbit. $\qquad\square$

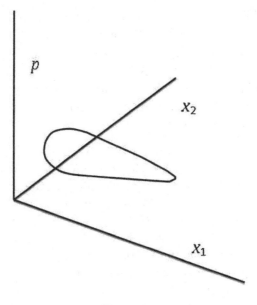

Fig. 10.4

10.4   Two Species Competition Models

In this section, we shall introduce an abstract model for two species competition. The abstract model has wide applications to many two species competition models in the form of ODEs, PDEs. The basic setup is as follows.

For $i = 1, 2$, let X_i be ordered Banach spaces with positive cones X_i^+ such that $IntX_i \neq \phi$. We use the same symbol for the partial orders generated by the cones X_i^+. If x_i, $\bar{x}_i \in X_i$, then we write $x_i \leq \bar{x}_i$ if $\bar{x}_i - x_i \in X_i^+$, $x_i < \bar{x}_i$ if $x_i \leq \bar{x}_i$ and $x_i \neq \bar{x}_i$, and $x_i \ll \bar{x}_i$ if $\bar{x}_i - x_i \in IntX_i^+$. If $x_i, y_i \in X_i$ satisfy $x_i < y_i$, then the order interval $[x_i, y_i]$ is defined by $[x_i, y_i] = \{u \in X_i : x_i \leq u \leq y_i\}$. If $x_i \ll y_i$, then $[[x_i, y_i]] = \{u \in X_i : x_i \ll u \ll y_i\}$ is called an open order interval.

Let $X = X_1 \times X_2$, $X^+ = X_1^+ \times X_2^+$, and $K = X_1^+ \times (-X_2^+)$. X^+ is a cone in X with nonempty interior given by $IntX^+ = IntX_1^+ \times IntX_2^+$. It generates the order relations $\leq$, $<$, $\ll$ in the usual way. In particular if $x = (x_1, x_2)$ and $\bar{x} = (\bar{x}_1, \bar{x}_2)$, then $x \leq \bar{x}$ if and only if $x_i \leq \bar{x}_i$, for $i = 1, 2$. Let $IntK = IntX_1^+ \times (-IntX_2^+)$. It generates the partial order relations

$\leq_K, <_K, \ll_K$. In this case,

$$x \leq_K \bar{x} \Leftrightarrow x_1 \leq \bar{x}_1 \text{ and } \bar{x}_2 \leq x_2.$$

A similar statement holds with $\ll_K$ replacing $\leq_K$ and $\ll$ replacing $\leq$. Assume that $T : [0, \infty) \times X^+ \to X^+$ is a continuous semiflow. We write $T_t(x) = T(t, x)$. The semiflow properties are (i) $T_0(x) = x$ for all $x \in X^+$, and (ii) $T_t \circ T_s = T_{t+s}$ for $t, s \geq 0$, (iii) $T(t, x)$ is continuous in t and x. The hypotheses are given below.

(H1) T is strictly order-preserving with respect to $<_K$. That is, $x <_K \bar{x}$ implies $T_t(x) <_K T_t(\bar{x})$. For each $> 0, T_t : X^+ \to X^+$ is order compact.

(H2) $T_t(0) = 0$ for all $t \geq 0$ and 0 is a repelling equilibrium. There exists a neighborhood U of 0 in X^+ such that for each $x \in U, x \neq 0$, there is $t_0 > 0$ such that $T_{t_0}(x) \notin U$.

(H3) $T_t(X_1^+ \times \{0\}) \subset X_1^+ \times \{0\}$ for all $t \geq 0$. There exists $\hat{x}_1 \gg 0$ such that $T_t((\hat{x}_1, 0)) = (\hat{x}_1, 0)$ for all $t \geq 0$, and $T_t((x_1, 0)) \to (\hat{x}_1, 0)$ as $t \to \infty$ for all $x_1 \neq 0$. The symmetric conditions hold for T on $\{0\} \times X_2^+$ with equilibrium point $(0, \hat{x}_2)$.

(H4) If $x = (x_1, x_2) \in X^+$ satisfies $x_i \neq 0, i = 1, 2$, then $T_t(x) \gg 0$ for $t > 0$. If $x, y \in X^+$ satisfy $x <_K y$ and either x or y belongs to $IntX^+$, then $T_t(x) \ll_K T_t(y)$ for $t > 0$.

The boundary equilibria are $E_1 = (\hat{x}_1, 0)$ and $E_2 = (0, \hat{x}_2)$. We say that E_* is a positive equilibrium of T if it belongs to the interior of X^+. If $x \in X^+$, then $O(x) = \{T_t(x) : t \geq 0\}$ is called the positive orbit of T. Let $I = [0, \hat{x}_1] \times [0, \hat{x}_2]$. In [HSW] the author proved the following classification for the abstract model of two species competition. The interested readers can find the proofs there.

Theorem 10.4.1 *Let* (H1)–(H4) *hold. Then the omega limit set of every orbit is contained in I and exactly one of the following holds:*

(a) *There exists a positive equilibrium E_* of T in I.*

(b) *$T_t(x) \to E_1$ as $t \to \infty$ for every $x = (x_1, x_2) \in I$ with $x_i \neq 0$, $i = 1, 2$.*

(c) *$T_t(x) \to E_2$ as $t \to \infty$ for every $x = (x_1, x_2) \in I$ with $x_i \neq 0$, $i = 1, 2$.*

Finally, if (b) *or* (c) *hold, $x = (x_1, x_2) \in X^+ \setminus I$ and $x_i \neq 0, i = 1, 2$, then either $T_t(x) \to E_1$ or $T_t(x) \to E_2$ as $t \to \infty$.*

If case (a) of Theorem 10.4.1 holds, then we have two cases: stable coexistence and bistability. In [ST] the authors proved the following:

Theorem 10.4.2 (Stable Coexistence)

Let (H1)–(H4) *hold. Assume that* E_1 *is not locally attractive from below and* E_2 *is not locally attractive from above. Then there exist two (not necessarily different) equilibria* E_1^c *and* E_2^c *such that*

$$E_2 <_K E_2^c \leq_K E_1^c <_K E_1$$

E_1^c *is the largest and* E_2^c *is the smallest equilibrium in* $Int X^+$ *with respect to* $\leq_K$.

All orbits starting in $[E_1^c, E_1]_K \cap Int X^+$ *converges to* E_1^c *while all orbits starting in* $[E_2, E_2^c,]_K \cap Int X^+$ *converges to* E_2^c. *In particular, if the interior equilibrium is unique, then* $E_1^c = E_2^c = E$ *and* $\beta = \{x : \omega(x) = E\} = I$, *where* β *is the domain of attraction of equilibrium* E.

Theorem 10.4.3 (Bistability [JLZ])

Let (H1)–(H4) *hold. Suppose the interior equilibrium* E *is unique. If* E_1 *and* E_2 *are locally stable with basin of attraction* β_1 *and* β_2 *respectively, in* X^+ *and that the mapping* T_t *is continuously differentiable and* $D_x T_t(E)$ *is strongly positive and the spectral radius* $r(D_x T_t(E)) > 1$. *Then* $M = X^+ \setminus \{\beta_1 \cup \beta_2\}$ *is an unordered and positively invariant* C^1 *submanifold with codimension one.*

Example 10.4.1 Consider Lotka-Volterra two species competition (Example 4.2.4)

$$\frac{dx_1}{dt} = r_1 x_1 \left(1 - \frac{x_1}{k_1}\right) - \alpha_1 x_1 x_2$$

$$\frac{dx_2}{dt} = r_2 x_2 \left(1 - \frac{x_1}{k_1}\right) - \alpha_2 x_1 x_2$$

$$x_1(0) > 0, x_2(0) > 0.$$

Then $E_0 = (0,0)$, $E_1 = (K_1, 0)$, $E_2 = (0, K_2)$. We define $\begin{pmatrix} x_1 \\ x_2 \end{pmatrix} \leq_K \begin{pmatrix} y_1 \\ y_2 \end{pmatrix}$ iff $x_1 \leq y_1$, $x_2 \geq y_2$. Then we can apply Theorem 10.4.1, Theorem 10.4.2 and Theorem 10.4.3 to obtain the results of competitive exclusion, stable coexistence and bistability.

Example 10.4.2 ([Smi] p. 65)

Lotka-Volterra two species competition model in two patches with migration. Let $x_i(t)$, $y_i(t)$ be the population densities of two competing species at i-th patch, $i = 1, 2$. Let ϵ, δ be the migration rate for population x and y respectively. The model takes the form:

$$x_1' = \epsilon(x_2 - x_1) + r_1 x_1 \left(1 - \frac{x_1}{k_1} - a_1 y_1 \right)$$

$$x_2' = \epsilon(x_1 - x_2) + r_2 x_2 \left(1 - \frac{x_2}{k_2} - a_2 y_2 \right)$$

$$y_1' = \delta(y_1 - y_2) + s_1 y_1 \left(1 - \frac{y_1}{L_1} - b_1 x_1 \right) \tag{10.42}$$

$$y_2' = \delta(y_1 - y_2) + s_2 y_2 \left(1 - \frac{y_2}{L_2} - b_2 x_2 \right),$$

where $\epsilon, \delta > 0$. Let $K_m = \{z = (x, y)\} \in R_+^4 : x \geq 0, y \leq 0\}$, $z = (x, y) \leq_m \bar{z} = (\bar{x}, \bar{y})$ if $x \leq \bar{x}$ and $y \geq \bar{y}$.

Then the flow $\varphi_t(z)$ is monotone with respect to the order relation $\leq_m$ and is strongly monotone in the interior of R_+^4.

Consider single population model for species x with two patches:

$$x_1' = \epsilon(x_2 - x_1) + r_1 x_1 \left(1 - \frac{x_1}{k_1} \right)$$

$$x_2' = \epsilon(x_1 - x_2) + r_2 x_2 \left(1 - \frac{x_2}{k_2} \right). \tag{10.43}$$

It is easy to show there is unique non-zero equilibrium $\hat{x}, \hat{x} \gg 0$ and it attracts all nontrivial solution (4.2). Similarly for the single population model for species y:

$$y_1' = \delta(y_2 - y_1) + s_1 y_1 \left(1 - \frac{y_1}{L_1} \right)$$

$$y_2' = \delta(y_1 - y_2) + s_2 y_2 \left(1 - \frac{y_2}{L_2} \right) \tag{10.44}$$

there is a unique non-zero equilibrium $\hat{y}, \hat{y} \gg 0$ and it attracts all nontrivial solution of (4.3). Let $E_x = (\hat{x}, 0)$, $E_y = (0, \hat{y})$. Since $\varphi_t((0, y)) \to E_y$ and $\varphi_t((x, 0)) \to E_x$ as $t \to \infty$, it follows that all positive orbits are attracted to the set

$$R \equiv [0, \hat{x}] \times [0, \hat{y}] = \{z : E_y \leq_m z \leq_m E_x\}.$$

Then we can apply Theorem 10.4.1, Theorem 10.4.2 and Theorem 10.4.3 to obtain the results of competitive exclusion, stable coexistence and bistability depending on the stability properties of E_x, E_y and E_c (E_c is a positive equilibrium).

10.5   Exercises

Exercise 10.1 Discuss the uniform persistence for the system of two competing predators for one prey [HHW]

$$x' = rx\left(1 - \frac{x}{k}\right) - \frac{m_1 x}{a_1 + x}y_1 - \frac{m_2 x}{a_2 + x}y_2,$$

$$y_1' = \left(\frac{m_1 x}{a_1 + x} - d_1\right)y_1,$$

$$y_2' = \left(\frac{m_2 x}{a_2 + x} - d_2\right)y_2,$$

$$x(0) > 0, \quad y_1(0) > 0, \quad y_2(0) > 0.$$

Exercise 10.2 Discuss the following system of two species N_1, N_2 competing for two complementary resources S and R in a chemostat [HCH]

$$S' = (S^{(0)} - S)D - \frac{1}{y_{1s}}f_1(S, R)N_1 - \frac{1}{y_{2s}}f_2(S, R)N_2,$$

$$R' = (R^{(0)} - R)D - \frac{1}{y_{1r}}f_1(S, R)N_1 - \frac{1}{y_{2r}}f_2(S, R)N_2,$$

$$N_1' = (f_1(S, R) - D)N_1,$$

$$N_2' = (f_2(S, R) - D)N_2,$$

$$S(0) \geq 0, \quad R(0) \geq 0, \quad N_1(0) > 0, \quad N_2(0) > 0.$$

where

$$f_1(S, R) = \min\left\{\frac{m_{1s}S}{a_{1s} + S}, \frac{m_{1r}R}{a_{1r} + R}\right\},$$

$$f_2(S, R) = \min\left\{\frac{m_{2s}S}{a_{2s} + S}, \frac{m_{2r}R}{a_{2r} + R}\right\}.$$

(1) Show that

$$S + \frac{1}{y_{1s}}N_1 + \frac{1}{y_{2s}}N_2 = S^{(0)} + O(e^{-Dt}) \quad as\, t \to \infty,$$

$$R + \frac{1}{y_{1r}}N_1 + \frac{1}{y_{2r}}N_2 = R^{(0)} + O(e^{-Dt}) \quad as\, t \to \infty.$$

(2) Analyze the limiting system

$$N_1' = (f_1(S, R) - D)N_1,$$

$$N_2' = (f_2(S, R) - D)N_2.$$

where

$$S = S^{(0)} - \frac{1}{y_{1s}}N_1 - \frac{1}{y_{2s}}N_2 \geq 0,$$

$$R = R^{(0)} - \frac{1}{y_{1r}}N_1 - \frac{1}{y_{2r}}N_2 \geq 0.$$

Exercise 10.3 Consider the following simple gradostat model [SW] with two vessels,

$$S_1' = 1 - 2S_1 + S_2 - f_u(S_1)u_1 - f_v(S_1)v_1,$$
$$S_2' = S_1 - 2S_2 - f_u(S_2)u_2 - f_v(S_2)v_2,$$
$$u_1' = -2u_1 + u_2 + f_u(S_1)u_1,$$
$$u_2' = u_1 - 2u_2 + f_u(S_2)u_2,$$
$$v_1' = -2v_1 + v_2 + f_v(S_1)v_1,$$
$$v_2' = v_1 - 2v_2 + f_v(S_2)v_2,$$
$$S_i(0) \geq 0, \quad u_i(0) \geq 0, \quad v_i(0) \geq 0, \quad i = 1, 2,$$

where

$$f_u(S) = \frac{m_1 S}{a_1 + S},$$
$$f_r(S) = \frac{m_2 S}{a_2 + S}.$$

(1) Show that

$$\Sigma_1(t) = \frac{2}{3} - S_1(t) - u_1(t) - v_1(t) \to 0 \quad as\, t \to \infty,$$
$$\Sigma_2(t) = \frac{1}{3} - S_2(t) - u_2(t) - v_2(t) \to 0 \quad as\, t \to \infty.$$

(2) Analyze the single species system

$$u_1' = -2u_1 + u_2 + f_u\left(\frac{2}{3} - u_1\right)u_1,$$

$$u_2' = u_1 - 2u_2 + f_u\left(\frac{1}{3} - u_2\right)u_2,$$

$$u_i(0) \geq 0, \quad i = 1, 2.$$

(3) Analyze the two species competition system

$$u_1' = -2u_1 + u_2 + f_u\left(\frac{2}{3} - u_1 - v_1\right)u_1,$$

$$u_2' = u_1 - 2u_2 + f_u\left(\frac{1}{3} - u_2 - v_2\right)u_2,$$

$$v_1' = -2v_1 + v_2 + f_v\left(\frac{2}{3} - u_1 - v_1\right)v_1,$$

$$v_2' = v_1 - 2v_2 + f_v\left(\frac{1}{3} - u_2 - v_2\right)v_2.$$

Exercise 10.4 [SW] The following is a mathematical model for two species competing for a single nutrient with internal storage in a chemostat. Let x_1, x_2 be two populations competing for a single nutrient of concentration S in the chemostat. The average amount of stored nutrient per individual of population x_1 is denoted by Q_1, and for population x_2 by Q_2. Then the model takes the form:

$$x_1' = x_1(\mu_1(Q_1) - D),$$
$$Q_1' = \rho_1(S, Q_1) - \mu_1(Q_1)Q_1,$$
$$x_2' = x_2(\mu_2(Q_2) - D),$$
$$Q_2' = \rho_2(S, Q_2) - \mu_2(Q_2)Q_2,$$
$$S' = (S^{(0)} - S)D - x_1\rho_1(S, Q_1) - x_2\rho_2(S, Q_2).$$

The function $\mu_i(Q_i)$ and $\rho(S, Q_i)$ satisfy $\mu_i(Q_i) \geq 0$, $\mu_i'(Q_i) > 0$, $\mu_i(\rho_i) = 0$ for $Q_i \geq \rho_i \geq 0$ where ρ_i is the minimal cell quota; $\rho_i(0, Q_i) = 0$, $\frac{\partial \rho_i}{\partial S} > 0$, $\frac{\partial \rho_i}{\partial Q_i} \leq 0$.

(1) Analyze the single population system

$$x' = x(\mu(Q) - D),$$
$$Q' = \rho(S, Q) - \mu(Q)Q,$$
$$S' = (S^{(0)} - S)D - x\rho(S, Q).$$

(2) Prove the conservation properties

$$S + x_1Q_1 + x_2Q_2 = S^{(0)} + O(e^{-Dt}) \quad as\, t \to \infty.$$

(3) Analyze the limiting system

$$x_1' = x_1(\mu_1(Q_1) - D),$$
$$Q_1' = \rho(S, Q_1) - \mu_1(Q_1)Q_1,$$
$$x_2' = x_2(\mu_2(Q_2) - D),$$
$$Q_2' = \rho_2(S, Q_2) - \mu_2(Q_1)Q_1,$$

where

$$S = S^{(0)} - x_1Q_1 - x_2Q_2 \geq 0,$$

by converting to the equivalent system of $x_1 = x_1$, $U_1 = x_1Q_1$, $x_2 = x_2$, $U_2 = x_2Q_2$.

$$x_1' = x_1 \left(\mu_1 \left(\frac{U_1}{x_1} \right) - D \right),$$

$$U_1' = \rho_1 \left(S^{(0)} - U_1 - U_2, \frac{U_1}{x_1} \right) - DU_1,$$

$$x_2' = x_2 \left(\mu_2 \left(\frac{U_2}{x_2} \right) - D \right),$$ \hfill (10.45)

$$U_2' = \rho_2 \left(S^{(0)} - U_1 - U_2, \frac{U_2}{x_2} \right) - DU_2,$$

with appropriate domain

$$\triangle = \{(x_1, U_1, x_2, U_2) \in \mathbb{R}_+^4 : x_i > 0, U_1 + U_2 \leq 1, i = 1, 2.\}$$

Define partial ordering defined by

$$(x_1, U_1, x_2, U_2) \leq_K (\overline{x_1}, \overline{U_1}, \overline{x_2}, \overline{U_2})$$

if and only if $x_1 \leq \overline{x_1}$, $U_1 \leq \overline{U_1}$, $x_2 \geq \overline{x_2}$, $U_2 \geq \overline{U_2}$. Prove that the system $(\star)$ preserves $\leq_K$ and apply Theorems 10.4.1, 10.4.2, 10.4.3 to obtain the global asymptotic behavior of the solutions of (10.45).

Chapter 11

INTRODUCTION TO HAMILTONIAN SYSTEMS

11.1 Definitions and Classic Examples

In this chapter we consider systems of the form

$$\begin{cases} \dot{x} = \dfrac{\partial}{\partial y} H(t,x,y), \\[2mm] \dot{y} = -\dfrac{\partial}{\partial x} H(t,x,y). \end{cases} \tag{H}$$

Here $x = (x_1, \cdots, x_n) \in D \subset \mathbb{R}^n$, D is open, $y = (y_1, \cdots, y_n) \in \mathbb{R}^n$, and $H \in C^2(\mathbb{R} \times D \times \mathbb{R}^n)$. Systems of the form are called *Hamiltonian systems*, and the function H is called the *Hamiltonian*. We say the *degree of freedom* of the system is n if it is autonomous (i.e. $H = H(x,y)$), the degree of freedom is $n + 1/2$ if otherwise. We call $D \times \mathbb{R}^n$ the *phase space* of (H). The flow induced by (H) is called a *Hamiltonian flow*, a terminology frequently abused even when solutions of (H) do not exist for all time.

Let ∇ be the gradient with respect to $z = (x,y)$, $I = I_n$ be the $n \times n$ identity matrix. Then (H) can be written

$$\dot{z} = J\nabla H(t,z), \quad J = \begin{pmatrix} 0 & I \\ -I & 0 \end{pmatrix}. \tag{H'}$$

Sometimes we denote ∇ by ∇_z to emphasize which space variables were taken. Alternatively, we may also write ∇H as $\frac{\partial}{\partial z} H^T$, regarding $\frac{\partial}{\partial z} H$ as a row vector.

Hamiltonian systems are of special importance in physics because most models in classical mechanics can be formulated as Hamiltonian systems, and the Schrödinger equation in quantum mechanics can be formally obtained from Hamiltonian formulation by replacing the variable y with some differential operators.

General theories for Hamiltonian systems are formulated as flows in vector fields on symplectic manifolds (i.e. manifolds equipped with some closed nondegenerate 2-form, with the skew-symmetric bilinear form induced by J in (H') as special case). As a brief introduction, we will present some important features and methods for Hamiltonian systems on Euclidean spaces, and illustrate the theory through some classic examples. Proofs presented herein avoid usage of differential forms. We refer to [Arnold; HLW; MHO; Moser] for further topics and examples.

Example 11.1.1 Any second order equation of the form

$$\ddot{x} + g(x) = 0 \tag{11.1}$$

is a Hamiltonian system. Here g is assumed to be continuous on an open interval $I \subset \mathbb{R}$. To see this, pick $x_0 \in I$, let $y = \dot{x}$, $G(x) = \int_{x_0}^{x} g(t)dt$, then define

$$H(x, y) = \frac{1}{2}y^2 + G(x), \quad (x, y) \in I \times \mathbb{R}. \tag{11.2}$$

It is easy to see that the Hamiltonian system with this Hamiltonian H is equivalent to (11.1). This is a Hamiltonian system with one degree of freedom. We have encountered many examples of this type, see Examples 1.1.6, 1.1.7, 1.1.9, 5.2.1, 5.2.2, 5.3.1. In such examples, x and y represent position and momentum (in rectangular or polar coordinates), H represents the total energy.

With the presence of an external forcing term $f(t)$,

$$\ddot{x} + g(x) = f(t), \tag{11.3}$$

the equation becomes a non-autonomous Hamiltonian system with Hamiltonian

$$H(t, x, y) = \frac{1}{2}y^2 + G(x) - xf(t), \quad (t, x, y) \in \mathbb{R} \times I \times \mathbb{R}.$$

It is a Hamiltonian system with 3/2-degree of freedom.

Example 11.1.2 Consider a mechanical system consisting of identical masses $m_1 = m_2 = m$ and three springs, all aligned on the real axis. One spring connects these two masses and the other two springs connect masses to fixed positions. Let κ_{12} be the spring constant for the one linking m_1 and m_2, and $\kappa > 0$ be the spring constant for the other two, x_i be the displacement from the equilibrium position of m_i, $i = 1, 2$. See Fig. 11.1. Equations of motion are

$$\begin{cases} m\ddot{x}_1 = -\kappa x_1 + \kappa_{12}(x_2 - x_1), \\ m\ddot{x}_2 = -\kappa x_2 - \kappa_{12}(x_2 - x_1). \end{cases}$$

This system is called *linearly coupled oscillation*. Let $y_i = m\dot{x}_i$ be the linear momentum of m_i. The system can be expressed as a Hamiltonian system with two degrees of freedom:

$$\begin{cases} \dot{x}_1 = \dfrac{y_1}{m}, \\[2mm] \dot{x}_2 = \dfrac{y_2}{m}, \\[2mm] \dot{y}_1 = -(\kappa + \kappa_{12})x_1 + \kappa_{12}x_2 \\[2mm] \dot{y}_2 = -(\kappa + \kappa_{12})x_2 + \kappa_{12}x_1 \,. \end{cases} \tag{CO}$$

The Hamiltonian H is the sum of kinetic energy K and potential energy V:

$$K(y_1, y_2) = \frac{1}{2m}\left(y_1^2 + y_2^2\right), \quad V(x_1, x_2) = \frac{\kappa}{2}\left(x_1^2 + x_2^2\right) + \frac{\kappa_{12}}{2}\left(x_1 - x_2\right)^2.$$

System (CO) can be used to describe small oscillations of various mechanical systems, such as vibrations of molecules or connected pendulums.

Fig. 11.1   Linearly coupled oscillation.

Example 11.1.3 Consider the Hénon-Heiles Hamiltonian given by

$$H(x_1, x_2, y_1, y_2) = \frac{1}{2}(x_1^2 + x_2^2 + y_1^2 + y_2^2) + \mu x_1^2 x_2 + \frac{\lambda}{3}x_2^3.$$

This corresponds a Hamiltonian system with two degrees of freedom:

$$\begin{cases} \dot{x}_1 = y_1, \\[1mm] \dot{x}_2 = y_2, \\[1mm] \dot{y}_1 = -x_1 - 2\mu x_1 x_2 \\[1mm] \dot{y}_2 = -x_2 - \mu x_1^2 - \lambda x_2^2 \,. \end{cases} \tag{HH}$$

Here μ and λ are constants. The system was studied by Hénon and Heiles in 1964 to model the nonlinear motion of a star in a planar galactic disc. The Hamiltonian is the sum of the kinetic energy K and potential energy V, where

$$K(y_1, y_2) = \frac{1}{2}(y_1^1 + y_2^2), \quad V(x_1, x_2) = \frac{1}{2}(x_1^2 + x_2^2) + \mu x_1^2 x_2 + \frac{\lambda}{3}x_2^3.$$

Example 11.1.4 The N-body problem (Example 1.1.11) concerns the motion of N mass points $m_1, \cdots, m_N$ moving in $\mathbb{R}^d$ ($d \in \{1, 2, 3\}$) under the influence of gravitational attraction. Let x_k be the position of mass m_k. According to Newton's law of universal gravitation, the equations of motion are

$$m_k \ddot{x}_k = \sum_{i \neq k} \frac{m_i m_k (x_i - x_k)}{|x_i - x_k|^3}, \quad k = 1, \ldots, N. \tag{11.4}$$

Let K be the kinetic energy and V be the Newtonian potential defined by

$$K(\dot{x}) = \frac{1}{2} \sum_{k=1}^{N} m_k |\dot{x}_k|^2, \quad V(x) = - \sum_{1 \leq i < j \leq N} \frac{m_i m_j}{|x_i - x_j|}.$$

The system (11.4) is a Hamiltonian system with dN degrees of freedom, where the Hamiltonian H is given by

$$H(x, y) = K(\mathbf{M}^{-1} y) + V(x),$$

where

$$\mathbf{M} = diag(\underbrace{m_1, \cdots, m_1}_{d \text{ copies}}, \underbrace{m_2, \cdots, m_2}_{d \text{ copies}}, \cdots, \underbrace{m_N, \cdots, m_N}_{d \text{ copies}}).$$

In examples above, we see that Hamiltonian plays the role of energy for a mechanical system. Variables x and y represent respectively the position and momentum vectors.

11.2   Linear Hamiltonian Systems

Linear Hamiltonian systems arise naturally in linear approximations and variational equations of nonlinear Hamiltonian systems (recall §2.5 and §4.1). As the Hamiltonian H is assumed to be of class C^2, the linearization of (H) at an equilibrium point $z^* = (x^*, y^*)$ is

$$\dot{z} = JD^2 H(t, z^*)(z - z^*). \tag{11.5}$$

Here $D^2 = D_z^2$ is the second derivative with respect to z. The variational equation of (H) along the solution $\phi(t, \tau, \xi)$ satisfying initial condition $\phi(\tau, \tau, \xi) = \xi$ is

$$\dot{Z} = JD^2 H(t, \phi(t, \tau, \xi))Z. \tag{11.6}$$

The solution to (11.6) with initial condition $Z(\tau) = I_{2n}$ is $Z(t) = \frac{\partial}{\partial \xi} \phi(t, \tau, \xi)$. For this reason we consider linear systems of the form

$$\dot{z} = JS(t)z, \quad S(t) \in \mathbb{R}^{2n \times 2n} \quad \text{is symmetric and continuous.} \tag{LH}$$

This is indeed a linear Hamiltonian system with Hamiltonian $H(t, z) = \frac{1}{2}\langle S(t)z, z \rangle$.

Since $J^T = -J$, the coefficient matrix $A = JS$ of (LH) satisfies

$$A^T J + JA = 0.$$

A matrix $A \in \mathbb{C}^{2n \times 2n}$ satisfying this identity is called a *Hamiltonian matrix*. Indeed, if matrix $A \in \mathbb{C}^{2n \times 2n}$ is Hamiltonian, then $-JA$ is symmetric, so that $A = JS$ with $S = -JA$ symmetric. This observation is summarized in the lemma below:

Lemma 11.2.1 *The mapping $S \mapsto JS$ from the space of symmetric matrices to the space of Hamiltonian matrices is bijective.*

We will study fundamental matrices and stability of (LH). As preparation, we begin with some preliminaries in linear algebra, focusing on properties of skew-symmetric and symplectic matrices.

Here are some basic properties for which we omit the proof.

Theorem 11.2.1 *Assume $A, B \in \mathbb{R}^{2n \times 2n}$ are skew-symmetric, $\alpha, \beta \in \mathbb{R}$. Then*

(a) Eigenvalues of A are zero or pure imaginary.

(b) $\alpha A + \beta B$ and $AB - BA$ are skew-symmetric.

(c) The matrix exponential e^A is orthogonal and has determinant 1.

(d) There exists some orthogonal matrix Q such that $Q^T AQ$ is a block diagonal matrix of the form

$$\text{diag}\left(\begin{pmatrix} 0 & a_1 \\ -a_1 & 0 \end{pmatrix}, \cdots, \begin{pmatrix} 0 & a_k \\ -a_k & 0 \end{pmatrix}, 0_{2(n-k) \times 2(n-k)} \right),$$

where $a_1, \cdots, a_k \neq 0$ are real. If $A \in \mathbb{C}^{2n \times 2n}$, then $a_1, \cdots, a_k \neq 0$ are complex.

Definition 11.2.1 Given skew-symmetric matrix $A = (a_{ij}) \in \mathbb{C}^{2n \times 2n}$, define its *Pfaffian* by

$$\text{Pf}(A) = \frac{1}{2^n n!} \sum_{\sigma \in S_{2n}} \text{sgn}(\sigma) a_{\sigma(1)\sigma(2)} a_{\sigma(3)\sigma(4)} \cdots a_{\sigma(2n-1)\sigma(2n)}.$$

Here S_{2n} stands for the symmetric group on $\{1, \cdots, 2n\}$.
The following simple example will be useful.

Example 11.2.1 The Pfaffian of the matrix J in (H') is 1. This can be easily verified by using the definition. The Pfaffian of the matrix in Theorem 11.2.1(d) is $a_1 \cdots a_n$ if the last block of zero matrix is empty ($k = n$), and the Pfaffian is 0 if otherwise ($k < n$). In particular, this tells us that the Pfaffian of any $n \times n$ skew-symmetric matrix with odd n must be 0. Detailed verifications are left to readers (Exercise 11.5).

Theorem 11.2.2 *Let $A \in \mathbb{C}^{2n \times 2n}$ by skew-symmetric. For any $B = (b_{ij}) \in \mathbb{C}^{2n \times 2n}$, we have*

$$\mathrm{Pf}(B^T AB) = \det(B)\mathrm{Pf}(A).$$

Proof. Clearly $B^T AB$ is also skew-symmetric.

Given $1 \leq \ell_i, k_i \leq n$, $i = 1, \cdots, n$. If $\{\ell_1, k_1, \ell_2, k_2, \cdots, \ell_n, k_n\} = \{1, 2, \cdots, 2n\}$, then $\mathrm{sgn}(\ell_1, k_1, \ell_2, k_2, \cdots, \ell_n, k_n)$ means the sign of the permutation

$$\begin{pmatrix} 1 & 2 & 3 & 4 & \cdots & 2n-1 & 2n \\ \ell_1 & k_1 & \ell_2 & k_2 & \cdots & \ell_n & k_n \end{pmatrix}$$

(i.e. it is 1 if the permutation is even, -1 if the permutation is odd).

If $\{\ell_1, k_1, \ell_2, k_2, \cdots, \ell_n, k_n\}$ is a proper subset of $\{1, 2, \cdots, 2n\}$, then we set $\mathrm{sgn}(\ell_1, k_1, \ell_2, k_2, \cdots, \ell_n, k_n) = 0$. Then

$$\det(B)\,\mathrm{sgn}(\ell_1, k_1, \ell_2, k_2, \cdots, \ell_n, k_n)$$

$$= \sum_{\sigma \in S_{2n}} \mathrm{sgn}(\sigma)\, b_{1\sigma(1)} b_{2\sigma(2)} \cdots b_{(2n)\sigma(2n)}\, \mathrm{sgn}(\ell_1, k_1, \ell_2, k_2, \cdots, \ell_n, k_n)$$

$$= \sum_{\sigma \in S_{2n}} \mathrm{sgn}(\sigma)\, b_{\ell_1\sigma(1)} b_{k_1\sigma(2)} \cdots b_{\ell_n\sigma(2n-1)} b_{k_n\sigma(2n)}$$

$$= \sum_{\sigma \in S_{2n}} \mathrm{sgn}(\sigma) \prod_{s=1}^{b} b_{\ell_s\sigma(2s-1)} b_{k_s\sigma(2s)}.$$

Using this observation, we find

$$
\begin{aligned}
\mathrm{Pf}(B^T A B) &= \frac{1}{2^n n!} \sum_{\sigma \in S_{2n}} \mathrm{sgn}(\sigma) \prod_{s=1}^{n} (B^T A B)_{\sigma(2s-1)\sigma(2s)} \\
&= \frac{1}{2^n n!} \sum_{\sigma \in S_{2n}} \mathrm{sgn}(\sigma) \sum_{\substack{1 \le \ell_1, \cdots, \ell_n \le 2n \\ 1 \le k_1, \cdots, k_n \le 2n}} \prod_{s=1}^{n} b_{\ell_s \sigma(2s-1)} a_{\ell_s k_s} b_{k_s \sigma(2s)} \\
&= \frac{1}{2^n n!} \sum_{\substack{1 \le \ell_1, \cdots, \ell_n \le 2n \\ 1 \le k_1, \cdots, k_n \le 2n}} \left(\prod_{s=1}^{n} a_{\ell_s k_s} \right) \sum_{\sigma \in S_{2n}} \mathrm{sgn}(\sigma) \prod_{s=1}^{n} b_{\ell_s \sigma(2s-1)} b_{k_s \sigma(2s)} \\
&= \frac{\det(B)}{2^n n!} \sum_{\substack{1 \le \ell_1, \cdots, \ell_n \le 2n \\ 1 \le k_1, \cdots, k_n \le 2n}} \left(\prod_{s=1}^{n} a_{\ell_s k_s} \right) \mathrm{sgn}(\ell_1, k_1, \ell_2, k_2, \cdots, \ell_n, k_n) \\
&= \det(B) \mathrm{Pf}(A)
\end{aligned}
$$

as desired. □

The following theorem due to Cayley in 1847 shows the close connection between Pfaffian and determinants of skew-symmetric matrices.

Theorem 11.2.3 *If $A \in \mathbb{C}^{2n \times 2n}$ is skew-symmetric, then $\det(A) = \mathrm{Pf}(A)^2$.*

Proof. By Theorem 11.2.1(d), A is orthogonally equivalent to a tridiagonal matrix

$$
D = \mathrm{diag}\left(\begin{pmatrix} 0 & a_1 \\ -a_1 & 0 \end{pmatrix}, \cdots, \begin{pmatrix} 0 & a_k \\ -a_k & 0 \end{pmatrix}, 0_{2(n-k) \times 2(n-k)} \right).
$$

Here $a_1, \cdots, a_k$ are nonzero. Then, by Theorem 11.2.2, we have $\mathrm{Pf}(A) = \mathrm{Pf}(D)$.

The Pfaffian $\mathrm{Pf}(D)$ is zero if the last zero block matrix is nonempty (i.e. $k < n$), for which case A is singular and we have $\det(A) = 0 = \mathrm{Pf}(D)^2 = \mathrm{Pf}(A)^2$. If the last zero block matrix is empty (i.e. $k = n$), then $\mathrm{Pf}(D) = a_1 \cdots a_n$ (by Example 11.2.1), so

$$
\mathrm{Pf}(A)^2 = (a_1 \cdots a_n)^2 = \det(D) = \det(A).
$$

This completes the proof. □

Definition 11.2.2 We say $M \in \mathbb{C}^{2n \times 2n}$ is *symplectic* if

$$
M^T J M = J. \tag{11.7}
$$

Denote the set of $2n \times 2n$ symplectic matrices by $\mathrm{Sp}(2n, \mathbb{C})$, called the *complex symplectic group*. The *real symplectic group* $\mathrm{Sp}(2n, \mathbb{R})$ consists of real matrices in $\mathrm{Sp}(2n, \mathbb{C})$.

In what follows we let $\mathbb{F}$ be either the complex field $\mathbb{C}$ or real field $\mathbb{R}$. It is easy to see that the *general linear group* $\mathrm{GL}(2n, \mathbb{F})$ which consists of invertible matrices over $\mathbb{F}$ is a group, and the *special linear group* $\mathrm{SL}(2n, \mathbb{F})$ consisting of matrices with determinant 1 is a subgroup of $\mathrm{GL}(2n, \mathbb{F})$. The following theorem justifies the choice of terminology "group" for the collection of symplectic matrices, and shows that every symplectic matrix has determinant 1.

Theorem 11.2.4 *The symplectic group* $\mathrm{Sp}(2n, \mathbb{F})$ *is a subgroup of* $\mathrm{SL}(2n, \mathbb{F})$.

Proof. Given any $M \in \mathrm{Sp}(2n, \mathbb{F})$, we have

$$\det(J) = \det(M^T J M) = \det(M^T) \det(J) \det(M) = \det(M)^2 \det(J).$$

This shows that symplectic matrices are invertible and have determinant ± 1. From (11.7) we have $M^{-1} \in \mathrm{Sp}(2n, \mathbb{F})$, because

$$(M^{-1})^T J M^{-1} = -(M^{-1})^T J^{-1} M^{-1} = -(M^T J M)^{-1} = -J^{-1} = J.$$

Given another $N \in \mathrm{Sp}(2n, \mathbb{F})$, clearly $MN \in \mathrm{Sp}(2n, \mathbb{F})$ since

$$(MN)^T J(MN) = N^T (M^T J M) N = N^T J N = J.$$

Therefore $\mathrm{Sp}(2n, \mathbb{F})$ is indeed a group.

Clearly $M^T J M$ is skew-symmetric since J is skew-symmetric. The theorem now follows by setting $A = J$, $M = B$ in Theorem 11.2.2. $\square$

It is a simple exercise to show that $\mathrm{Sp}(2, \mathbb{F}) = \mathrm{SL}(2, \mathbb{F})$ but $\mathrm{Sp}(2n, \mathbb{F}) \neq \mathrm{SL}(2n, \mathbb{F})$ for $n \geq 2$ (Exercise 11.9).

Theorem 11.2.5 *If* λ *is an eigenvalue of* $M \in \mathrm{Sp}(2n, \mathbb{R})$, *then so are* $\bar{\lambda}$, $1/\lambda$, $1/\bar{\lambda}$.

Proof. By Theorem 11.2.4 we know that $\lambda \neq 0$. Since we are dealing with real matrices, it is sufficient to show that $1/\lambda$ is an eigenvalue. Let I be the $2n \times 2n$ identity matrix. Then

$$
\begin{aligned}
\det(M - \lambda I) &= \det(J^{-1}(M^T)^{-1}J - \lambda I) \\
&= \det(J^{-1}((M^T)^{-1} - \lambda I)J) \\
&= \det((M^T)^{-1} - \lambda I) \\
&= \det(M^{-1} - \lambda I) \\
&= \det(I - \lambda M) \quad \text{(since } \det(M) = 1, \text{ by Theorem 11.2.4)} \\
&= \det(\lambda M - I) \quad \text{(since the space is even dimensional)} \\
&= \lambda^{2n} \det\left(M - \frac{1}{\lambda}I\right).
\end{aligned}
$$

Thus λ is an eigenvalue if and only if $1/\lambda$ is an eigenvalue. $\quad\square$

Remark 11.2.1 Real symplectic matrices has a simple geometric interpretation. Let ω^2 be the skew-symmetric bilinear form on $\mathbb{R}^{2n}$ defined by

$$
\omega^2(z, w) = \langle z, Jw \rangle = z^T Jw. \tag{11.8}
$$

Then $M \in \mathrm{Sp}(2n, \mathbb{R})$ if and only if $\omega^2(z, w) = \omega^2(Mz, Mw)$ for all z, $w \in \mathbb{R}^{2n}$. Note that

$$
\omega^2(z, w) = \sum_{k=1}^{n}(z_k w_{n+k} - z_{n+k}w_k) = \sum_{k=1}^{n} \det\begin{pmatrix} z_k & w_k \\ z_{n+k} & w_{n+k} \end{pmatrix}
$$

is the sum of oriented areas of parallelograms obtained by projecting the parallelogram spanned by z and w to $x_k y_k$-planes. Therefore, real symplectic matrices can be interpreted as those real matrices which preserve the sum of oriented areas of projected parallelograms on all $x_k y_k$-planes. Putting x_k in front of y_k is a choice of orientation; one may project to $y_k x_k$-planes instead and change the sign of these areas. Clearly the "parallelogram" in $\mathbb{R}^{2n}$ can be replaced by planar regions which can be approximated by disjoint union of parallelograms.

Alternatively, in terms of exterior algebra, real symplectic matrices are matrices which preserve the 2-form

$$
\omega^2 := dx \wedge dy = dx_1 \wedge dy_1 + dx_2 \wedge dy_2 + \cdots + dx_n \wedge dy_n. \tag{11.9}
$$

Here x_k, y_k are coordinate functions, $dx_k \wedge dy_k$ is the bilinear form on $\mathbb{R}^{2n}$:

$$
(dx_k \wedge dy_k)(z, w) = \det\begin{pmatrix} x_k(z) & x_k(w) \\ y_k(z) & y_k(w) \end{pmatrix} = \det\begin{pmatrix} z_k & w_k \\ z_{n+k} & w_{n+k} \end{pmatrix}.
$$

Same notation ω^2 were selected in (11.8) and (11.9) since they are clearly identical. One may also conclude real symplectic matrices have determinant

1 by observing their preservation of the $2n$-form $(\omega^2)^n = \omega^2 \wedge \cdots \wedge \omega^2$ on $\mathbb{R}^{2n}$, and that $(\omega^2)^n$ is the standard volume form. See [Arnold] for details about algebraic and geometric aspects of exterior algebra.

We have finished preparations on skew-symmetric and symplectic matrices, and is now ready for main results in this section regarding fundamental matrices and stability of (LH).

Theorem 11.2.6 *Let $\Phi(t)$ be the fundamental matrix of the linear Hamiltonian system (LH) with $\Phi(0) = I$, then $\Phi(t) \in \mathrm{Sp}(2n, \mathbb{R})$ for any t. Conversely, if $\Phi : \mathbb{R} \to \mathrm{Sp}(2n, \mathbb{R})$ is of class C^1, then $\Phi(t)$ is a fundamental matrix of (LH) for some symmetric and continuous $S(t)$.*

Proof. Note that $J^T = J^{-1} = -J$. Therefore,

$$\frac{d}{dt}\left(\Phi^T J \Phi\right) = \dot{\Phi}^T J \Phi + \Phi^T J \dot{\Phi} = \left(\Phi^T S J^T\right) J \Phi + \Phi^T J \left(JS\Phi\right) = 0.$$

Since $\Phi(0)^T J \Phi(0) = J$, we conclude that $\Phi(t)^T J \Phi(t) = J$ for any t.

For the converse, assume $\Phi : \mathbb{R} \to \mathrm{Sp}(2n, \mathbb{R})$ is of class C^1. Define $S = -J\dot{\Phi}\Phi^{-1}$, then $S(t)$ is continuous and $\dot{\Phi} = JS\Phi$. By Theorem 3.2.4 and the fact that $\Phi(t)$ is invertible for any t, $\Phi(t)$ is a fundamental matrix for the system $\dot{z} = JSz$. By differentiating the identity $\Phi(t)^T J \Phi(t) = J$ with respect to t, we find

$$\dot{\Phi}^T J \Phi + \Phi^T J \dot{\Phi} = 0,$$

and thus

$$S^T = (\Phi^T)^{-1}\dot{\Phi}^T J = (\Phi^T)^{-1}\left(\dot{\Phi}^T J \Phi\right)\Phi^{-1}$$

$$= -(\Phi^T)^{-1}\left(\Phi^T J \dot{\Phi}\right)\Phi^{-1} = S.$$

This shows that S is symmetric, and this completes the proof. $\square$

Recall that 0 is asymptotically stable for linear system $x' = Ax$ with constant coefficient if all eigenvalues of A have negative real parts (Theorem 3.3.5). More generally, an equilibrium x^* of autonomous system $x' = f(x)$ is asymptotically stable if all eigenvalues of the variational matrix $D_x f(x^*)$ have negative real parts (Theorem 4.2.1). Such occasion never occur for (LH). To see this, consider (LH) with constant symmetric matrix S, then e^{JSt} is a fundamental matrix for (LH). By Theorem 11.2.6, e^{JSt} is symplectic for any t. Note that an eigenvalue of the Hamiltonian matrix

JS with negative (resp. positive) real part corresponds an eigenvalue of the symplectic matrix e^{JS} inside (resp. outside) the unit disc. Therefore, by Theorem 11.2.5, JS has an eigenvalue with negative real part if and only if it has an eigenvalue with positive real part.

From Theorems above and Chapter 3, we are now ready to obtain the following stability criteria for linear Hamiltonian systems.

Theorem 11.2.7 *Consider the linear Hamiltonian system (LH).*

(a) *Assume S is constant. Then $z \equiv 0$ is stable if and only if eigenvalues of $A = JS$ are pure imaginary and have simple elementary divisors. Moreover, $z \equiv 0$ can't be asymptotically stable.*

(b) *Assume S is T-periodic. Then $z \equiv 0$ is stable if and only if characteristic multipliers have unit modulus and have simple elementary divisors. Moreover, $z \equiv 0$ can't be asymptotically stable.*

Proof. First observe that the characteristic polynomial of A is even, since

$$\det(A - \lambda I) = \det(A^T - \lambda I) = \det(-SJ - \lambda I) = \det(SJ + \lambda I)$$
$$= \det(JS + \lambda I) = \det(A + \lambda I).$$

This implies, in particular, that A has an eigenvalue with negative real part if and only if it has an eigenvalue with positive real part. With this observation, part (a) follows easily from Theorem 3.3.6 and Theorem 11.2.5.

Now consider part (b). Let $\Phi(t)$ be a fundamental matrix of (LH). By the Floquet theorem, there exists a T-periodic nonsingular matrix $P(t) \in \mathbb{C}^{2n \times 2n}$, and a matrix $R \in \mathbb{C}^{2n \times 2n}$, such that $\Phi(t) = P(t)e^{tR}$. Recall that the change of variables $z = P(t)w$ allows us to rewrite (LH) as $\dot{w} = Rw$ (see the proof of Theorem 3.5.5). By (a) and Theorem 3.5.5, we see that the solution $w \equiv 0$, which is equivalent to $z \equiv 0$, is stable if and only if every characteristic multiplier λ has simple elementary divisor and satisfies $|\lambda| = 1$. $\square$

Remark 11.2.2 Assuming $\Phi(t)$ is a fundamental matrix of (LH) with T-periodic S. A Hamiltonian version of the Floquet theorem states that, $\Phi(t + T)$ is also a fundamental matrix of (LH), and there exist $P : \mathbb{R} \to \mathrm{Sp}(2n, \mathbb{C})$ and Hamiltonian matrix $R \in \mathbb{C}^{2n \times 2n}$ such that P is T-periodic and

$$\Phi(t) = P(t)e^{tR}.$$

The proof of this version of Floquet theorem requires finding logarithm for the symplectic matrix $\Phi(t+T)\Phi(t)^{-1}$ in the space of Hamiltonian matrices. It is a nontrivial task which requires further preparations. A relatively simple approach is based on analytic theory for matrices (c.f. [YS]). We skip it for brevity as it is not needed for the stability criterion Theorem 11.2.7.

11.3   First Integrals and Poisson Bracket

One simple feature of autonomous Hamiltonian system is the conservation of Hamiltonian (energy); that is, for any solution $(x(t), y(t))$ of (H) with $H = H(x, y)$, we have

$$\frac{d}{dt}H(x(t), y(t)) = \frac{\partial H}{\partial x} \cdot \dot{x} + \frac{\partial H}{\partial y} \cdot \dot{y} = 0.$$

Conserved quantities can be defined for more general systems as follows.

Definition 11.3.1 Consider a system of differential equations

$$\frac{dz}{dt} = f(t, z), \quad f \in C^1(I \times D), \ I \subset \mathbb{R}, \ D \subset \mathbb{R}^n \ \text{are open}.$$

We say a C^1 function $F(z)$ is a *first integral* (or *integral of motion*, or *constant of motion*) of the system if $F(z(t))$ is constant along any solution curve $z(t)$.

Many people exclude constant functions from the collection of first integrals, but the definition adopted here include them as trivial first integrals. For simplicity we consider $I = \mathbb{R}$ and solutions which exist globally in time.

First integrals allow us to reduce the system to lower dimensional spaces; if $F_1, \cdots, F_k$ is a family of non-constant first integrals, then any solution of the system are confined to move on level sets of $F_1, \cdots, F_k$. If their differentials $dF_1, \cdots, dF_k$ at some point z_0 are linearly independent, let $c_i = F_i(z_0)$, then by the implicit function theorem, the level set $F_1^{-1}(c_1) \cap \cdots \cap F_k^{-1}(c_k)$ is locally an $(n-k)$-dimensional manifold. In case $k = n-1$ this level set is a curve along which the solution resides. We say the system in $\mathbb{R}^n$ is an *integrable system* if there exist $n-1$ independent first integrals.

Roughly speaking, more first integrals were identified, the better we know about behavior of solutions. Integrability is an indication that the system is regular, and non-integrability is an indication of complexity or chaos. For Hamiltonian systems, there is a special structure known as *Poisson structure* (or *Lie-Poisson structure*) which reveals algebraic nature of first integrals. This section is an introduction to this connection.

Example 11.3.1 Consider the second order equation (11.1) in Example 11.1.1. By conservation of the Hamiltonian $H(x, y)$ in (11.2), solutions of the corresponding Hamiltonian system move along level sets of H. Two typical examples are

$$H(x, y) = \frac{k}{2}x^2 + \frac{1}{2m}y^2 \quad \text{(harmonic oscillator, Example 5.2.1)}$$

$$H(\theta, \psi) = \frac{1}{2}\psi^2 - k^2 \cos\theta \quad \text{(simple pendulum, Example 5.3.1)}.$$

Here k and m are positive constants. Both of them are integrable.

Example 11.3.2 Consider linearly coupled oscillators (CO) in Example 11.1.2. Apart from the Hamiltonian, it can be easily verified that the followings are also first integrals:

$$F_1(x_1, x_2, y_1, y_2) = \frac{\kappa}{2}(x_1 + x_2)^2 + \frac{1}{2m}(y_1 + y_2)^2,$$

$$F_2(x_1, x_2, y_1, y_2) = \frac{\kappa + 2\kappa_{12}}{2}(x_1 - x_2)^2 + \frac{1}{2m}(y_1 - y_2)^2.$$

They are not independent from the Hamiltonian H; clearly $F_1 + F_2 = 2H$. These two first integrals suggest new variables $\xi_1 = x_1 + x_2$, $\xi_2 = x_1 - x_2$ to rewrite the system (CO):

$$m\ddot{\xi}_1 = -\kappa\xi_1, \quad m\ddot{\xi}_2 = -(\kappa + 2\kappa_{12})\xi_2.$$

This becomes uncoupled harmonic oscillators.

Definition 11.3.2 A *Lie algebra* $\mathfrak{g}$ (over $\mathbb{R}$ or $\mathbb{C}$) is a vector space with a skew-symmetric bilinear map $[\cdot, \cdot] : \mathfrak{g} \times \mathfrak{g} \to \mathfrak{g}$, called the *(Lie) bracket*, which satisfies the *Jacobi identity*:

$$[x, [y, z]] + [y, [z, x]] + [z, [x, y]] = 0 \quad \text{for any } x, y, z \in \mathfrak{g}. \quad (11.10)$$

A *subalgebra* of $\mathfrak{g}$ is a linear subspace which is closed under the bracket.

Example 11.3.3 Here are some simple examples of Lie algebras. Verification of the Jacobi identity is an easy exercise left to readers.

(1) $\mathbb{R}^3$ with the bracket $[x, y] = x \times y$, where $\times$ is the cross product.
(2) The space $\mathfrak{gl}(n) = \mathbb{R}^{n \times n}$ with bracket (or commutator) $[A, B] = AB - BA$.
(3) The space $\mathfrak{so}(n)$ consisting of real skew-symmetric matrices is a subalgebra of $\mathfrak{gl}(n)$. This is a reiteration of Theorem 11.2.1(b).

(4) The space $\mathfrak{sp}(2n)$ of $2n \times 2n$ real Hamiltonian matrices is a subalgebra of $\mathfrak{gl}(2n)$.

Examples (2), (3), (4) are matrix algebra. Conventional symbols $\mathfrak{gl}$, $\mathfrak{so}$, $\mathfrak{sp}$ were chosen because of their correspondence with matrix groups GL (general linear group), SO (special orthogonal group), Sp (symplectic group), through the matrix exponential.

Definition 11.3.3 Let $D \subset \mathbb{R}^n$ be an open set. Consider the space $C^\infty(\mathbb{R} \times D \times \mathbb{R}^n)$ of real-valued smooth functions. Let $(x_1, \cdots, x_n, y_1, \cdots, y_n)$ be coordinates of $D \times \mathbb{R}^n$. The *Poisson bracket* $\{\cdot, \cdot\}$ is the map on $C^\infty(\mathbb{R} \times D \times \mathbb{R}^n) \times C^\infty(\mathbb{R} \times D \times \mathbb{R}^n)$ defined by

$$\{F, G\} = \sum_{k=1}^{n} \left(\frac{\partial F}{\partial x_k} \frac{\partial G}{\partial y_k} - \frac{\partial F}{\partial y_k} \frac{\partial G}{\partial x_k} \right). \tag{11.11}$$

Writing the gradient ∇ of a function with respect to $z = (x, y)$ as a column vector, then (11.11) can be alternatively written

$$\{F, G\} = \nabla F^T J \nabla G. \tag{11.12}$$

Remark 11.3.1 Clearly $\{F, G\}$ in (11.12) is well-defined for $F, G \in C^1(\mathbb{R} \times D \times \mathbb{R}^n)$. We confine the domain of definition to smooth functions so as to match the domain with the image space. Furthermore, we can allow Poisson bracket to be defined on vector valued functions. In particular, consider $z = (x, y)$ as the identity map from (x, y) to (x, y), we can rewrite (H') as

$$\dot{z} = \{z, H\}. \tag{H''}$$

Example 11.3.4 The space $C^\infty(\mathbb{R} \times D \times \mathbb{R}^n)$ together with the Poisson bracket $\{\cdot, \cdot\}$ is a Lie algebra. Clearly the Poisson bracket is a bilinear skew-symmetric map. Verification of the Jacobi identity is a routine exercise left to readers.

The next theorem shows the connection between first integrals of Hamiltonian systems and the Lie algebra $C^\infty(\mathbb{R} \times D \times \mathbb{R}^n)$ with Poisson bracket.

Theorem 11.3.1 *Consider an autonomous Hamiltonian system with Hamiltonian function $H \in C^\infty(D \times \mathbb{R}^n)$, $D \subset \mathbb{R}^n$ is open. Let $\{\cdot, \cdot\}$ be the Poisson bracket on $C^\infty(D \times \mathbb{R}^n)$. Then for any $F, G \in C^\infty(D \times \mathbb{R}^n)$, we have*

(a) F is a first integral if and only if $\{F, H\} = 0$.

(b) (Poisson Theorem) If F and G are first integrals, then so is $\{F, G\}$.

Proof. By (11.12), for any $F \in C^\infty(\mathbb{R} \times D \times \mathbb{R}^n)$, and any solution z of (H"), we have

$$\frac{d}{dt}F(t, z(t)) = \frac{\partial}{\partial t}F(t, z(t)) + \{F, H\}. \tag{11.13}$$

Part (a) follows trivially from (11.13). Part (b) follows from (a) and the Jacobi identity

$$\{F, \{G, H\}\} + \{G, \{H, F\}\} + \{H, \{F, G\}\} = 0.$$

$\square$

Remark 11.3.2 The Poisson theorem can be alternatively stated as: For a fixed smooth Hamiltonian, the set of first integrals for the Hamiltonian system is a subalgebra of the Lie algebra $(C^\infty(D \times \mathbb{R}^n), \{\cdot, \cdot\})$. This suggests an algebraic classification of Hamiltonian systems using Lie algebras.

Example 11.3.5 Consider the Hénon-Heiles system in Example 11.1.3 and let H be its Hamiltonian. If $\mu = \lambda = 1$, then

$$F(x_1, x_2, y_1, y_2) = x_1 x_2 + y_1 y_2 + x_1 x_2^2 + \frac{1}{3}x_1^3$$

is a first integral of the system since $\{F, H\} = 0$. If $\mu = -1$, $\lambda = -6$, then

$$G(x_1, x_2, y_1, y_2) = x_1^4 + 4x_1^2 x_2^2 + 4y_1(y_1 x_2 - y_2 x_1) - 4x_1^2 x_2 + 3(x_1^2 + y_1^2)$$

is a first integral of the system since $\{G, H\} = 0$. We leave these straightforward verifications to readers. The system (HH) is chaotic for generic choices of parameters μ, λ. Figure 11.2 shows a chaotic trajectory for the case $\mu = 1$, $\lambda = -1$.

Example 11.3.6 Assume the motion of a mechanical system consisting of N particles $q_1, \cdots, q_N$ in $\mathbb{R}^3$ is governed by an autonomous Hamiltonian system. This includes the N-body problem (Example 1.1.11) and charged particles moving in a fixed magnetic field as special cases. Let p_k be the linear momentum of particle q_k. The angular momentum of the whole system is

$$\mathbf{L} = (L_1, L_2, L_3) = \sum_{k=1}^{N} q_k \times p_k.$$

Fig. 11.2   Poincaré section for the Hénon-Heiles system with $\mu = 1$, $\lambda = -1$. The one on the left is the section at $x_1 = y_1 = 0$, and the one on the right is the section at $x_1 = y_2 = 0$.

Suppose two components of the angular momentum, say L_1 and L_2, are known to be first integrals. Let $q_k = (x_k, y_k, z_k)$, $p_k = (u_k, v_k, w_k)$. Then

$$\{L_1, L_2\} = \left\{\sum_{i=1}^{N}(y_i w_i - z_i v_i), \sum_{j=1}^{N}(z_j u_j - x_j w_j)\right\}$$

$$= \sum_{k=1}^{N}\left((0, w_k, -v_k) \cdot (z_k, 0, -x_k) - (0, -z_k, y_k) \cdot (-w_k, 0, u_k)\right)$$

$$= \sum_{k=1}^{N}(x_k v_k - y_k u_k) \;=\; L_3.$$

The Poisson theorem tells us that the third component L_3 of the angular momentum is also a first integral. By rotating coordinate systems, we may conclude from this example that if two components of the angular momentum are conserved, then the angular momentum is conserved. Note that discussions above are valid for arbitrary potential functions.

Example 11.3.7 The equation of motion for the Kepler problem (the two-body problem) is given by

$$\ddot{q} \;=\; -\frac{\mu q}{|q|^3}. \tag{11.14}$$

Here $\mu > 0$ is a constant, and $q \in \mathbb{R}^3$ is the vector from one mass to the other. This formula fixes the mass center at the origin and has taken conservation of linear momentum into consideration. Kepler's first law of planetary motions states that the trajectory of any solution q of (11.14) is a conic section. We will show it by identifications of first integrals.

Equation (11.14) can be expressed as a Hamiltonian system with 3 degrees of freedom. By setting $p = \dot{q}$, the Hamiltonian H is

$$H(q,p) = \frac{1}{2}|p|^2 - \frac{\mu}{|q|}.$$

It is straightforward to verify the conservation of the angular momentum $\mathbf{L} = q \times p$:

$$\frac{d}{dt}\mathbf{L} = \dot{q} \times p + q \times \dot{p} = \frac{\partial H}{\partial p} \times p - q \times \frac{\partial H}{\partial q} = p \times p + \mu q \times \frac{q}{|q|^3} = 0.$$

This provides 3 first integrals since it a vector in $\mathbb{R}^3$. Together with the energy integral H, we have a total of 4 first integrals. Another set of first integrals is given by the *eccentricity vector*:

$$\mathbf{A} = \frac{1}{\mu}p \times \mathbf{L} - \frac{q}{|q|}. \tag{11.15}$$

The eccentricity vector is more commonly, but improperly, known as the *Laplace-Runge-Lenz vector*. Its conservation can be also verified directly, but here we appeal to Theorem 11.3.1.

Recall the notation ∇_x for gradient with respect to x, D_x for the total derivative with respect to x. Gradient vectors are written as column vectors, so the total derivative $D_x f$ of $f : \mathbb{R}^n \to \mathbb{R}^m$ as an $m \times n$ matrix has $\nabla_x f_1^T, \cdots, \nabla_x f_m^T$ as its row vectors. To show that $\mathbf{A} = (A_1, A_2, A_3)$ is conserved, by Theorem 11.3.1(a) and (11.12) (Remark 11.3.1) all we need is showing

$$\{\mathbf{A}, H\} = D\mathbf{A}J\nabla H = 0.$$

Here $D\mathbf{A} = D_{q,p}\mathbf{A}$ is a 3×6 matrix whose rows are ∇A_i^T, $i \in \{1,2,3\}$.

Three elementary identities from vector analysis, proofs for which are easy exercises left to readers (Exercise 11.7), will be convenient (q, p, u, v are in $\mathbb{R}^3$):

$$D_q\left(\frac{q}{|q|}\right)p = \frac{1}{|q|}\left(p - \frac{q \cdot p}{|q|^2}q\right) \tag{i}$$

$$q \times (q \times p) = (q \cdot p)q - |q|^2 p \tag{ii}$$

$$D_p(p \times u)v = v \times u. \tag{iii}$$

Knowing that $\mathbf{L}$ is a constant vector, the eccentricity vector $\mathbf{A}$ is also conserved because

$$D\mathbf{A}J\nabla H = \left(-D_q\left(\frac{q}{|q|}\right), \frac{1}{\mu}D_p(p \times \mathbf{L})\right)\left(\begin{array}{c} p \\ \nabla_q(\mu/|q|) \end{array}\right)$$

$$= -D_q\left(\frac{q}{|q|}\right)p - D_p(p \times \mathbf{L})\frac{q}{|q|^3}$$

$$= -\frac{1}{|q|}\left(p - \frac{q \cdot p}{|q|^2}q\right) - \frac{q}{|q|^3} \times (q \times p) = 0.$$

Clearly $\mathbf{A}$ is perpendicular to $\mathbf{L}$, thus q, p, and $\mathbf{A}$ remain coplanar for all time. Let θ be the angle between $\mathbf{A}$ and q. Then

$$|\mathbf{A}||q|\cos\theta = \mathbf{A} \cdot q = \frac{1}{\mu}(p \times \mathbf{L}) \cdot q - |q| = \frac{|\mathbf{L}|^2}{\mu} - |q|.$$

Therefore,

$$|q| = \frac{|\mathbf{L}|^2/\mu}{1 + |\mathbf{A}|\cos\theta}.$$

This implies Kepler's first law of planetary motions — the trajectory of q is a conic section with a focus at the origin, eccentricity $|\mathbf{A}|$, and semi-latus rectum $|\mathbf{L}|^2/\mu$. It also justifies the name "eccentricity vector" for $\mathbf{A}$. The proof of Kepler's first law via conservation of the eccentricity vector is due to Hermann and Bernoulli in the 1710s, long before eras of Laplace, Runge, and Lenz.

11.4   Symplectic Transformations

We have seen the close relation between linear Hamiltonian systems and symplectic matrices. In this section, we explore similar connections for general Hamiltonian systems.

Definition 11.4.1 Given an open set $U \subset \mathbb{R}^{2n}$, we say $F \in C^1(U, \mathbb{R}^{2n})$ is *symplectic* (or *canonical*) if $DF(z) \in \mathrm{Sp}(2n, \mathbb{R})$ for any $z \in U$.

This definition includes symplectic matrices as special cases of symplectic transformations. By the inverse function theorem, a symplectic transformation F is locally a C^1 diffeomorphism and its local inverse is also symplectic. Clearly compositions of symplectic transformations are symplectic.

When dealing with time-dependent functions $F(t, z)$, being "symplectic" means symplectic as function of z. In this case, we still use the notation $DF = D_z F$ for the total derivative of F with respect to the space variable z.

We shall follow this conventional notation and denote partial differentiation with respect to the first (time) variable by $\partial/\partial t$, denote the total derivative with respect to time by using dot " $\cdot$ " or d/dt. It is also convenient to use $\partial F/\partial z$ for the total derivative of F with respect to z.

To clarify such notations, let us consider for instance a flow $\phi(t, \tau, \xi)$ induced by $\dot{z} = f(t, z)$. Then

$$\frac{\partial}{\partial \xi} F(t, \phi(t, \tau, \xi)) = DF(t, \phi(t, \tau, \xi)) \frac{\partial}{\partial \xi} \phi(t, \tau, \xi),$$

$$\frac{d}{dt} F(t, \phi(t, \tau, \xi)) = \frac{\partial}{\partial t} F(t, \phi(t, \tau, \xi)) + DF(t, \phi(t, \tau, \xi)) f(t, \phi(t, \tau, \xi)).$$

Assume the time-dependent symplectic transformation $F(t, z)$ is of class C^1 with respect to t. By the inverse function theorem again, the mapping $(t, z) \mapsto (t, F(t, z))$ is locally a C^1 diffeomorphism. Let $\zeta = F(t, z)$, then locally z is a function of (t, ζ), so it makes sense to write $\partial \zeta/\partial z$ and $\partial z/\partial \zeta$, which are inverses of each other.

The theorem below states that such symplectic changes of variables send Hamiltonian systems to *locally Hamiltonian systems*; i.e. systems which can be locally written as Hamiltonian systems. It is not hard to find locally Hamiltonian systems which are not Hamiltonian systems (Exercise 11.6).

Theorem 11.4.1 *Suppose $\zeta := F(t, z)$ is symplectic for z in an open set $U \subset \mathbb{R}^{2n}$ and is C^1 with respect to t. Then the Hamiltonian system (H') in terms of the new variable ζ is a locally Hamiltonian system.*

Proof. Since $\zeta = F$ is symplectic, we have identity

$$\frac{\partial \zeta}{\partial z}^T J \frac{\partial \zeta}{\partial z} = J.$$

Differentiate this identity with respect to t, we find

$$\frac{\partial^2 \zeta}{\partial t \partial z}^T J \frac{\partial \zeta}{\partial z} + \frac{\partial \zeta}{\partial z}^T J \frac{\partial^2 \zeta}{\partial t \partial z} = 0.$$

Let $G = -J\frac{\partial \zeta}{\partial t}$. Then

$$
\begin{aligned}
D_\zeta G(t, z(t, \zeta)) &= -J\frac{\partial^2 \zeta}{\partial t \partial z}\frac{\partial z}{\partial \zeta} \\
&= \left(\frac{\partial \zeta}{\partial z}^T\right)^{-1}\left(\frac{\partial^2 \zeta}{\partial t \partial z}^T J\frac{\partial \zeta}{\partial z}\right)\frac{\partial z}{\partial \zeta} \\
&= -\frac{\partial z}{\partial \zeta}^T\frac{\partial^2 \zeta}{\partial t \partial z}^T J^T \\
&= D_\zeta G(t, z(t, \zeta))^T.
\end{aligned}
$$

Therefore, $D_\zeta G$ is symmetric and, by Theorem 5.4.2, for each ζ_0 in its domain of the space variable, there is a neighborhood of ζ_0 and a function $K(t, \zeta)$ with ζ in this neighborhood such that G as a function of (t, ζ) is equal to $\nabla_\zeta K(t, \zeta)$.

Let z be solution of the Hamiltonian system (H'). Then

$$
\begin{aligned}
\dot\zeta &= \frac{\partial}{\partial t}F(t, z) + DF(t, z)J\frac{\partial}{\partial z}H(t, z)^T \\
&= \frac{\partial \zeta}{\partial t} + \frac{\partial \zeta}{\partial z}J\left(\frac{\partial}{\partial \zeta}H(t, z(t, \zeta))\frac{\partial \zeta}{\partial z}\right)^T \\
&= JG + J\frac{\partial}{\partial \zeta}H(t, z(t, \zeta))^T \\
&= J\nabla_\zeta K(t, \zeta) + J\nabla_\zeta H(t, z(t, \zeta)).
\end{aligned}
$$

Therefore, in terms of ζ, the system is a locally Hamiltonian system with Hamiltonian $K(t, \zeta) + H(t, z(t, \zeta))$. $\square$

Recall the Poisson bracket defined in (11.12). To emphasize the space variable in use, we add the space variable to the subscript of $\{\cdot, \cdot\}$. Then

$$
\begin{aligned}
\{F, G\}_\zeta &= \nabla_\zeta F^T J\nabla_\zeta G \\
&= \nabla_z F^T\left(\frac{\partial z}{\partial \zeta}J\frac{\partial z}{\partial \zeta}^T\right)\nabla_z G \\
&= \nabla_z F^T J\nabla_z G = \{F, G\}_z.
\end{aligned}
$$

This simple observation is summarized in the theorem below.

Theorem 11.4.2 *Under the assumption of Theorem 11.4.1, the Poisson bracket is preserved by the symplectic change of variable $\zeta = F(t, z)$; i.e.* $\{F, G\}_\zeta = \{F, G\}_z.$

Coordinates functions obtained from symplectic change of variables are called *symplectic coordinates*. By the above theorem and the Poisson theorem, first integrals F, G, $\{F, G\}$ for the Hamiltonian system (H) are still first integrals for the locally Hamiltonian system in terms of any other symplectic coordinates.

The next theorem can be regarded as a nonlinear and local version of Theorem 11.2.6. The idea of proof has appeared in the proofs of Theorem 11.2.6 and Theorem 11.4.1.

Theorem 11.4.3 *Given open set $D \subset \mathbb{R}^{2n}$ and $f \in C^1(\mathbb{R} \times D, \mathbb{R}^{2n})$. Consider*

$$\dot{z} = f(t, z), \quad (t, z) \in \mathbb{R} \times D.$$

Let φ^t be the flow induced by the system. Then the system is a locally Hamiltonian system if and only if φ^t is symplectic for sufficiently small $|t|$.

Proof. Assume the system is a locally Hamiltonian system. For any $z_0 \in D$, there exists neighborhood U of z_0, neighborhood $(-\tau, \tau)$ of 0, and a scalar function H on $(-\tau, \tau) \times U$ such that $f = J\nabla_z H$ on $(-\tau, \tau) \times U$. Moreover, H is of class C^2 with respect to z, so that $D^2 H$ is symmetric.

Fix $\xi \in U$ and let $Z(t) = \frac{\partial}{\partial \xi}\varphi^t(\xi)$, then $Z(0) = I_{2n}$ and Z solves the variational equation

$$\dot{Z} = J D^2 H(t, \varphi^t(\xi))Z.$$

This is valid for $|t|$ sufficiently small and ξ sufficiently close to z_0. Therefore,

$$\frac{d}{dt}\left(Z^T J Z\right) = \dot{Z}^T J Z + Z^T J \dot{Z}$$

$$= \left(Z^T D^2 H(t, \varphi^t(\xi))J^T\right) J Z + Z^T J \left(J D^2 H(t, \varphi^t(\xi))Z\right) = 0.$$

Again, we have used $J^T = J^{-1} = -J$.

Since $Z^T J Z = J$ at $t = 0$, we conclude that the identity $Z^T J Z = J$ holds for $|t|$ sufficiently small. Thus φ^t is symplectic for small $|t|$.

To prove the converse, assume φ^t is symplectic for $t \in (-\tau, \tau)$. Let $Z(t) = \frac{\partial}{\partial \xi}\varphi^t(\xi)$ be as above. Then $Z(t) \in \mathrm{Sp}(2n, \mathbb{R})$ for small $|t|$, and it solves the variational equation

$$\dot{Z} = Df(t, \varphi^t(\xi))Z.$$

By differentiating the identity

$$Z^T J Z = J$$

with respect to t, we find

$$\dot{Z}^T J Z + Z^T J \dot{Z} = 0,$$

and thus

$$Z^T Df(t, \varphi^t(\xi))^T JZ + Z^T JDf(t, \varphi^t(\xi))Z = 0.$$

This implies $JDf(t, \varphi^t(\xi))$ is symmetric. Since $\xi \in D$ is arbitrary, we conclude that $JDf(t, z)$ is symmetric for arbitrary $z \in D$ and small $|t|$. By Theorem 5.4.2, for any $z_0 \in D$, there is a neighborhood of z_0 and a function $H(t, z)$ with z in this neighborhood such that $-Jf(t, z) = \nabla H(t, z)$. This shows that the system $\dot{z} = f(t, z) = J\nabla H(t, z)$ is a locally Hamiltonian system.   □

Example 11.4.1 Consider the harmonic oscillator in Example 5.2.1 and Example 11.3.1 with Hamiltonian $H(x, y) = \frac{k}{2}x^2 + \frac{1}{2m}y^2$: $(k, m > 0)$

$$\dot{x} = \frac{y}{m},$$
$$\dot{y} = -kx.$$

Define $\Phi : \mathbb{R}^2 \setminus \{0\} \to \mathbb{R}_+ \times \mathbb{R}/2\pi\mathbb{Z}$ by $\Phi(x, y) = (I, \phi)$, where

$$I = \frac{1}{2}\left(\sqrt{km}\,x^2 + \frac{y^2}{\sqrt{km}}\right), \quad \phi = \tan^{-1}\left(\frac{1}{\sqrt{km}}\frac{y}{x}\right).$$

This transformation is symplectic because

$$\frac{\partial(I, \phi)}{\partial(x, y)} = \begin{pmatrix} \sqrt{km}\,x & \frac{1}{\sqrt{km}}y \\ \frac{-\sqrt{\frac{k}{m}}y}{kx^2 + y^2/m} & \frac{\sqrt{\frac{k}{m}}x}{kx^2 + y^2/m} \end{pmatrix} \in \mathrm{SL}(2, \mathbb{R}) = \mathrm{Sp}(2, \mathbb{R}).$$

According to Theorem 11.4.1, the system in terms of new variables (I, ϕ) is a locally Hamiltonian system. Let $\tilde{H}$ be the Hamiltonian H in terms of I, ϕ, then

$$\tilde{H} = \tilde{H}(I) = \sqrt{\frac{k}{m}}I$$

depends only on I. Thus the original system becomes

$$\dot{I} = \frac{\partial \tilde{H}}{\partial \phi} = 0,$$
$$\dot{\phi} = -\frac{\partial \tilde{H}}{\partial I} = -\sqrt{\frac{k}{m}}.$$

Therefore solutions are $I = \text{constant}$ and $\phi(t) = \phi(0) - \sqrt{k/m}\,t$.

The symplectic change of variables $(x, y) \to (I, \phi)$ is the composition of a linear symplectic transformation Φ_1 and the *symplectic polar coordinates* Φ_2:

$$\begin{pmatrix} x \\ y \end{pmatrix} \xrightarrow{\Phi_1} \begin{pmatrix} X \\ Y \end{pmatrix} = \begin{pmatrix} \sqrt{km}\, x \\ \frac{1}{\sqrt{km}}\, y \end{pmatrix} \xrightarrow{\Phi_2} \begin{pmatrix} I \\ \phi \end{pmatrix} = \begin{pmatrix} \frac{1}{2}(X^2 + Y^2) \\ \tan^{-1}(Y/X) \end{pmatrix}.$$

In general, if symplectic coordinates $(I_1, \cdots, I_n, \phi_1, \cdots, \phi_n)$ were selected for the Hamiltonian system (H) such that $I_1, \cdots, I_n$ are first integrals, then solutions can be obtained by direct integration. We call such coordinates *action-angle variables*. Coordinates $I_1, \cdots, I_n$ are action variables and $\phi_1, \cdots, \phi_n$ are angle variables.

Let $\mathbb{T}^n = (\mathbb{R}/2\pi\mathbb{Z})^n$ be the n-dimensional torus. If in terms of the action-angle variables $(I, \phi) = (I_1 \cdots, I_n, \phi_1, \cdots, \phi_n) \in \mathbb{R}^n \times \mathbb{T}^n$ the Hamiltonian is a smooth function of the form $H_0(I)$, then as in Example 11.4.1 we have equations of motion

$$\dot{I}_i = 0, \quad \dot{\phi}_i = -\frac{\partial H_0}{\partial I_i} =: \omega_i(I).$$

Solution with initial data (I^0, ϕ^0) is then given by

$$I = I^0, \quad \phi(t) = \phi^0 + \omega(I^0)t.$$

Tori $\{I^0\} \times \mathbb{T}^n$ are invariant and the dynamics is completely determined by the *frequency vector* $\omega(I) = (\omega_1(I), \cdots, \omega_n(I))$.

Adding a smooth small perturbation term $H_\varepsilon(I, \phi)$ to the Hamiltonian would typically destroy many invariant tori, resulting chaotic regions in the phase space. A classic result by Kolmogorov (1954), Arnold (1963), and Moser (1962) states that tori with "strongly non-resonant" frequency ω will persist. More precisely, it means that for some $\alpha > 0$, $\tau > n - 1$, the frequency vector ω satisfies

$$|\langle k, \omega \rangle| \geq \frac{\alpha}{|k|^\tau} \quad \text{for all } k \in \mathbb{Z}^n \setminus \{0\}.$$

This is also known as the *Diophantine condition*. Figure 11.2 exhibits how invariant tori ("islands" in the figure) coexist with chaotic regions. The study of this phenomenon has been developed into an important branch in dynamical systems known as the *KAM theory*. Interested readers are referred to [L; P].

11.5   Generating Functions and Hamilton-Jacobi's Method

We have seen how a Hamiltonian system can be simplified by choosing proper symplectic coordinates. However, it is in general a nontrivial task to find ideal symplectic coordinates. In this section we introduce Hamilton-Jacobi's method for finding such coordinates.

Let us begin with the simplest case: $n = 1$. By writing $\mathbb{R}^2 = \mathbb{R}_x \times \mathbb{R}_y$, we mean the first coordinate is x and the second is y. Given a simply connected region $D \subset \mathbb{R}^2$ with positively oriented boundary ∂D. By Green's theorem,

$$\text{Area}(D) = \oint_{\partial D} x \, dy = -\oint_{\partial D} y \, dx = \iint_D dx dy.$$

The sign of area changes when the orientation is reversed.

As differential of coordinate functions, dx is the linear map sending (ξ, η) to ξ, dy is the linear map sending (ξ, η) to η. More generally, for any smooth function $f : \mathbb{R}^2 \to \mathbb{R}$, its differential $df_{(x_0,y_0)}$ at (x_0, y_0) is the linear mapping

$$df_{(x_0,y_0)}(\xi, \eta) = \frac{\partial f}{\partial x}(x_0, y_0)\xi + \frac{\partial f}{\partial y}(x_0, y_0)\eta$$

$$= \frac{\partial f}{\partial x}(x_0, y_0)dx(\xi, \eta) + \frac{\partial f}{\partial y}(x_0, y_0)dy(\xi, \eta).$$

For brevity, (x_0, y_0) it is often suppressed and the above equation is written

$$df = \frac{\partial f}{\partial x} \, dx + \frac{\partial f}{\partial y} \, dy.$$

We may also write the linear map $df_{(x_0,y_0)}$ as $Df(x_0, y_0)$ (the row vector $(\partial f/\partial x, \partial f/\partial y)$) or $\nabla f(x_0, y_0)$ (the column vector $(\partial f/\partial x, \partial f/\partial y)^T$) product with $\binom{dx}{dy}$.

For any continuous real-valued function g and smooth path $\gamma : [a, b] \to \mathbb{R}^2$, we have the Riemann-Stieltjes integral over γ

$$\int_\gamma g \, df = \lim_{\max_i |t_{i+1}-t_i| \to 0} \sum_k g(\gamma(t_k))df_{\gamma(t_k)}(\gamma(t_{k+1}) - \gamma(t_k))$$

$$= \lim_{\max_i |t_{i+1}-t_i| \to 0} \sum_k g(\gamma(t_k))df_{\gamma(t_k)}(\gamma'(t_k))(t_{k+1} - t_k).$$

Here $\{a = t_0 < t_1 < \cdots < t_n = b\}$ is a partition of $[a, b]$. Linear combinations of mappings of the form $g \, df$ are called (*differential*) 1-*forms*. The Riemann-Stieltjes integral above defines the integration of 1-forms over paths.

If $\Phi : \mathbb{R}^2 = \mathbb{R}_x \times \mathbb{R}_y \to \mathbb{R}^2 = \mathbb{R}_X \times \mathbb{R}_Y$ is symplectic, denote it by $\Phi(x, y) = (X, Y)$, then area-preserving property of $D\Phi$ implies

$$\text{Area}(\Phi(D)) = \oint_{\partial\Phi(D)} X\, dY = -\oint_{\partial\Phi(D)} Y\, dX = \iint_{\Phi(D)} dX dY$$
$$= \iint_D \det\left(D\Phi^{-1}\right) dx dy = \iint_D dx dy = \text{Area}(D).$$

Take $\oint_{\partial D} x\, dy$ and $\oint_{\partial\Phi(D)} X\, dY$, for example, we may rewrite the later in terms of x, y:

$$\oint_{\partial\Phi(D)} X\, dY = \oint_{\partial D} X(x, y) \frac{\partial Y}{\partial x} dx + X(x, y) \frac{\partial Y}{\partial y} dy.$$

Combining with $\oint_{\partial D} x\, dy$, we find the line integral

$$\int_\gamma x\, dy - X\, dY$$

is independent of path, which implies the local existence of function $S(y, Y)$ such that

$$dS = x\, dy - X\, dY.$$

Symplectic coordinates $\Phi = (X, Y)$ can be generated by various choices of S. For example,

$$S(y, Y) = yY \Rightarrow (x, -X) = (Y, y)$$
$$\Rightarrow \Phi(x, y) = (-y, x),$$
$$S(y, Y) = \frac{1}{2} y^2 \cot Y \Rightarrow (x, -X) = \left(y \cot Y, -\frac{y^2}{2} \csc^2 Y\right)$$
$$\Rightarrow \Phi(x, y) = \left(\frac{x^2 + y^2}{2}, \tan^{-1}\left(\frac{y}{x}\right)\right).$$

The second one is exactly the symplectic polar coordinates.

We may take other area integrals in discussions above, say $\oint_{\partial D} x\, dy$ and $\oint_{\partial\Phi(D)} Y\, dX$, for which case we have

$$\frac{\partial S}{\partial X} dX + \frac{\partial S}{\partial y} dy = dS = Y\, dX + x\, dy.$$

Symplectic coordinates $\Phi = (X, Y)$ can be also generated by such S. For example,

$$S(X, y) = Xy \Rightarrow (Y, x) = (y, X) \Rightarrow \Phi(x, y) = (x, y),$$

$$S(X, y) = \frac{y}{2}\sqrt{2X - y^2} + X \tan^{-1}\left(\frac{y}{\sqrt{2X - y^2}}\right)$$

$$\Rightarrow (Y, x) = \left(\tan^{-1}\left(\frac{y}{\sqrt{2X - y^2}}\right), \sqrt{2X - y^2}\right)$$

$$\Rightarrow \Phi(x, y) = \left(\frac{x^2 + y^2}{2}, \tan^{-1}\left(\frac{y}{x}\right)\right).$$

The first one is simply the identity map, and the second one is again the symplectic polar coordinates. It is not apparent how the second S was obtained and what advantages we gain by finding such an S. In below, we will introduce the Hamilton-Jacobi method and answer these questions.

For more general n, write $\mathbb{R}^{2n} = \mathbb{R}_{x_1} \times \cdots \mathbb{R}_{x_n} \times \mathbb{R}_{y_1} \times \cdots \mathbb{R}_{y_n}$, we can fix k and apply discussion above to $x_k y_k$-planes $\mathbb{R}^2 = \mathbb{R}_{x_k} \times \mathbb{R}_{y_k}$. Coordinates (x, y) and (X, Y) above are replaced with (x_k, y_k) and (X_k, Y_k), Area(D) in our discussions are replaced with projections of a parallelogram in $\mathbb{R}^{2n}$ to $x_k y_k$-planes. We can then generalize arguments above to surfaces in $\mathbb{R}^{2n}$ that can be approximated by disjoint union of parallelograms.

Coordinate functions $x = (x_1, \cdots, x_n)^T$, $y = (y_1, \cdots, y_n)^T$ in (H) are written as column vectors, so are their differentials $dx = (dx_1, \cdots, dx_n)^T$, $dy = (dy_1, \cdots, dy_n)^T$. The 1-form $x_1 dy_1 + \cdots + x_n dy_n$ is supposed to be written $x^T dy$ as matrix multiplication, but it is conventionally written $x\, dy$. Same for $y\, dx$.

Theorem 11.5.1 *A C^2 transformation $\Phi(x, y) = (X, Y)$ is symplectic if and only if any one of the followings holds:*

(a) *There exists locally a C^2 function $S_1(x, X)$ such that*

$$dS_1 = y\, dx - Y\, dX.$$

(b) *There exists locally a C^2 function $S_2(x, Y)$ such that*

$$dS_2 = y\, dx + X\, dY.$$

(c) *There exists locally a C^2 function $S_3(X, y)$ such that*

$$dS_3 = Y\, dX + x\, dy.$$

(d) *There exists locally a C^2 function $S_4(y, Y)$ such that*

$$dS_4 = x\, dy - X\, dY.$$

Real-valued functions $S_1, \cdots, S_4$ in this theorem are called *generating functions* for the symplectic transformation Φ.

Proof. We only prove the equivalence of symplecticity of Φ and (a) since all others are similar.

Write the Jacobian matrix of Φ as four $n \times n$ block matrices:

$$\frac{\partial(X,Y)}{\partial(x,y)} = \begin{pmatrix} \frac{\partial X}{\partial x} & \frac{\partial X}{\partial y} \\ \frac{\partial Y}{\partial x} & \frac{\partial Y}{\partial y} \end{pmatrix}.$$

It is symplectic if and only if

$$\frac{\partial(X,Y)}{\partial(x,y)}^T J \frac{\partial(X,Y)}{\partial(x,y)} = J,$$

or equivalently,

$$\begin{pmatrix} \frac{\partial X}{\partial x}^T \frac{\partial Y}{\partial x} - \frac{\partial Y}{\partial x}^T \frac{\partial X}{\partial x} & \frac{\partial X}{\partial x}^T \frac{\partial Y}{\partial y} - \frac{\partial Y}{\partial x}^T \frac{\partial X}{\partial y} \\ \frac{\partial X}{\partial y}^T \frac{\partial Y}{\partial x} - \frac{\partial Y}{\partial y}^T \frac{\partial X}{\partial x} & \frac{\partial X}{\partial y}^T \frac{\partial Y}{\partial y} - \frac{\partial Y}{\partial y}^T \frac{\partial X}{\partial y} \end{pmatrix} = \begin{pmatrix} 0 & I_n \\ -I_n & 0 \end{pmatrix}.$$

It is easy to see that this matrix identity holds if and only if the $2n \times 2n$ matrix

$$\begin{pmatrix} \frac{\partial X}{\partial x}^T \frac{\partial Y}{\partial x} & \frac{\partial X}{\partial x}^T \frac{\partial Y}{\partial y} - I_n \\ \frac{\partial X}{\partial y}^T \frac{\partial Y}{\partial x} & \frac{\partial X}{\partial y}^T \frac{\partial Y}{\partial y} \end{pmatrix}$$

is symmetric.

Since Φ is of class C^2, the Jacobian matrix $\frac{\partial^2 X_i}{\partial(x,y)^2}$ of each X_i is a $2n \times 2n$ symmetric matrix. Therefore, the matrix above is symmetric if and only if

$$D \begin{pmatrix} \frac{\partial X}{\partial x}^T Y - y \\ \frac{\partial X}{\partial y}^T Y \end{pmatrix} = \begin{pmatrix} \frac{\partial X}{\partial x}^T \frac{\partial Y}{\partial x} & \frac{\partial X}{\partial x}^T \frac{\partial Y}{\partial y} - I_n \\ \frac{\partial X}{\partial y}^T \frac{\partial Y}{\partial x} & \frac{\partial X}{\partial y}^T \frac{\partial Y}{\partial y} \end{pmatrix} + \sum_{i=1}^{n} Y_i \frac{\partial^2 X_i}{\partial(x,y)^2} \tag{11.16}$$

is symmetric. We leave it as exercise to verify this identity (Exercise 11.11).

By Theorem 5.4.2, the matrix above is symmetric if and only if there exists locally a C^2 function $S_1(x,y)$ such that

$$\frac{\partial S_1}{\partial x} = -\left(\frac{\partial X}{\partial x}^T Y - y \right)^T, \quad \frac{\partial S_1}{\partial y} = -\left(\frac{\partial X}{\partial y}^T Y \right)^T.$$

This can be alternatively written

$$dS_1 = \left(y^T - Y^T \frac{\partial X}{\partial x} \right) dx - \left(Y^T \frac{\partial X}{\partial y} \right) dy$$

$$= y\,dx - Y^T \left(\frac{\partial X}{\partial x}\,dx + \frac{\partial X}{\partial y}\,dy \right)$$

$$= y\,dx - Y\,dX.$$

This completes the proof. □

As observed in the previous section, if symplectic coordinates (X, Y) are action-angle variables, where $X_1, \cdots, X_n$ are action variables, then the new Hamiltonian $\tilde{H}$ is a function of X, and solutions can be obtained by direct integration. Locally (x, y) are functions of (X, Y) and vice versa. The idea is to find a generating function of the form $S(x, X)$ so that

$$H\left(t, x, \frac{\partial S}{\partial x} \right) = \tilde{H}(X). \qquad (11.17)$$

This partial differential equation is called a *Hamilton-Jacobi equation*. One the left-hand side, variable x is expressed as $x(X, Y)$ after differentiating S. If the generating function is of the form $S(X, y)$, then

$$H\left(t, \frac{\partial S}{\partial y}, y \right) = \tilde{H}(X). \qquad (11.18)$$

If succeed, the Hamiltonian in terms of new coordinates do not depend on Y (call such coordinates *cyclic*), then the Hamiltonian system become

$$\dot{X} = 0,$$

$$\dot{Y} = -\frac{\partial \tilde{H}}{\partial X}.$$

Then $X = (X_1, \cdots, X_n)$ are first integrals, and we can solve the system as in Section 11.4.

Hamilton-Jacobi's method is to find action-angle variables for the Hamiltonian system (H) by finding suitable generating functions through solving the Hamilton-Jacobi equation. This approach is somewhat surprising because solving the Hamilton-Jacobi equation does not sound easier than solving the original system directly. However, in many practical applications, finding solutions for the Hamilton-Jacobi equation are indeed easier. We illustrate Hamilton-Jacobi's method in the following examples.

Example 11.5.1 Consider Hamiltonian system with one degree of freedom and with Hamiltonian

$$H(x, y) = \frac{1}{2}y^2 + V(x).$$

The Hamiltonian is assumed to be of class C^2 on open domain $D \subset \mathbb{R}^2$. We seek action-angle variables (X, Y), with action X and angle Y. Suppose the Hamiltonian H expressed in terms of (X, Y) is $\tilde{H} = \tilde{H}(X)$.

Consider generating function of the form $S(x, X)$, case (a) in Theorem 11.5.1. From $y = \partial S / \partial x$ and the Hamilton-Jacobi equation (11.17), we have

$$\frac{1}{2} \left(\frac{\partial S}{\partial x} \right)^2 + V(x) = \tilde{H}(X),$$

$$\dot{x} = \frac{\partial H}{\partial y} = y = \pm\sqrt{2}\sqrt{\tilde{H}(X) - V(x)}.$$

Thus, with fixed X_0 and suitable x_0 which depends on X_0,

$$S(x, X_0) = \pm\sqrt{2} \int_{x_0}^{x} \sqrt{\tilde{H}(X_0) - V(s)} \, ds, \tag{11.19}$$

$$dt = \frac{dx}{\pm\sqrt{2}\sqrt{\tilde{H}(X) - V(x)}}. \tag{11.20}$$

Relate X_0 and x_0 by $\tilde{H}(X_0) = h_0 = H(x_0, 0) = V(x_0)$ and assume $\frac{\partial}{\partial x} H(x_0, 0) \neq 0$. By the implicit function theorem, the x-intercept $(\bar{x}, 0)$ of H's level curve is locally a function of X satisfying $\bar{x}(X_0) = x_0$. Then the integral (11.19) yields a locally well-defined generating function S:

$$S(x, X) = \pm\sqrt{2} \int_{\bar{x}}^{x} \sqrt{\tilde{H}(X) - V(s)} \, ds. \tag{11.21}$$

The angle variable Y is then obtained by

$$Y = -\frac{\partial S}{\partial X} = \pm\sqrt{2}\sqrt{\tilde{H}(X) - V(\bar{x})} \, \bar{x}'(X) \mp \frac{1}{\sqrt{2}} \int_{\bar{x}}^{x} \frac{\tilde{H}'(X)}{\sqrt{\tilde{H}(X) - V(s)}} \, ds.$$

The first term on the right-hand side is zero by the definition of $\bar{x}$. Thus

$$Y = \mp\frac{1}{\sqrt{2}} \int_{\bar{x}}^{x} \frac{\tilde{H}'(X)}{\sqrt{\tilde{H}(X) - V(s)}} \, ds = -\tilde{H}'(X) t. \tag{11.22}$$

The last identity uses (11.20) and t is the transfer time from $\bar{x}$ to x.

Consider the case of harmonic oscillator, with $V(x) = x^2/2$. Define $X = (x^2 + y^2)/2$ (the action variable), then $\tilde{H}$ is simply the identity map and $\bar{x} = \pm\sqrt{2X}$. Note that $\frac{\partial}{\partial x}H(x_0, 0) \neq 0$ whenever $x_0 \neq 0$. Consider (x, y) in the first quadrant, for instance, then S in (11.21) is nonnegative and $\bar{x} = \sqrt{2X}$. A closed form for the generating function $S(x, X)$ in (11.21) can be found

$$S(x, X) = \int_{\sqrt{2X}}^{x} \sqrt{2X - s^2}\, ds = \frac{x}{2}\sqrt{2X - x^2} + X \tan^{-1}\left(\frac{x}{\sqrt{2X - x^2}}\right).$$

The integral (11.22) becomes

$$Y = -\frac{1}{\sqrt{2}} \int_{\sqrt{2X}}^{x} \frac{1}{\sqrt{X - s^2/2}}\, ds$$

$$= -\tan^{-1}\left(\frac{x}{\sqrt{2X - x^2}}\right) - \frac{\pi}{2} = \tan^{-1}\left(\frac{y}{x}\right)$$

as expected. We have seen in Example 11.4.1 that the angle variable $Y(t)$ is linear in t, which also follows from (11.22).

For more general potential V, integrals (11.21) and (11.22) may not have closed forms, but we can still describe generating functions and angle-angle variables in integral form as outlined in this example.

Example 11.5.2 Consider Hamiltonian system with two degrees of freedom and with Hamiltonian

$$H(x_1, x_2, y_1, y_2) = \frac{x_2(x_1^2 + y_1^2)}{x_1\, y_1\, y_2}, \quad x_1, y_1, y_2 \neq 0.$$

By observing symmetry of the Hamiltonian, it is natural to use new variables $u_1 = x_1/y_1$ and $u_2 = x_2/y_2$ in place of either (x_1, x_2) or (y_1, y_2). Using Hamilton-Jacobi's method is a less direct approach, but we shall use this simple example to illustrate Hamilton-Jacobi's method for systems with higher degrees of freedom.

As in the previous example, we seek action-angle variables (X, Y), with action variables $X = (X_1, X_2)$ and angle variables $Y = (Y_1, Y_2)$. In terms of these new variables, the Hamiltonian H becomes $\tilde{H} = \tilde{H}(X_1, X_2)$.

Consider generating function of the form $S(X_1, X_2, y_1, y_2)$, case (c) in Theorem 11.5.1. From $x_i = \partial S/\partial y_i$, $i \in \{1, 2\}$, and the Hamilton-Jacobi equation (11.18), we have

$$\frac{\partial S}{\partial y_2}\left(\left(\frac{\partial S}{\partial y_1}\right)^2 + y_1^2\right) = \tilde{H}(X_1, X_2)\left(\frac{\partial S}{\partial y_1}\right)y_1 y_2.$$

A trick of finding S is by separating variables:

$$\frac{1}{y_2}\frac{\partial S}{\partial y_2} = \tilde{H}(X_1, X_2)\frac{\left(\frac{\partial S}{\partial y_1}\right)y_1}{\left(\frac{\partial S}{\partial y_1}\right)^2 + y_1^2}.$$

Equations

$$\frac{1}{y_2}\frac{\partial S}{\partial y_2} = c_2, \qquad \left(\frac{\partial S}{\partial y_1}\right)^2 + y_1^2 = c_1 y_1 \frac{\partial S}{\partial y_1}$$

suggest setting S to be quadratic in both y_1 and y_2.

Setting

$$S(X_1, X_2, y_1, y_2) = y_1^2 X_1 + y_2^2 X_2,$$

then

$$x_1 = \frac{\partial S}{\partial y_1} = 2y_1 X_1,$$

$$x_2 = \frac{\partial S}{\partial y_2} = 2y_2 X_2,$$

$$Y_1 = \frac{\partial S}{\partial X_1} = y_1^2,$$

$$Y_1 = \frac{\partial S}{\partial X_2} = y_2^2.$$

Therefore, the symplectic transformation $\Phi = (X_1, X_2, Y_1, Y_2)$ is given by

$$(X_1, X_2, Y_1, Y_2) = \left(\frac{x_1}{2y_1}, \frac{x_2}{2y_2}, y_1^2, y_2^2\right).$$

In terms of these symplectic coordinates, we find

$$\tilde{H} = \tilde{H}(X_1, X_2) = 2X_2\left(2X_1 + \frac{1}{2X_1}\right).$$

From the Hamiltonian system given by $\tilde{H}$, we find X_1 and X_2 are constants, say $X_1 = c_1$ and $X_2 = c_2$, and that

$$x_1 = 2c_1 y_1, \qquad x_2 = 2c_2 y_2.$$

Also, y_1 and y_2 are given by $y_1^2 = Y_1$, $y_2^2 = Y_2$, and from equations for $\dot{Y}_1$, $\dot{Y}_2$:

$$y_1^2 = -2c_2\left(2 - \frac{1}{2c_1^2}\right)t + d_1, \qquad y_2^2 = -\left(2c_1 + \frac{1}{2c_1}\right)t + d_2$$

for some constants d_1, d_2.

Example 11.5.3 We finish the chapter by considering the *two-center problem*, also known as *Euler's three-body problem*. It concerns the motion of a point mass moving in space under the Newtonian gravitational attraction due to two fixed bodies, or a point charge in the space under the electrostatic force according to Coulomb's law. In celestial mechanics, this problem has been used to model the two-body problem with one oblate primary.

For simplicity, we consider the planar problem and fixed bodies (called centers) with masses (or charges) μ_1 and μ_2 placed at $\mathbf{c}_1 = (-c, 0)$ and $\mathbf{c}_2 = (c, 0)$, respectively, where $c > 0$ is constant. Let (x_1, x_2) be the position of the moving unit mass (or charge). Equations of motion for the free particle is the Hamiltonian system with Hamiltonian

$$H(x_1, x_2, y_1, y_2) = \frac{1}{2}(y_1^2 + y_2^2) - \frac{\mu_1}{r_1} - \frac{\mu_2}{r_2},$$

where

$$r_1 = \sqrt{(x_1 + c)^2 + x_2^2}, \quad r_2 = \sqrt{(x_1 - c)^2 + x_2^2}$$

are distances from the free particle to fixed centers. When one of μ_1, μ_2 is zero, it becomes the Kepler problem and we have seen in Example 11.3.7 that conic sections with a focus at the nonzero μ_i are trajectories of solutions. Both branches of hyperbola are trajectories if μ_i is allowed to be negative, which occurs when charged particles are been considered.

We begin by choosing the "correct" coordinates for this problem. It is motivated from the following curious observation: for arbitrary μ_1, $\mu_2 \in \mathbb{R}$, conic sections with foci $(\pm c, 0)$ are trajectories of solutions for the two-center problem.

We divide this observation into two parts. Let $q_1(t)$ be a solution for the force field $\mathbf{F}_1$ due to μ_1 and $q_2(t)$ be a solution for the force field $\mathbf{F}_2$ due to μ_2, and we assume that they parametrize the same curve γ. Assume q_1 and q_2 have the same orientation. Consider a new parametrization q for γ whose speed $|\dot{q}|$ at time t is given by

$$|\dot{q}(t)|^2 = |\dot{q}_1(t_1)|^2 + |\dot{q}_2(t_2)|^2,$$

where t_1 and t_2 are chosen so that $q(t) = q_1(t_1) = q_2(t_2)$.

(I). Assume $\alpha(t)$ is a regular parametrization for γ and parameter s denotes the arc length, then $s(t) = \int_{t_0}^{t} |\dot{\alpha}(t)| dt$ for some t_0, and so

$$\frac{1}{2} \frac{d}{ds} |\dot{\alpha}(t)|^2 = \ddot{\alpha}(t) \cdot \vec{\tau},$$

where $\vec{\tau} = \dot{\alpha}/|\dot{\alpha}|$ is the unit tangent vector in the direction of $\dot{\alpha}(t)$. Therefore, since q_1 and q_2 have the same orientation, with $q(t) = q_1(t_1) = q_2(t_2)$ we have

$$\ddot{q}(t) \cdot \vec{\tau} = (\ddot{q}_1(t_1) + \ddot{q}_2(t_2)) \cdot \vec{\tau}$$

$$= -\left(\frac{\mu_1(q_1(t_1) - \mathbf{c}_1)}{|q_1(t_1) - \mathbf{c}_1|^3} + \frac{\mu_2(q_2(t_2) - \mathbf{c}_2)}{|q_2(t_2) - \mathbf{c}_2|^3} \right) \cdot \vec{\tau}$$

$$= (\mathbf{F}_1 + \mathbf{F}_2)(q(t)) \cdot \vec{\tau}.$$

(II). Let α be as above, denote by R the radius of curvature, then R is given by $|\dot{\alpha}|^3/|\ddot{\alpha} \times \dot{\alpha}|$, which is independent of parametrizations. Thus at $q(t) = q_1(t_1) = q_2(t_2)$ we have

$$|\ddot{q}(t) \times \vec{\tau}| = \frac{|\ddot{q}(t) \times \dot{q}(t)|}{|\dot{q}(t)|} = \frac{1}{R}|\dot{q}(t)|^2 = \frac{1}{R}\left(|\dot{q}_1(t_1)|^2 + |\dot{q}_2(t_2)|^2\right)$$

$$= \frac{|\ddot{q}_1(t_1) \times \dot{q}_1(t_1)|}{|\dot{q}_1(t_1)|} + \frac{|\ddot{q}_2(t_2) \times \dot{q}_2(t_2)|}{|\dot{q}_2(t_2)|}$$

$$= |\ddot{q}_1(t_1) \times \vec{\tau}| + |\ddot{q}_2(t_2) \times \vec{\tau}|$$

$$= |(\mathbf{F}_1 + \mathbf{F}_2)(q(t)) \times \vec{\tau}|.$$

The last line follows since q_1 and q_2 have the same orientation.

Putting together (I) and (II), we conclude that $q(t)$ is a solution for the combined forced field $\mathbf{F}_1 + \mathbf{F}_2$. In case one center is attractive and the other is repulsive, one may modify discussions above by considering q_1 and q_2 with opposite orientations.

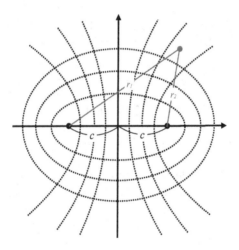

Fig. 11.3   The two-center problem.

Observation above is a special case of Bonnet's theorem in classical mechanics (see Exercise 11.17). For the two-center problem, it suggests that we rewrite the Hamiltonian in terms of elliptic coordinates (see Fig. 11.3):

$$\xi_1 = \frac{r_1 + r_2}{2c}, \quad \xi_2 = \frac{r_1 - r_2}{2c}.$$

It is routine to check the followings (Exercise 11.16):

(1) $|\xi_2| \le 1 \le \xi_1$.

(2) $x_1 = c\xi_1\xi_2$, $x_2 = \pm c\sqrt{(\xi_1^2 - 1)(1 - \xi_2^2)}$.

(3) $\dot{x}_1 = c(\dot{\xi}_1\xi_2 + \xi_1\dot{\xi}_2)$, $\dot{x}_2 = \pm c\dfrac{\xi_1\dot{\xi}_1(1 - \xi_2^2) - \xi_2\dot{\xi}_2(\xi_1^2 - 1)}{\sqrt{(\xi_1^2 - 1)(1 - \xi_2^2)}}$.

By (3), we can write the kinetic energy in terms of elliptic coordinates ξ_1, ξ_2:

(4) $\dfrac{1}{2}(\dot{x}_1^2 + \dot{x}_2^2) = \dfrac{c^2}{2}\left(\dfrac{(\xi_1^2 - \xi_2^2)\dot{\xi}_1^2}{\xi_1^2 - 1} + \dfrac{(\xi_1^2 - \xi_2^2)\dot{\xi}_2^2}{1 - \xi_2^2}\right)$.

Let L be the *Lagrangian*; i.e. kinetic energy minus the potential energy (see Exercise 11.2),

$$L(\xi_1, \xi_2, \dot{\xi}_1, \dot{\xi}_2) = \frac{c^2}{2}\left(\frac{(\xi_1^2 - \xi_2^2)\dot{\xi}_1^2}{\xi_1^2 - 1} + \frac{(\xi_1^2 - \xi_2^2)\dot{\xi}_2^2}{1 - \xi_2^2}\right) + \frac{\mu_1}{c(\xi_1 + \xi_2)} + \frac{\mu_2}{c(\xi_1 - \xi_2)}.$$

The corresponding Hamiltonian is obtained by the *Legendre transformation* (see also Exercise 11.2) $\eta_i = \partial L/\partial\dot{\xi}_i$. Then we have

$$\eta_1 = c^2\left(\frac{\xi_1^2 - \xi_2^2}{\xi_1^2 - 1}\right)\dot{\xi}_1, \quad \eta_2 = c^2\left(\frac{\xi_1^2 - \xi_2^2}{1 - \xi_2^2}\right)\dot{\xi}_2,$$

and the Hamiltonian function $H(\xi_1, \xi_2, \eta_1, \eta_2)$ is

$$H = \frac{1}{2c^2(\xi_1^2 - \xi_2^2)}\left[(\xi_1^2 - 1)\eta_1^2 + (1 - \xi_2^2)\eta_2^2\right] - \frac{\mu_1}{c(\xi_1 + \xi_2)} - \frac{\mu_2}{c(\xi_1 - \xi_2)}.$$

In order to find symplectic transformation $\Phi(\xi_1, \xi_2, \eta_1, \eta_2) = (X_1, X_2, Y_1, Y_2)$ so that $H = \tilde{H}(X_1, X_2)$, we may search for generating function of the form $S(\xi_1, \xi_2, X_1, X_2)$.

From the Hamilton-Jacobi equation,

$$2c^2(\xi_1^2 - \xi_2^2)\tilde{H}(X_1, X_2)$$
$$= (\xi_1^2 - 1)\left(\frac{\partial S}{\partial \xi_1}\right)^2 + (1 - \xi_2^2)\left(\frac{\partial S}{\partial \xi_2}\right)^2 - 2c\mu_1(\xi_1 - \xi_2) - 2c\mu_2(\xi_1 + \xi_2).$$

As in Example 11.5.2, using the trick of separating variables,

$$(\xi_1^2 - 1)\left(\frac{\partial S}{\partial \xi_1}\right)^2 - 2c(\mu_1 + \mu_2)\xi_1 - 2c^2\tilde{H}\xi_1^2$$
$$= -(1 - \xi_2^2)\left(\frac{\partial S}{\partial \xi_2}\right)^2 + 2c(\mu_1 - \mu_2)\xi_2 - 2c^2\tilde{H}\xi_2^2.$$

The first equation is independent of ξ_2 and the second equation is independent of ξ_1, so they are equal to the same constant. Setting both equations $= c_1$ and $\tilde{H} = c_2$, we obtain S expressed as elliptic integrals of the form

$$S(\xi_1, \xi_2, c_1, c_2) = \int_{\bar{\xi}_1}^{\xi_1} \sqrt{\frac{c_1 + 2c(\mu_1 + \mu_2)\xi_1 + 2c^2 c_2 \xi_1^2}{\xi_1^2 - 1}} \, d\xi_1$$
$$+ \int_{\bar{\xi}_2}^{\xi_2} \sqrt{\frac{-c_1 + 2c(\mu_1 - \mu_2)\xi_2 - 2c^2 c_2 \xi_2^2}{1 - \xi_2^2}} \, d\xi_2.$$

Taking partial derivatives with respect to ξ_1 and ξ_2 and putting back to the Hamilton-Jacobi equation, then

$$H\left(\xi_1, \xi_2, \frac{\partial S}{\partial \xi_1}, \frac{\partial S}{\partial \xi_2}\right)$$

is indeed independent of ξ_1 and ξ_2, as desired in (11.17).

As our final remark, this two-center problem can be also solved by using another set of elliptic coordinates (ξ, η), which are defined by

$$x_1 = c \cosh \xi \cos \eta, \quad x_2 = c \sinh \xi \sin \eta.$$

Using the corresponding Lagrangian and the same trick of separating variables, ξ and η can be expressed as functions of a parameter given by some elliptic integrals. See [Wh] for details.

11.6 Exercises

Exercise 11.1 The double pendulum consists of pendulum attached to the end of another pendulum. Assume for simplicity that both pendulums have unit length and a unit mass is placed at the end of each pendulum. Consider planar motions and let (x_1, y_1) be the position of the first pendulum, (x_2, y_2) be the position of the second pendulum attached to the first pendulum. The kinetic and potential energies are

$$K = \frac{1}{2}(\dot{x}_1^2 + \dot{y}_1^2) + \frac{1}{2}(\dot{x}_2^2 + \dot{y}_2^2), \quad V = y_1 + y_2.$$

Let θ_1 be the angle of the first pendulum from equilibrium position, and θ_2 be the angle of the second pendulum from equilibrium position.

(1) Write equations of motion as a Hamiltonian system using variables θ_1, θ_2.
(2) Show that the system can be written as a linearly coupled oscillation when the oscillation is small (i.e. θ_1, θ_2 are close to 0).

Exercise 11.2 Let $L(q, \dot{q}, t)$ be a smooth real-valued function on $D \times \mathbb{R}^n \times \mathbb{R}$, $D \subset \mathbb{R}^d$ is open. The *Euler-Lagrange equation* for L is

$$\frac{d}{dt}\frac{\partial L}{\partial \dot{q}} = \frac{\partial L}{\partial q}.$$

We call L the *Lagrangian* for this equation.

(1) Find suitable Lagrangians to rewrite examples in Section 11.1 as Euler-Lagrangian equations.

(2) Assume the second derivative $\partial^2 L / \partial \dot{q}^2$ is positive definite. Transform the Euler-Lagrange equation to a Hamiltonian system by using new variables (q, p) and Hamiltonian function $H = p \cdot q - L$, where $p = \partial L / \partial \dot{q}$, called *generalized momentum*. This change of variables is called the *Legendre transformation*.

(3) The condition on $\partial^2 L / \partial \dot{q}^2$ in (2) implies L is strictly convex in $\dot{q}$. Extend the definition of Legendre transformations to general convex functions.

Exercise 11.3 The motion $z = (x, y)$ of a small asteroid or satellite in a gravitational field generated by two celestial bodies moving along circular orbits can be written

$$\begin{cases} \ddot{x} - 2\dot{y} = -\dfrac{\partial V}{\partial x}, \\ \ddot{y} + 2\dot{y} = -\dfrac{\partial V}{\partial y}, \end{cases}$$

were

$$V(x, y) = \frac{x^2 + y^2}{2} + \frac{1 - \mu}{\sqrt{(x + \mu)^2 + y^2}} + \frac{\mu}{\sqrt{(x - 1 + \mu)^2 + y^2}}.$$

Here μ, $1 - \mu > 0$ are masses of the two celestial bodies (e.g. Sun and Jupiter).

(1) Write the system as an Euler-Lagrange equation by finding suitable Lagrangian, and write the system as a Hamiltonian system by finding suitable Hamiltonian.

(2) Verify that the following is a first integral (known as the *Jacobi integral*)

$$h = \frac{1}{2}(\dot{x}^2 + \dot{y}^2) - V(x, y).$$

Then use it to sketch regions in $\mathbb{R}^2$ on which $h \geq -V(x, y)$ for different values of h. This is known as the *Hill region*. What is the physical meaning of this region?

(3) Find critical points of the system, and show that critical points located on the x-axis are unstable.

Hint: Check $V_{xx}V_{yy} < V_{xy}^2$, and show that this condition implies eigenvalues of the linearized matrix at these critical points cannot be pure imaginary.

Exercise 11.4 Prove Theorem 11.2.1.

Exercise 11.5 Verify details of Example 11.2.1.

Exercise 11.6 Find a system of the form $\dot{z} = f(z)$, $z \in D$ and $D \subset \mathbb{R}^{2n}$ is open, such that the system is not a Hamiltonian system but is a locally Hamiltonian system.

Exercise 11.7 Verify identities (i), (ii), (iii) in Example 11.3.7.

Exercise 11.8 In Example 11.3.6 we consider a mechanical system consisting of N particles $q_1, \cdots, q_N \in \mathbb{R}^3$, and showed that conservation of two components of the angular momentum implies the conservation of the angular momentum. Assume $m_1, \cdots, m_N > 0$ and the Hamiltonian is

$$H(q_1, \cdots, q_N, p_1, \cdots, p_N) = \frac{1}{2} \sum_{k=1}^{N} \frac{p_k^2}{m_k} + V(q_1, \cdots, q_N).$$

Here the potential function V is arbitrary, $p_k = m_k \dot{q}_k$ is the linear momentum of the particle q_k. Consider linear momentum $\mathbf{P}$ and angular momentum $\mathbf{L}$ of the system:

$$\mathbf{P} = (P_1, P_2, P_3) = \sum_{k=1}^{N} p_k, \quad \mathbf{L} = (L_1, L_2, L_3) = \sum_{k=1}^{N} q_k \times p_k.$$

Generalize result in Example 11.3.6 by finding minimal subsets of $\{P_1, P_2, P_3, L_1, L_2, L_3\}$ such that conservations of terms in this subset imply conservation of all other terms.

Hint: Try $\{P_2, P_3, L_1, L_2\}$.

Exercise 11.9 Consider the symplectic group $\mathrm{Sp}(2n, \mathbb{F})$, $\mathbb{F} = \mathbb{R}$ or $\mathbb{C}$.

(1) Show that $M \in \mathrm{Sp}(2n, \mathbb{F})$ if and only if $M^T \in \mathrm{Sp}(2n, \mathbb{F})$.
(2) Show that $\mathrm{Sp}(2n, \mathbb{F})$ is a proper subgroup of $\mathrm{SL}(2n, \mathbb{F})$ if and only if $n \geq 2$.

(3) Consider $n = 2$. Let $M \in \mathbb{C}^{4 \times 4}$ be a matrix consisting of four 2×2 block matrices A, B, C, D. Namely,

$$M = \begin{pmatrix} A & B \\ C & D \end{pmatrix}.$$

Determine necessary and sufficient conditions on A, B, C, D for $M \in$ Sp$(4, \mathbb{C})$.

(4) Assume that all eigenvalues of $M \in$ Sp$(4, \mathbb{R})$ have non-zero real parts. Determine possible Jordan forms for M.

(5) Consider orthogonal matrices in Sp$(2n, \mathbb{R})$. Discuss its relation with the group of unitary matrices in $\mathbb{C}^{n \times n}$.

Exercise 11.10 An alternative proof for the fact that symplectic matrices have determinant 1 (Theorem 11.2.4), without using Pfaffian, can be obtained as follows. Given $M \in$ Sp$(2n, \mathbb{R})$.

(1) Find orthogonal matrix U and positive definite matrix P such that $M = UP$. Show that both U and P must be symplectic.
(2) Write the orthogonal symplectic matrix U as four $n \times n$ block matrices. Show that U must be of the form

$$U = \begin{pmatrix} A & B \\ -B & A \end{pmatrix}.$$

(3) Show that real matrices of the form in (b) have nonnegative determinants. Use it to conclude that $\det(M) = 1$.

Exercise 11.11 Verify the identity (11.16), and prove equivalence of sympleticity of Φ and (b), (c), (d) of Theorem 11.5.1.

Exercise 11.12 We have seen generating functions $S(x, X)$, $S(y, Y)$, and $S(X, y)$ for the symplectic polar coordinates

$$X = \frac{x^2 + y^2}{2}, \quad Y = \tan^{-1}\left(\frac{y}{x}\right).$$

Determine the generating function $S(x, Y)$ for the symplectic polar coordinates.

Exercise 11.13 Find action-angle variables for coupled oscillators (CO) given in Example 11.1.2.
Hint: Adjust variables in Example 11.3.2.

Exercise 11.14 Denote a symplectic transformation $\Phi : \mathbb{R}^4 \to \mathbb{R}^4$ by

$$\Phi(x_1, x_2, y_1, y_2) = (X_1, X_2, Y_1, Y_2).$$

Determine symplectic transformations Φ generated by the following generating functions:

(1) $S(x_1, x_2, X_1, X_2) = x_1 X_2^2 + x_2^2 X_1$.
(2) $S(X_1, X_2, y_1, y_2) = X_1 y_1 \cos y_2 + X_2 y_1 \sin y_2$.

Exercise 11.15 Consider the Hamiltonian system with Hamiltonian

$$H(x_1, x_2, y_1, y_2) = y_1^2(1 + x_1^2) + y_2^2(1 + x_2^2) + x_1.$$

Use Hamilton-Jacobi's method to find a generating function of the form $S(x_1, x_2, Y_1, Y_2)$, and determine the corresponding symplectic coordinates (X_1, X_2, Y_1, Y_2).

Exercise 11.16 Verify (1)–(4) of Example 11.5.3.

Exercise 11.17 (Bonnet's theorem) In Example 11.5.3 we proved that, for arbitrary $\mu_1, \mu_2 \in \mathbb{R}$, given a solution q_1 due to the force field $\mathbf{F}_1$ generated by μ_1 and a solution q_2 due to the force field $\mathbf{F}_2$ generated by μ_2, if q_1, q_2 parametrize the same geometric orbit, then a solution q due to the combined force field $\mathbf{F}$ parametrize the same geometric orbit. Generalize it to arbitrary number of centers and more general law of forces (not necessarily inverse square law, like Newton's law or Coulomb's law).

APPENDIX A

A.1

The proof for the local existence of I.V.P. (2.1) [CL] has two steps. First we construct a sequence of "approximate" solutions to the I.V.P. (2.1). Then we apply Ascoli-Arzela's Theorem to extract a convergent subsequence which converges to a solution of (2.1).

Definition A.1.1 *We say a piecewise C^1 function $\varphi(t)$ defined on interval I is an ε-approximate solution of I.V.P. (2.1) if*

$$|\varphi'(t) - f(t, \varphi(t))| < \varepsilon \text{ for all } t \in I$$

whenever $\varphi'(t)$ exists.

Since $(t_0, x_0) \in D$ and D is open in $\mathbb{R} \times \mathbb{R}^n$, there exist $a, b > 0$ such that

$$S = \{(t, x) : |t - t_0| \leq a, \ |x - x_0| \leq b\} \subseteq D.$$

From the assumption $f \in C(D)$, we obtain

$$|f(t, x)| \leq M \text{ on the compact set } S.$$

Let $c = \min\{a, b/M\}$.

Theorem A.1.1 *For any $\varepsilon > 0$, there exists an ε-approximate solution of I.V.P. (2.1) on the interval $I = \{t : |t - t_0| \leq c\}$.*

Proof. From the fact that $f(t,x)$ is continuous on the compact set S, it follows that $f(t,x)$ is uniformly continuous on S. Then, given $\varepsilon > 0$, there exists $\delta = \delta(\varepsilon) > 0$ such that

$$|f(t,x) - f(s,y)| \; < \varepsilon \text{ whenever } |(t,x) - (s,y)| < \delta. \qquad (A.1)$$

Partition the interval $[t_0, t_0 + c]$ into m subintervals, $t_0 < t_1 < \cdots < t_m = t_0 + c$ with $t_{j+1} - t_j < \min\left\{\frac{\delta}{2}, \frac{\delta}{2M}\right\}$ for $j = 0, \cdots, m-1$. Construct an ε-approximate solution $\varphi(t)$ by Euler's method: Let $\varphi(t_0) = x_0$, and define $\varphi(t)$ by the following:

$$\begin{aligned} \varphi(t) &= \varphi(t_0) + f(t_0, \varphi(t_0))(t - t_0), \text{on } [t_0, t_1] \\ \varphi(t) &= \varphi(t_j) + f\left(t_j, \varphi(t_j)\right)(t - t_j), \text{on } [t_j, t_{j+1}], 1 \le j \le m-1. \end{aligned} \qquad (A.2)$$

First we check that $(t, \varphi(t)) \in S$ for all $t_0 \le t \le t_0 + c$. Obviously $|t - t_0| \le c \le a$. For $t_0 \le t \le t_1$, $|\varphi(t) - x_0| = |f(t_0, x_0)|(t - t_0) \le Mc \le M \cdot b/M = b$. By induction we assume $(t, \varphi(t)) \in S$ for $t_0 \le t \le t_j$. Then for $t_j \le t \le t_{j+1}$,

$$\begin{aligned} |\varphi(t) - x_0| &\le |\varphi(t) - \varphi(t_j)| + |\varphi(t_j) - \varphi(t_{j-1})| + \cdots \\ &+ \cdots + |\varphi(t_0) - x_0| \le M\left((t - t_j) + (t_j - t_{j-1}) + \cdots + (t_1 - t_0)\right) \quad (A.3) \\ &= M(t - t_0) \le Mc \le M \cdot b/M = b. \end{aligned}$$

Now we verify that $\varphi(t)$ is an ε-approximate solution. Let $t \in (t_j, t_{j+1})$. Then $|t - t_j| \le |t_{j+1} - t_j| < \delta/2$ and from (A.2) $|\varphi(t) - \varphi(t_j)| = |f(t_j, \varphi(t_j))(t - t_j)| \le M|t - t_j| \le M \cdot \delta/2M = \delta/2$. Hence it follows that $|(t, \varphi(t)) - (t_j, \varphi(t_j))| < \delta$ and from (A.1) we have that

$$|\varphi'(t) - f(t, \varphi(t))| = |f(t_j, \varphi(t_j)) - f(t, \varphi(t))| < \varepsilon.$$

$\square$

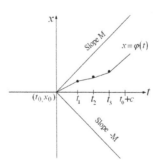

Fig. A.1

Theorem A.1.2 *Let $f \in C(D)$, $(t_0, x_0) \in D$. Then the I.V.P. (2.1) has a solution on the interval $I = [t_0 - c, t_0 + c]$ for some $c > 0$.*

Proof. Let $\varepsilon_m = \frac{1}{m} \downarrow 0$ as $m \to +\infty$ and let $\varphi_m(t)$ be an ε_m-approximate solution on I. Claim: $\{\varphi_m\}$ is uniformly bounded and equicontinuous.

<u>Uniform boundedness</u>: From (A.3)

$$|\varphi_m(t)| \le |\varphi_m(t_0)| + |\varphi_m(t) - \varphi_m(t_0)|$$
$$\le |x_0| + M(t - t_0) \le |x_0| + Mc$$

for all t and m.

<u>Equicontinuity of $\{\varphi_m\}$</u>: For any $s < t$ with $s < t_j < \cdots < t_k < t$ for some j and k,

$$|\varphi_m(t) - \varphi_m(s)| \le |\varphi_m(t) - \varphi_m(t_k)| + |\varphi_m(t_k) - \varphi_m(t_{k-1})|$$
$$+ \cdots + |\varphi_m(t_{j+1}) - \varphi_m(t_j)| + |\varphi_m(t_j) - \varphi_m(s)|$$
$$\le M(t - t_k) + M(t_k - t_{k-1}) + \cdots + M(t_j - s)$$
$$= M|t - s|.$$

Hence for any $\varepsilon > 0$ we choose $\delta = \varepsilon/M$, then

$$|\varphi_m(t) - \varphi_m(s)| < \varepsilon \text{ whenever } |t - s| < \delta.$$

From Ascoli-Arzela Theorem, we can extract a convergent subsequence $\{\varphi_{m_k}\}$. Let $\varphi_{m_k} \to \varphi \in C(I)$. Then $f(t, \varphi_{m_k}(t)) \to f(t, \varphi(t))$ uniformly on I. Let $E_m(t) = \varphi'_m(t) - f(t, \varphi_m(t))$. Then E_m is piecewise continuous on I and $|E_m(t)| \le \varepsilon_m$ on I. It follows that

$$\varphi_m(t) = x_0 + \int_{t_0}^t [f(s, \varphi_m(s)) + E_m(s)] \, ds,$$
$$\downarrow \qquad \qquad \downarrow$$
$$\varphi(t) = x_0 + \int_{t_0}^t f(s, \varphi(s)) ds.$$

From Lemma 2.2.1 we complete the proof Theorem A.1.2. $\qquad \square$

A.2

Let $f : D \subseteq R \times R^n \times R^k \to R^n$ be a C^1 function. In Appendix A.2 we prove the continuous dependence of the solution $\varphi(t; \tau, \xi, \lambda)$ of (A.4)

$$x' = f(t, x, \lambda)$$
$$x(\tau) = \xi \tag{A.4}$$

on (τ, ξ, λ) where τ is the initial time; ξ is the initial value; λ is the parameter. First we consider the discrete version of the continuous dependence property.

Theorem A.2.1 *Let $\{f_n(t, x)\}$ be a sequence of continuous functions on D and $\lim\limits_{n \to \infty} f_n = f$ uniformly on any compact set in D. Let $(\tau_n, \xi_n) \in D$, $(\tau_n, \xi_n) \to (\tau, \xi)$ and $\varphi_n(t)$ be any noncontinuable solution of $x' = f_n(t, x)$, $x(\tau_n) = \xi_n$. If $\varphi(t)$ is the unique solution of $x' = f(t, x)$, $x(\tau) = \xi$, defined on $[a, b]$. Then $\varphi_n(t)$ is defined on $[a, b]$ for n large and $\varphi_n \to \varphi$ uniformly on $[a, b]$.*

Proof. It suffices to show that $\varphi_n \to \varphi$ uniformly on $[\tau, b]$. The proof is split into two parts.

Part 1: Show that there exists $t^1 > \tau$ such that $\varphi_n \to \varphi$ uniformly on $[\tau, t^1]$. Since $f \in C(D)$ and $(\tau, \xi) \in D$, we let E be a compact subset of D such that Int $E \supseteq$ graph $\varphi = \{(t, \varphi(t)) : a \le t \le b\}$. Let $|f| < M$ in E. Then, $|f_n| < M$ for $n \ge n_0$ for some large n_0. Choose $\delta > 0$ sufficiently small such that

$$R = \{(t, x) : |t - \tau| \le \delta, \ |x - \xi| \le 3M\delta\} \subseteq E.$$

Fig. A.2

Choose $n_1 \geq n_0$ large such that

$$|\tau_n - \tau| < \delta, \ |\xi_n - \xi| < M\delta \quad \text{for} \quad n \geq n_1.$$

Let $n \geq n_1$, then $(\tau_n, \xi_n) \in R$. For $|t - \tau| \leq \delta$, as long as $(t, \varphi_n(t))$ stays in E, we have

$$|\varphi_n(t) - \xi_n| = \left| \int_{\tau_n}^{t} f_n(s, \varphi_n(s)) ds \right| \leq M|t - \tau_n|$$

and

$$\begin{aligned}
|\varphi_n(t) - \xi| &< |\xi - \xi_n| + |\varphi_n(t) - \xi_n| \\
&\leq M\delta + M|t - \tau_n| \\
&\leq M\delta + M(|t - \tau| + |\tau - \tau_n|) \\
&\leq M\delta + M\delta + M\delta = 3M\delta.
\end{aligned}$$

Then $\varphi_n(t)$ is defined and $\{(t, \varphi_n(t)) : |t - \tau| \leq \delta\} \subseteq R$. Obviously $\{\varphi_n\}$ is uniformly bounded on $I_\delta = \{t : |t - \tau| \leq \delta\}$. From

$$|\varphi_n(t) - \varphi_n(s)| = \left| \int_{t}^{s} f_n(s', \varphi_n(s')) ds' \right| \leq M|t - s| \quad \text{for} \quad s > t \ \ s, t \in I_\delta,$$

it follows that φ_n is equicontinuous on I_δ. By Ascoli-Arzela Theorem, we can extract a convergent subsequence $\{\varphi_{n_k}\}$ such that $\varphi_{n_k} \to \bar{\varphi}$ uniformly on I_δ.

Since

$$\varphi_n(t) = \xi_n + \int_{\tau_n}^{t} f_n(s, \varphi_n(s)) ds,$$

letting $n_k \to \infty$ yields that

$$\bar{\varphi}(t) = \xi + \int_{\tau}^{t} f(s, \bar{\varphi}(s)) ds.$$

By uniqueness of $\varphi(t)$, we have $\bar{\varphi}(t) \equiv \varphi(t)$ on I_δ. Obviously by the uniqueness property every convergent subsequence of $\{\varphi_n\}$ converges to φ. Now we claim: $\varphi_k \to \varphi$ on I_δ. If not, then there exists $\varepsilon > 0$ and a subsequence φ_{n_ℓ} such that $\| \varphi_{n_\ell} - \varphi \| > \varepsilon$ for each $\ell = 1, 2, \cdots$. Apply Ascoli-Arzela Theorem to $\{\varphi_{n_\ell}\}$, we obtain a subsequence $\varphi_{n_{\ell_k}} \to \varphi$ which is a desired contradiction. Hence $\varphi_k \to \varphi$ uniformly on $[\tau, t^1]$, $t^1 = \tau + \delta$.

Part 2: To show $\varphi_n \to \varphi$ uniformly on $[\tau, b]$. Let $t^* = \sup\{t^1 \leq b : \varphi_n \to \varphi$ uniformly on $[\tau, t^1]\}$ we claim that $t^* = b$. If not, $t^* < b$. Choose $\delta^1 > 0$ small such that

$$R^1 = \left\{ (t, x) : |t - t^*| \leq 2\delta^1, \ |x - \varphi(t^*)| \leq 4M\delta^1 \right\} \subseteq E.$$

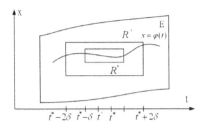

Fig. A.3

Choose t^1, $t^* - \delta^1 < t^1 < t^*$. Then

$$|\varphi(t^*) - \varphi(t^1)| \leq \int_{t^1}^{t^*} |f(s, \varphi(s))| ds \leq M\delta^1.$$

Hence

$$R'' = \{(x,t) : |t - t^1| \leq \delta^1, |x - \varphi(t^1)| \leq 3M\delta^1\} \subseteq R^1.$$

Let $\xi_n^1 = \varphi_n(t^1)$, $\xi^1 = \varphi(t^1)$. Then $(t^1, \xi_n^1) \to (t^1, \xi^1)$.

Apply part 1 with δ replaced by δ^1, we obtain that

$$\varphi_n(t) \to \varphi(t) \quad \text{uniformly on} \quad [t^1, t^1 + \delta^1].$$

But $t^1 + \delta^1 > t^*$. This contradicts to the definition of t^*.   □

Now we state the continuous version of the continuous dependence property.

Theorem A.2.2 *Let $f(t, x, \lambda)$ be a continuous function of (t, x, λ) for all (t, x) in open set D and for all λ near λ_0 and let $\varphi(t; \tau, \xi, \lambda)$ be any non-continuable solution of*

$$x' = f(t, x, \lambda),$$
$$x(\tau) = \xi.$$

If $\varphi(t; t_0, \xi_0, \lambda_0)$ is defined on $[a, b]$ and is unique, then $\varphi(t; \tau, \xi, \lambda)$ is defined on $[a, b]$ for all (τ, ξ, λ) near (t_0, ξ_0, λ_0) and is a continuous function of (t, τ, ξ, λ).

Proof. By Theorem A.2.1 we have

$$|\varphi(t; \tau, \xi, \lambda) - \varphi(t_1; t_0, \xi_0, \lambda_0)|$$
$$\leq |\varphi(t; \tau, \xi, \lambda) - \varphi(t; t_0, \xi_0, \lambda_0)| + |\varphi(t, t_0, \xi_0, \lambda_0) - \varphi(t_1, t_0, \xi_0, \lambda_0)|$$
$$\leq \varepsilon + |\varphi(t; t_0, \xi_0, \lambda_0) - \varphi(t_1; t_0, \xi_0, \lambda_0)| < 2\varepsilon,$$

for $a \leq t \leq b$ if $|(\tau, \xi, \lambda) - (t_0, \xi_0, \lambda_0)| < \delta$ and $|t - t_1| < \delta_1$.   □

A.3

In this Appendix A.3, we prove Theorem A.3.1.

Theorem A.3.1 [Cop] *Let $\varphi(t, \lambda_0)$ be the unique solution of*

$$\frac{dx}{dt} = f(t, x, \lambda_0),$$
$$x(t_0) = \xi_0,$$

on compact interval $J = [a, b]$. Assume $f \in C^1$ in x and λ at all points $(t, \varphi(t, \lambda_0), \lambda_0)$, $t \in J$. Then for λ sufficiently near λ_0, the system $E_\lambda : \frac{dx}{dt} = f(t, x, \lambda)$, $x(t_0) = \xi_0$ has a unique solution $\varphi(t, \lambda)$ defined on J. Moreover, $\varphi_\lambda(t, \lambda_0)$ exists and satisfies

$$y' = f_x(t, \varphi(t, \lambda_0), \lambda_0)y + f_\lambda(t, \varphi(t, \lambda_0), \lambda_0),$$
$$y(t_0) = 0.$$

Proof. Since f_x, f_λ are continuous in (t, x, λ) and $\varphi(t, \lambda_0)$ is continuous in t, then given any $\varepsilon > 0$ and for each $s \in J$, there exists $\delta = \delta(s, \varepsilon) > 0$ such that

$$|f_x(t, x, \lambda) - f_x(t, \varphi(t, \lambda_0), \lambda_0)| \leq \varepsilon,$$
$$\text{and} \hspace{5cm} (A.5)$$
$$|f_\lambda(t, x, \lambda) - f_\lambda(t, \varphi(t, \lambda_0), \lambda_0)| \leq \varepsilon,$$

if $|t - s| \leq \delta$, $|x - \varphi(t, \lambda_0)| \leq \delta$, $|\lambda - \lambda_0| \leq \delta$.

Since $\bigcup\limits_{s \in [a,b]} B(s, \delta(s, \varepsilon)) \supseteq [a, b]$, from Heine-Borel Theorem, there exist $N > 0$ such that $\bigcup_{i=1}^{N} B(s_i, \delta(s_i, \varepsilon)) \supseteq [a, b]$. Let $\delta^1 = \delta^1(\varepsilon) = \min(\delta(s_1, \varepsilon), \cdots, \delta(s_N, \varepsilon))$. For any $t \in J$, if

$$|x - \varphi(t, \lambda_0)| \leq \delta^1, \quad |\lambda - \lambda_0| \leq \delta^1, \hspace{2cm} (A.6)$$

then $|t - s_i| \leq \delta_i(s_i, \varepsilon) = \delta_i$ for some i and then $|x - \varphi(t, \lambda_0)| \leq \delta^1 < \delta_i$, $|\lambda - \lambda_0| \leq \delta^1 < \delta_i$. Hence (A.5) holds for any $t \in J$ satisfying (A.6). Let

$$R = \{(t, x, \lambda) : |x - \varphi(t, \lambda_0)| \leq \delta^1, \ |\lambda - \lambda_0| \leq \delta^1, \ t \in J\}.$$

Then, f_x, f_λ are bounded on R, say $|f_x| \leq A$, $|f_\lambda| \leq B$. Hence f satisfies a Lipschitz condition in x and, if $|\lambda - \lambda_0| \leq \delta^1$, the I.V.P. (E_λ) has a unique solution passing through any point in the region $\Omega : |x - \varphi(t, \lambda_0)| \leq \delta^1$, $t \in J$. Moreover, the solution $\varphi(t, \lambda)$ passing through the point (t_0, ξ_0) satisfies

$$|\varphi'(t, \lambda) - f(t, \varphi(t, \lambda), \lambda_0)| \leq B|\lambda - \lambda_0|,$$

as long as $(t, \varphi(t, \lambda))$ stays in Ω. Therefore, by Theorem 2.2.4, we have

$$|\varphi(t, \lambda) - \varphi(t, \lambda_0)| \leq C|\lambda - \lambda_0|, \tag{A.7}$$

where $C = B[e^{Ah} - 1]/A$ and $h = b - a$. It follows that $\varphi(t, \lambda)$ is defined on J for all λ sufficiently near λ_0. Furthermore by (A.5), (A.7) and the mean value theorem, if λ is sufficiently near λ_0, then for all $t \in J$,

$$\begin{aligned}&|f(t, \varphi(t, \lambda), \lambda) - f(t, \varphi(t, \lambda_0), \lambda_0) - f_x(t, \varphi(t, \lambda_0), \lambda_0)\\&\cdot [\varphi(t, \lambda) - \varphi(t, \lambda_0)] - f_\lambda(t, \varphi(t, \lambda_0), \lambda_0)(\lambda - \lambda_0)| \leq \varepsilon|\lambda - \lambda_0|.\end{aligned} \tag{A.8}$$

Put

$$\psi(t, \lambda) = \varphi(t, \lambda) - \varphi(t, \lambda_0) - y(t)(\lambda - \lambda_0).$$

Then $\psi(t_0, \lambda) = 0$ and (A.8) can be rewritten as

$$|\psi'(t, \lambda) - f_x(t, \varphi(t, \lambda_0), \lambda_0)\psi(t, \lambda)| \leq \varepsilon|\lambda - \lambda_0|.$$

Since $y(t) \equiv 0$ is the unique solution of the differential equation

$$y' = f_x(t, \varphi(t, \lambda_0), \lambda_0)y$$
$$y(t_0) = 0$$

then, from Theorem 2.2.4 it follows that

$$|\psi(t, \lambda)| \leq \varepsilon|\lambda - \lambda_0| \left(\frac{e^{Ah} - 1}{A} \right).$$

Therefore $\varphi_\lambda(t, \lambda_0)$ exists and is equal to $y(t)$. $\qquad\square$

APPENDIX B

We say that $W^u(0, U)$ is a Lipschitz graph over $P\mathbb{R}^n$ if there is a neighborhood $V \subseteq P\mathbb{R}^n$ of 0 and a Lipschitz continuous function $h : V \to Q\mathbb{R}^n$ such that
$$W^u(0, U) = \{(\xi, \eta) \in \mathbb{R}^n : \eta = h(\xi),\ \xi \in V\}.$$
The set $W^u(0, U)$ is said to be tangent to $P\mathbb{R}^n$ at 0 if $\frac{|h(\xi)|}{|\xi|} \to 0$ as $\xi \to 0$ in $W^u(0, U)$. We say that $W^u(0, U)$ is a C^k (or analytic) graph over $P\mathbb{R}^n$ if the above function h is C^k (or analytic). Similar definitions hold for $W^s(0, U)$.

Theorem B.1 *(Stable Manifold Theorem)*
If $g\ Re(\sigma(A)) \neq 0$, then there is a neighborhood U of 0 in $\mathbb{R}^n$ such that $W^u(0, U)$ (or $W^s(0, U)$) is a Lipschitz graph over $E^u := P\mathbb{R}^n$ (or $E^s := Q\mathbb{R}^n$) which is tangent to E^u (or E^s) at 0. Also, there are positive constants K_1, α_1 such that $x \in W^u(0, U)$ (resp. $W^s(0, U)$) then the solution $\varphi^t(x)$ of (4.8) satisfies
$$|\varphi^t(x)| \leq K_1 e^{-\alpha_1 |t|}|x|, \quad t \leq 0\ (resp.\ t \geq 0).$$
Furthermore if g is a C^k function (or an analytic function) in a neighborhood U of 0, then so are $W^u(0, U)$ and $W^s(0, U)$.

Then local stable and unstable manifold can be extended to global stable and unstable manifolds. Define
$$W^s(0) = \text{the global stable manifold of } 0$$
$$= \bigcup_{t \leq 0} \{\varphi_t(x), x \in W^s_{loc}(0)\},$$

and

$$W^u(0) = \text{the global unstable manifold of } 0$$
$$= \bigcup_{t \geq 0} \{\varphi_t(x), x \in W^u_{loc}(0)\}.$$

Proof. We shall apply the contraction mapping principle. With the function ρ in (4.9) and K, α in (4.14), we choose $\delta > 0$ so that

$$4K\rho(\delta) < \alpha, \quad 8K^2\rho(\delta) < \alpha. \tag{B.1}$$

Let $S(\delta)$ be the set of continuous functions $x : (-\infty, 0] \to \mathbb{R}^n$ such that $|x| = \sup_{-\infty \leq t \leq 0} |x(t)| \leq \delta$. Then the set $S(\delta)$ is a complete metric space with the metric induced by the uniform topology.

For any $y(\cdot) \in S(\delta)$ and any $\xi \in P\mathbb{R}^n$ with $|\xi| \leq \delta/2K$, we define, for $t \leq 0$

$$(T(y, \xi))(t) = e^{At}\xi + \int_0^t e^{A(t-s)} Pg(y(s))ds + \int_{-\infty}^t e^{A(t-s)} Qg(y(s))ds. \tag{B.2}$$

Next we want to show that

$$T : S(\delta) \times \{\xi \in P\mathbb{R}^n : |\xi| \leq \delta/2K\} \to S(\delta)$$

is Lipschitz continuous and $T(\cdot, \xi)$ is a contraction mapping with contraction constant $1/2$ for all $\xi \in P\mathbb{R}^n$ with $|\xi| \leq \frac{\delta}{2K}$. From (4.9), (4.13), (4.14), (B.1) and (B.2) for $t \leq 0$, we have

$$|T(y, \xi)(t)| \leq Ke^{\alpha t}|\xi| + \int_t^0 Ke^{\alpha(t-s)}\rho(\delta)|y(s)|ds$$

$$+ \int_{-\infty}^t Ke^{-\alpha(t-s)}\rho(\delta)|y(s)|ds$$

$$\leq K|\xi| + \rho(\delta) \cdot \delta \cdot K\frac{1}{\alpha} + \rho(\delta)\delta\frac{K}{\alpha}$$

$$< \delta/2 + \delta/2 = \delta.$$

Thus $T(\cdot, \xi) : S(\delta) \to S(\delta)$. Furthermore from (4.9), (4.13), (4.14), (B.1) and (B.2), we have

$$|T(y_1, \xi)(t) - T(y_2, \xi)(t)|$$

$$\leq \int_t^0 K e^{\alpha(t-s)} |g(y_1(s)) - g(y_2(s))| ds$$

$$+ \int_{-\infty}^t K e^{\alpha(t-s)} |g(y_1(s)) - g(y_2(s))| ds$$

$$\leq \int_t^0 K e^{\alpha(t-s)} \rho(\delta) |y_1(s) - y_2(s)| ds$$

$$+ \int_{-\infty}^t K e^{\alpha(t-s)} \rho(\delta) |y_1(s) - y_2(s)| ds$$

$$\leq \frac{1}{2} |y_1 - y_2|.$$

Therefore $T(\cdot, \xi)$ has a unique fixed point $x^*(\cdot, \xi)$ in $S(\delta)$. The fixed point satisfies (4.15) and thus is a solution of (4.8) by Lemma 4.3.1.

The function $x^*(\cdot, \xi)$ is continuous in ξ. Also

$$|x^*(t, \xi)| \leq K e^{\alpha t} |\xi| + K \rho(\delta) \int_t^0 e^{\alpha(t-s)} |x^*(s, \xi)| ds \qquad \text{(B.3)}$$
$$+ K \rho(\delta) \int_{-\infty}^t e^{-\alpha(t-s)} |x^*(s, \xi)| ds.$$

From inequality (B.3) and the following Lemma 4.3.2, one can prove that

$$|x^*(t, \xi)| \leq 2K e^{\alpha t/2} |\xi|, \quad t \leq 0. \qquad \text{(B.4)}$$

This estimate shows $x^*(\cdot, 0) = 0$ and $x^*(0, \xi) \in W^u(0)$.

The similar estimate as above shows that

$$|x^*(t, \xi) - x^*(t, \tilde{\xi})| \leq 2K e^{\alpha t/2} |\xi - \tilde{\xi}|, \quad t \leq 0. \qquad \text{(B.5)}$$

In particular, $x^*(\cdot, \xi)$ is Lipschitz in ξ. Since $x^*(t, \xi)$ satisfies

$$x^*(t, \xi) = e^{At} \xi + \int_0^t e^{A(t-s)} Pg(x^*(s, \xi)) ds \qquad \text{(B.6)}$$
$$+ \int_{-\infty}^t e^{A(t-s)} Qg(x^*(s, \xi)) ds.$$

Set $t = 0$ in (B.6), we have

$$x^*(0, \xi) = \xi + \int_{-\infty}^0 e^{A(-s)} Qg(x^*(s, \xi)) ds. \qquad \text{(B.7)}$$

From (B.7) and $\xi \in P\mathbb{R}^n$ we have $Px^*(0, \xi) = \xi$ (see Fig. B.1).

Thus we define

$$h(\xi) = Qx^*(0, \xi) = \int_{-\infty}^0 e^{A(-s)} Qg(x^*(s, \xi)) ds. \qquad \text{(B.8)}$$

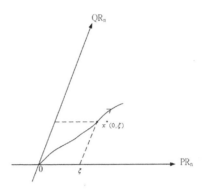

Fig. B.1

Now we prove that $W^u(0, U)$ is tangent to $P\mathbb{R}^n$ at 0. From (4.9) and $x^*(t, 0) \equiv 0$, let $\delta = |x^*(s, \xi)|$, it follows that

$$|g(x^*(s, \xi)) - g(x^*(s, 0))| \le \rho(|x^*(s, \xi)|) \cdot |x^*(s, \xi) - x^*(s, 0)|,$$

or

$$|g(x^*(s, \xi))| \le \rho(|x^*(s, \xi)|) \cdot |x^*(s, \xi)|.$$

From (B.4) and (B.8) we have

$$
\begin{aligned}
|h(\xi)| &= \left| \int_{-\infty}^0 e^{-As} Q g(x^*(s, \xi)) ds \right| \\
&\le \int_{-\infty}^0 \left\| e^{-As} Q \right\| \cdot |g(x^*(s, \xi))| \, ds, \\
&\le \int_{-\infty}^0 K e^{\alpha s} \rho \left(|x^*(s, \xi)| \right) |x^*(s, \xi)| ds, \quad \left(|x^*(s, \xi)| \le 2K e^{\frac{\alpha s}{2}} |\xi|, s \le 0 \right) \\
&\le \int_{-\infty}^0 K e^{\alpha s} \rho \left(2K e^{\frac{\alpha s}{2}} |\xi| \right) \cdot 2K e^{\frac{\alpha s}{2}} |\xi| ds, \\
&\le 2K^2 \rho(2K|\xi|) \left(\int_{-\infty}^0 e^{\frac{3}{2}\alpha s} ds \right) |\xi|, \\
&= \frac{4K^2}{3\alpha} \rho(2K|\xi|)|\xi|.
\end{aligned}
$$

Therefore from (B.1)

$$\frac{|h(\xi)|}{|\xi|} \le \frac{4K^2}{3\alpha} \rho(2K|\xi|) \to 0 \text{ as } |\xi| \to 0.$$

$\square$

Lemma B.1 *Suppose that $\alpha > 0, \gamma > 0, K, L, M$ are nonnegative constants and u is a nonnegative bounded continuous function satisfying one of the inequalities*

$$u(t) \leq Ke^{-\alpha t} + L \int_0^t e^{-\alpha(t-s)}u(s)ds + M \int_0^\infty e^{-\gamma s}u(t+s)ds, \ t \geq 0 \ \ (B.9)$$

$$u(t) \leq Ke^{\alpha t} + L \int_t^0 e^{\alpha(t-s)}u(s)ds + M \int_{-\infty}^0 e^{\gamma s}u(t+s)ds, \ t \leq 0. \ \ (B.10)$$

If $\beta \equiv L/\alpha + \frac{M}{\gamma} < 1$, then

$$u(t) \leq (1-\beta)^{-1}Ke^{-[\alpha - (1-\beta)^{-1}L]|t|}. \tag{B.11}$$

Proof. It suffice to prove (B.9). We obtain (B.10) directly from (B.9) by changing variables $t \to -t$, $s \to -s$. Let $\delta = \limsup_{t\to\infty} u(t)$. Since u is nonnegative, bounded, then it follows that $\delta < \infty$. We claim that $\delta = 0$, i.e. $\lim_{t\to\infty} u(t) = 0$. If not, $\delta > 0$. Choose θ, $\beta < \theta < 1$, then there exists $t_1 > 0$ such that $u(t) \leq \theta^{-1}\delta$ for $t \geq t_1$. For $t \geq t_1$

$$u(t) \leq Ke^{-\alpha t} + L \int_0^{t_1} e^{-\alpha(t-s)}u(s)ds + L \int_{t_1}^t e^{-\alpha(t-s)}u(s)ds$$

$$+ M\theta^{-1}\delta \int_0^\infty e^{-\gamma s}ds$$

$$\leq Ke^{-\alpha t} + Le^{-\alpha t}\int_0^{t_1} e^{\alpha s}u(s)ds + \left(\frac{L}{\alpha} + \frac{M}{\gamma}\right)\theta^{-1}\delta.$$

Let $t \to \infty$. Then we have a contradiction:

$$\limsup_{t\to\infty} u(t) \leq \left(\frac{L}{\alpha} + \frac{M}{\gamma}\right)\theta^{-1}\delta = \beta\theta^{-1}\delta < \delta.$$

Let $v(t) = \sup_{s\geq t} u(s)$. Since $u(t) \to 0$ as $t \to \infty$, for any $t \in [0, \infty)$, there exists $t_1 \geq t$ such that

$$v(t) = v(s) = u(t_1), \quad t \leq s \leq t_1,$$
$$v(s) < v(t_1) \quad \text{for} \ s > t_1.$$

Fig. B.2

Then

$$v(t) = u(t_1) \le Ke^{-\alpha t_1} + L \int_0^{t_1} e^{-\alpha(t_1-s)} u(s)ds$$

$$+ M \int_0^\infty e^{-\gamma s} u(t_1 + s)ds$$

$$\le Ke^{-\alpha t_1} + L \left(\int_0^t e^{-\alpha(t_1-s)} u(s)ds \right) + L \left(\int_t^{t_1} e^{-\alpha(t_1-s)} u(s)ds \right)$$

$$+ M \int_0^\infty e^{-\gamma s} v(t+s)ds$$

$$\le Ke^{-\alpha t} + L \int_0^t e^{-\alpha(t-s)} v(s)ds + Lv(t) \int_t^{t_1} e^{-\alpha(t_1-s)} ds$$

$$+ v(t)M \int_0^\infty e^{-\gamma s} ds$$

$$\le Ke^{-\alpha t} + Le^{-\alpha t} \int_0^t e^{\alpha s} v(s)ds + \beta v(t).$$

Let $z(t) = e^{\alpha t} v(t)$. Then

$$z(t) \le K + L \int_0^t z(s)ds + \beta z(t)$$

or

$$z(t) \le (1-\beta)^{-1} K + (1-\beta)^{-1} L \int_0^t z(s)ds.$$

Then from Gronwall's inequality, it follows that

$$z(t) \le (1-\beta)^{-1} K \exp \left((1-\beta)^{-1} Lt \right).$$

Thus the estimate (B.11) follows. □

For the regularity of $W^u(0, U)$, we observe from (B.7), (4.14) and (4.9),

$$\left| x^*(0, \xi) - x^*(0, \bar{\xi}) \right| \ge |\xi - \bar{\xi}| - \int_{-\infty}^0 K\rho(\delta) e^{\alpha s} \cdot \left| x^*(s, \xi) - x^*(s, \bar{\xi}) \right| ds$$

$$\ge |\xi - \bar{\xi}| \left[1 - \frac{4K^2 \rho(\theta)}{3\alpha} \right] \ge 1/2 |\xi - \bar{\xi}|.$$

Thus the mapping $\xi \longmapsto x^*(0, \xi)$ is 1-1 with continuous inverse.

Hence $W^u(0, U)$ is a Lipschitz manifold. It is not easy to prove that $W^u(0, U)$ is C^k-manifold if $f \in C^k$. The reader may consult [Hen]. The stable and unstable manifold are defined as follows

$$W^s(0) = \bigcup_{t \le 0} \varphi^t \left(W^s_{loc}(0, U) \right),$$

$$W^u(0) = \bigcup_{t \ge 0} \varphi^t \left(W^u_{loc}(0, U) \right).$$

Bibliography

[ASY]   Alligood, K., Sauer, T. and Yorke, J. (1996). *Chaos, An Introduction to Dynamical Systems*, Springer-Verlag.

[AP]   A. Ambrosetti and G. Prodi, *A primer of Nonlinear Analysis*, Cambridge University Press.

[Apo]   Apostol, T. M. (1975). *Mathematical Analysis*, (Addison-Wesley).

[Arnold] Arnold, V. (1989), *Mathematical Methods of Classical Mechanics*, translated by K. Vogtmann and A. Weinstein, 2nd edition, Springer.

[AW]   Aronson, D. and Weinberg, H. (1975). *Nonlinear diffusion in population genetics, combustion, and nerve conduction, in Partial Differential Equations and Related Topics*, ed., J. A. Goldstein, Lecture Note in Mathematics 446, 5–49, New York: Springer.

[BB]   Berger, M. and Berger, M. (1968). *Perspective in Nonlinearity. An Introduction to Nonlinear Analysis*, W. A. Benjamin, (Inc. New York-Amsterdam).

[BDiP]   Boyce, W. E. and DiPrima, R. C. (1986). *Elementary Differential Equations and Boundary Value Problems*, John Wiley & Sons.

[Che]   Cheng, K. S. (1981). *Uniqueness of a limit cycle for a predator prey system*, SIAM Math. Anal. 12, pp. 541–548.

[CHW]   Chi, C. W., Hsu, S. B. and Wu, L. I. (1998). *On the asymmetric May-Leonard model of three competing species*, SIAM J. Appl. Math., pp. 211–226.

[CL]   Coddington, E. A. and Levinson, N. (1955). *Ordinary Differential Equations*, New York, McGraw-Hill.

[Cole]   Cole, J. D. (1968). *Perturbation Methods in Applied Mathematics*, Waltham, Mass., Blaisdell Pub. Co.

[Cop]   Coppel, W. A. (1965). *Stability and Asymptotic Behavior of Differential Equations*, Heath, Boston.

[E]   Evans, L. C. (1998). *Partial Differential Equations*, Graduate Studies in Mathematics, Vol. 19 p. 234.

[FMcL]   Fife, P. C. and McLeod, J. B. (1977). *The Approach of Solutions of Nonlinear Diffusion Equations to Travelling Front Solutions*, Archiv. Rat. Mech. Anal. 65, pp. 335–361.

355

[FW]   Freedman, H. I. and Waltman, P. (1977). *Mathematical analysis of some three-species food-chain models*, Math. Biosci. 33(3-4), pp.

[HLW]   Hairer, E., Lubich, C. and Wanner, G. (2002). *Geometric Numerical Integration — Structure-Preserving Algorithms for Ordinary Differential Equations*, Springer Series in Comput. Mathematics, Vol. 31, Springer-Verlag.

[H1]   Hale, J. (1969). *Ordinary Differential Equations*, New York, Wiley-Interscience.

[H2]   Hale, J. (1988). *Asymptotic Behavior of Dissipative Systems*, American Mathematical Society.

[HK]   Hale, J. and Kocak, H. (1991). *Dynamics and Bifurcations*, Springer-Verlag.

[Ha]   Hartman, P. (1964). *Ordinary Differential Equations*, Wiley, New York.

[Hen]   Henry, D. (1981). *Geometric Theory of Semilinear Parabolic Equations*, Lecture Notes in Mathematics, Vol. 840, Springer-Verlag.

[Hir1]   Hirsch, M. (1982). *Systems of differential equations which are competitive or cooperative I; limit sets*, SIAM J. Math. Anal. Vol. 13, pp. 167–179.

[Hir2]   Hirsch, M. (1984). *The dynamical systems approach to differential equations*, Bull. A.M.S. 11, pp. 1–64.

[HS1]   Hirsch, M. and Smale, S. (1974). *Differential Equations, Dynamical Systems and Linear Algebra*, New York, Academic Press.

[HS2]   Hirsch, M. and Smith, H. (2005). *Monotone Dynamical Systems*, Handbook of Differential Equations, Ordinary Differential Equations (Second Volume), eds., A. Canada, P. Drabek, A. Fonda, Elsevier, pp. 239–357.

[Hsu]   Hsu, S. B. (1978). *Limiting behavior of competing species*, SIAM J. Appl. Math. 34, pp. 760–763.

[HCH]   Hsu, S. B., Cheng, K. S. and Hubbell, S. P. (1981). *Exploitative competition of microorganisms for two complementary nutrients in continuous cultures*, SIAM J. Appl. Math., 41, pp. 422–444.

[HHW]   Hsu, S. B., Hubbel, S. and Waltman, P. (1978). *Competing Predators*, SIAM J. Appl. Math. Vol. 35, No. 4, pp. 617–625.

[HH]   Hsu, S. B. and Hwang, S. F. (1988). *Analysis of large deformation of a heavy cantilever*, SIAM J. Math. Anal. Vol. 19, No. 4, pp. 854–866.

[HN]   Hsu, S. B. and Ni, W. M. (1988). *On the asymptotic behavior of solutions of $v''(x)+x\sin v(x) = 0$*, Bulletin of Institute of Math. Academia Sinica, pp. 109–114.

[HSW]   Hsu, S. B., Smith, H. L. and Waltman, P. (1996). *Competitive exclusion and coexistence for competitive systems on ordered Banach space*, Trans. A.M.S. 348, pp. 4083–4094.

[HW]   Hsu, S. B. and Waltman, P. (1991). *Analysis on a model of two competitors in a Chemostat with an external inhibitor*, SIAM J. Appl. Math., 52, pp. 528–540.

[Hw1]   Hwang, T. W. (2003). *Global analysis of the predator-prey system with Beddington-DeAngelis functional response*, J. Math. Anal. Appl. 281(1), pp. 395–401.

[Hw2]   Hwang, T. W. (2004). *Uniqueness of limit cycles of the predator-prey system with Beddington-DeAngelis functional response*, J. Math. Anal. Appl. 290(1), pp. 113–122.

[Hxc]   Huang, X. C. (1988). *Uniqueness of limit cycles of generalised Lienard systems and Predator-prey systems*, J. Phys. A: Math. Gen. 21(1988)L685-L691.

[IK]   Isaacson, E. and Keller, H. (1966). *Analysis of Numerical Method*, New York, Wiley.

[JLZ]   Jiang, J. F., Liang, X. and Zhao, X. Q. (2004). *Saddle-point behavior for monotone semiflows and reaction-diffusion models*, J. Differential Equations, 203, pp. 313–330.

[JS]   Jordan, D. W. and Smith, P. (1977). *Nonlinear Ordinary Differential Equations*, Oxford, New York, Clarendon Press.

[KF]   Kuang, Y. and Freedman, H.I. (1988) *Uniqueness of limit cycle in Gauss-type model of predator-prey system*, Math. Biosience 88, 67–84(1988).

[Kee1]   Keener, J. (1988). *Principles of Applied Mathematics*, Addison Wiley.

[Kee2]   Keener, J. P. (1998). *Mathematical Physiology*, Springer-Verlag.

[KH]   Klebanoff, A. and Hastings, A. (1994). *Chaos in three species food chains*, J. Math. Biol., 32, pp. 427–451.

[KLY]   Kellogg R.B., T.Y. Li and J.A. Yorke (1976). *A constructive proof of the Brouwer Fixed Point Theorem and Computational results*, SIAM J. Numerical Analysis Vol. 13, pp. 473–483.

[L]   De La Llave, R. (2001). *A tutorial on KAM theory*, Proc. Symp. Pure Math. 69, 175–292.

[Lef]   Lefschetz, S. (1943). *Existence of periodic solutions for certain differential equations*, Prof. Nat. Acad. Sci. 29(1), pp. 29–32.

[LS]   Lin, C. C. and Segel, L. A. (1974). *Mathematics Applied to Deterministic Problems in the Natural Science*, Macmillan, NY.

[ML]   May, R. and Leonard, W. J. (1975). *Nonlinear aspects of competition between three species*, SIAM Appl. Math. 29, pp. 243–253.

[M-S]   Maynard-Smith, J. (1974). *Models in Ecology*, Cambridge University Press.

[M-PS]   Mallet-Paret, J. and Smith, H. (1990). *The Poincaré-Bendixson Theorem for monotone cyclic feedback systems*, J. Dyn. Diff. Eq. 2, pp. 367–421.

[MY]   Markus, L. and Yamabe, H. (1960). *Global stability criteria for differential systems*, Osaka Math. J. 12, pp. 305–317.

[MHO]   Meyer, K., Hall, G. and Offin, D. (2009), *Introduction to Hamiltonian Dynamical Systems and the N-Body Problem*, 2nd edition, Springer.

[MM]   Miller, R. K. and Michel, A. (1982). *Ordinary Differential Equations*, New York, Academic Press.

[Moser]   Moser, J. (1968), *Lectures on Hamiltonian Systems*, Memoirs of the American Mathematical Society, Number 81, American Mathematical Society.

[Mur]   Murdock, J. A. (1991). *Perturbation Theory and Methods*, John Wiley & Sons, Inc. New York.

[Murr]   Murray, J. (1989). *Mathematical Biology*, Springer-Verlag.

[NS]   Nemitskii, V. and Stepanov, V. (1960). *Qualitative Thoery of Differential Equations*, Princeton University Press.

[P]   Pöschel, J. (2001). *A lecture on the classical KAM theorem*, Proc. Symp. Pure Math. 69, 707–732.

[O'M]   O'Malley, R. (1974). *Introduction to Singular Perturbation*, New York, Academic Press.

[R]   Robinson, C. (1994). *Dynamical Systems, Stability, Symbolic Dynamics, and Chaos*, CRC Press.

[Sel]   Selgrade, J. F. (1982/83). *A Hopf bifurcation in single-loop positive-feedback systems*, Quart. Appl. Math. 40(3), pp. 347–351.

[Sma]   Smale, S. (1980). *The Mathematics of Time: Essays on Dynamical Systems*, Economic Process and Related Topics, Springer-Verlag.

[Smi]   Smith, H. (1995). *Monotone Dynamical Systems*, AMS Monographs, Vol. 41.

[ST]   [ST] Smith, H. L. and Thieme, H. R. (2001). *Stable coexistence and bistability for competitive systems on ordered Banach space*, J. Differential Equations, 176, pp. 195–222.

[SW]   Smith, H. and Waltman, P. (1995). *The Theory of the Chemostat*, Cambridge University Press.

[Stak]   Stakgold, I. (1972). *Boundary Value Problems of Mathematical Physics*, Vol. I, The Macmillan Company.

[Str]   Strogatz, S. H. (1994). *Nonlinear Dynamics and Chaos*, Addison Wiley.

[V]   Viana, M. (2000). *What's new on Lorenz strange attractors?*, The Mathematical Intelligence, Vol. 22, No. 3, pp. 6–19.

[Wal]   Waltman, P. (1983). *Competition Models in Population Biology*, CBMS Vol. 45, SIAM.

[WHK]   Wan, Y. H., Hassard, B. D. and Kazarinoff, N. D. (1981). *Theory and Application of Hopf Bifurcation*, Cambridge University Press.

[Wh]   Whittaker, E. T. (1937). *A Treatise on the Analytical Dynamics of Particles and Rigid Bodies, with an Introduction to the Problem of Three Bodies (4th ed.)*, New York: Dover Publications.

[YS]   Yakubovich, V. A. and Starzhinskii, V. M. (1975). *Linear Differential Equations with Periodic Coefficients*, Vols. 1 & 2, John-Wiley & Sons.

[Z]   [Z] Zhao, X. Q. (2003). *Dynamical Systems in Population Biology*, Springer.

[Zh]   Zhang, Z. (1986). *Proof of the uniqueness theorem of limit cycle of generalized Lienard equations*, Applicable Analysis 23: 1-2, 63-76(1986).

Index

Printed in the United States
by Baker & Taylor Publisher Services